At last — if you want to truly understand distributed ledger technology and its implications for payments, here finally is an authoritative and accessible reference book. No polemics, breathless praise, or impenetrable prose — the authors impartially explain how DLT works and how it can be used to upgrade our outdated payment systems. Along the way, the authors delight the reader with fascinating historical details and many practical applications. This book has instantly become the definitive source.

Prof. Darrell Duffie, The Adams Distinguished Professor of Management and
Professor of Finance, Stanford University

An amazing tour-de-force, and a must-read for anyone hoping to build distributed ledger systems for financial transactions. Good material for both students and experts alike.

Prof. Alex Pentland, Toshiba Professor, MIT

This is an excellent introduction to DLT which is not just credible and insightful for advanced students trained in quantitative fields — but also very accessible for the general reader with limited knowledge of math or cryptography. As the book so aptly explains, there are at least two big challenges for anybody who seeks to understand the key principles of DLT: the field is dominated by "zealots or their arch enemies" and the subject requires a good knowledge of cryptography, game theory and economics — three separate skills rarely taught in combination.

However, the book aims for a "balanced and well-reasoned discourse" — which it delivers in style — and it succinctly and effectively covers all three of the key elements of knowledge that need to be grasped to understand how DLT works.

It is thus a tour de force! It should be required reading for regulators, finance practitioners, investors and economists, particularly given that DLT will almost certainly reshape the future financial system.

Dr. Gillian Tett
Chair Editorial Board, Editor-at-large, US
The Financial Times

Regardless of your views on cryptocurrencies, any modern practitioner of finance should be familiar with distributed ledger technology (DLT). Professor Alex Lipton and Adrien Treccani have written a fascinating tractatus that answers fundamental questions about the nature of money. They effectively demonstrate the intrinsic value offered by DLT, as the solution to many of the problems that have plagued financial systems for centuries. If there is one book that everyone should read on DLT, this is the one.

Dr. Marcos Lopez de Prado, Global Head — Quantitative Research and Development,
Strategy & Planning, Abu Dhabi Investment Authority

We are just at the tip of the iceberg when it comes to the disruptive potential of DLT. Trillions of dollars of value will be created over the coming decades as entrepreneurs find ways to capitalize on the technology to democratize finance and wealth management. This comprehensive book from Alex Lipton and Adrien Treccani provides a powerful place to start exploring the DLT opportunity as well as a definitive reference to those already active in the space. The book is unique in its balanced and well-rounded approach.

Francisco Fernandez, Avaloq Founder and serial entrepreneur

A very well-researched and rounded tour d'horizon of the brave new world of Blockchain, crypto and the technology's impact on how we will transfer ownership and value in the future. Alex and Adrien strike the right balance between academic rigor and practical relevance. A highly recommended read for both industry newcomers as well as veterans.

Dr. Mathias Imbach, Co-Founder & Group CEO, Sygnum

This book provides comprehensive coverage of the cryptocurrency ecosystem, with lucid explanations of the underlying math, technology, and economics. Technical material is seamlessly integrated with the authors' broad perspective and deep understanding of the past and future of blockchains, digital currencies, and decentralized finance. The book is an essential resource for practitioners and academics.

Prof. Paul Glasserman, Jack R. Anderson Professor of Business, Columbia University

Blockchain and Distributed Ledgers is a wonderful book. It is at once what the Russians call a brick, for its size, and what the English call a gem, for its clarity. And it is timely.

The astronomer Carl Sagan says: "We live in a society exquisitely dependent on science and technology, in which hardly anyone knows anything about science and technology."

It is this problem Lipton and Treccani are attempting to confront. A comfort, then, that some of our leading mathematicians are among our more discerning philosophers too.

Edward Price, Adjunct Instructor, New York University (NYU)
Interdisciplinary Journal of Economics and Business Law, 2021, Volume 10, Issue 2,
pp. 141–143

Blockchain and Distributed Ledgers

Mathematics, Technology, and Economics

Blockchain
and
Distributed
Ledgers

Mathematics, Technology, and Economics

Alexander Lipton

Sila Money, USA & Hebrew University of Jerusalem, Israel

Adrien Treccani

METACO, Switzerland

World Scientific

NEW JERSEY · LONDON · SINGAPORE · BEIJING · SHANGHAI · HONG KONG · TAIPEI · CHENNAI · TOKYO

Published by

World Scientific Publishing Co. Pte. Ltd.

5 Toh Tuck Link, Singapore 596224

USA office: 27 Warren Street, Suite 401-402, Hackensack, NJ 07601

UK office: 57 Shelton Street, Covent Garden, London WC2H 9HE

Library of Congress Cataloging-in-Publication Data

Names: Lipton, Alexander, author. | Treccani, Adrien, author.
Title: Blockchain and distributed ledgers : mathematics, technology, and economics /
    Alexander Lipton, Adrien Treccani.
Description: New Jersey : World Scientific, [2022] | Includes bibliographical references and index.
Identifiers: LCCN 2021015351 | ISBN 9789811221514 (hardcover) |
    ISBN 9789811221521 (paperback) | ISBN 9789811221538 (ebook) |
    ISBN 9789811221545 (ebook other)
Subjects: LCSH: Blockchains (Databases) | Blockchains (Databases)--Industrial applications. |
    Finance--Data processing. | Cryptocurrencies.
Classification: LCC QA76.9.B56 L57 2022 | DDC 005.74--dc23
LC record available at https://lccn.loc.gov/2021015351

British Library Cataloguing-in-Publication Data
A catalogue record for this book is available from the British Library.

For any available supplementary material, please visit
https://www.worldscientific.com/worldscibooks/10.1142/11857#t=suppl

Desk Editor: Jiang Yulin

Typeset by Diacritech Technologies Pvt. Ltd.
Chennai - 600106, India

Printed in Singapore

I dedicate this book to Marsha, my love, soulmate, and lodestar, and our fabulous daughter Rachel for all her love and support.

Alexander Lipton

I dedicate this book to my love and guide Jessica and our cherished son Henri.

Adrien Treccani

"The madness of passion soon passes, and what remains are chains and common sense that tells us that these chains are unbreakable."

A. N. Ostrovsky, Without a Dowry, A drama in four acts.

About the Authors

Alexander Lipton is Co-Founder and Chief Information Officer of Sila, Partner at Numeraire, Visiting Professor and Dean's Fellow at the Hebrew University of Jerusalem, and Connection Science Fellow at MIT. Alex is a board member of Sila and an advisory board member of several fintech companies worldwide. In 2006–2016, Alex was Co-Head of the Global Quantitative Group and Quantitative Solutions Executive at Bank of America. Earlier, he was a senior manager at Citadel, Credit Suisse, Deutsche Bank, and Bankers Trust. At the same time, Alex held visiting professorships at EPFL, NYU, Oxford University, Imperial College, and the University of Illinois. Before becoming a banker, Alex was a Full Professor of Mathematics at the University of Illinois and a Consultant at Los Alamos National Laboratory. In 2000 Alex was awarded the Inaugural Quant of the Year Award and in 2021 the Buy-side Quant of the Year Award by *Risk Magazine*. Alex authored/edited 10 other books and more than a hundred scientific papers. Alex is an Associate Editor of several journals, including *Finance and Stochastics*, *Journal of FinTech*, *International Journal of Theoretical and Applied Finance*, and *Quantitative Finance*. He is a frequent keynote speaker at Quantitative Finance and FinTech conferences and forums worldwide.

Founder and CEO of METACO, a leading provider of security infrastructure for digital assets, **Adrien Treccani** is a software engineer specialized in high performance computing and financial engineering. He has been an active member of the fintech community since 2012 and advised numerous banks and financial institutions globally on distributed ledger technology, cryptocurrencies and decentralized finance. Adrien lectures at University of Lausanne and Ecole Polytechnique Fédérale de Lausanne and has published in top peer-reviewed journals including *Management Science* and the *Journal of Financial Econometrics*.

Adrien holds a Bachelor degree in computer science and a Master degree in financial engineering at the Ecole Polytechnique Fédérale de Lausanne. He obtained a Ph.D. in mathematical finance at the Swiss Finance Institute and completed a post doctorate in high performance computing at University of Zürich. He worked in the hedge fund industry as a quantitative analyst before founding METACO in 2015.

Contents

List of Figures

List of Tables

List of Abbreviations

AD	Anderson–Darling (test)
AES	Advanced Encryption Standard
AIC	Akaike Information Criterion
AML	Anti-Money Laundering
AMM	Automated Market Maker
AS	Autonomous System
ASIC	Application-Specific Integrated Circuit
BaaS	Blockchain as a Service
BB	Bulletin Board
BE	Big-Endian
BFT	Byzantine Fault Tolerance
BGP	Byzantine Generals Problem
BIC	Bayesian Information Criterion
BIP	Bitcoin Improvement Proposal
BIS	Bank of International Settlements
BTC	Bitcoin
CaaS	Consensus as a Service
CBDC	Central Bank Digital Currency
CCAR	Comprehensive Capital Analysis and Review
CCP	Central Clearing Counterparty
cdf	Cumulative Density Function
CDO	Collateralized Debt Obligation
CKD	Child Key Derivation
CLS	Continuous Linked Settlement
CME	Chicago Mercantile Exchange
CMS	Centralized Mixing Service
CSD	Central Securities Depository
CSPRNG	Cryptographically Secure Pseudo Random Number Generator

DAC	Decentralized Autonomous Corporation
DAG	Direct Acyclic Graph
DAO	Distributed Autonomous Organization
DDO	DID Document Object
DeFi	Decentralized Finance
DER	Distinguished Encoding Rules
DES	Data Encryption Standard
DEX	Decentralized Exchange
DH	Diffie-Hellman (key exchange)
DID	Digital Identity
DIF	Decentralized Identity Foundation
DLS	Distributed Ledger System
DLT	Distributed Ledger Technology
DoS	Denial of Service
DPKI	Decentralized Public Key Infrastructure
DPoS	Delegated Proof-of-Stake
DRE	Direct-Recording Electronic (voting machine)
DSA	Digital Signature Algorithm
DTC	Digital Trade Coin
E2E	End-To-End
ECB	European Central Bank
ECC	Elliptic Curve Cryptography
ECDH	Elliptic Curve Diffie–Hellman (key exchange)
ECDSA	Elliptic Curve Digital Signature Algorithm
EDA	Encryption-Decryption Algorithm
EDTC	Environmental Digital Trade Coin
EHR	Electronic Health Record
ELB	Economic Lower Bound
EOA	Externally Owned Account
ERC	Ethereum Request for Comments
ETH	Ether
EVM	Electronic Voting Machine
FBA	Federated Byzantine Agreement
FDIC	Federal Deposit Insurance Corporation
FFC	Finite Field Cryptography
FFT	Fast Fourier Transform
FRC	Finite Ring Cryptography

GCD	Greatest Common Divisor
GCHQ	Government Communications Headquarters
GFC	Global Financial Crisis
GH	Generalized Hyperbolic (distribution)
GPM	Green Programmable Money
HMLR	Her Majesty's Land Registry
HQIC	Hannan–Quinn Information Criterion
IAM	Identity and Access Management
IC	Interactive Consistency
ICO	Initial Coin Offering
IF	Information Flow
IoT	Internet of Things
IP	Internet Protocol
IPO	Initial Public Offering
ISP	Internet Service Provider
IT	Information Technology
KGB	Committee for State Security
KS	Kolmogorov–Smirnov (test)
KYC	Know Your Customer
KYT	Know Your Transaction
LC	Letter of Credit
LCG	Linear Congruential Generator
LE	Little-Endian (format)
LFSR	Linear Feedback Shift Register
Libor	London Interbank Offered Rate
LIFO	Last-In-First-Out
LPoS	Leased Proof-of-Stake
M2M	Machine-to-Machine (payment)
MPC	Multi-Party Computation
MTT	Maria Theresa Thaler
NB	Narrow Bank
NBS	National Bureau of Standards
NFT	Non-Fungible Token
NIRP	Negative Interest-Rate Policy
NIST	National Institute of Standards and Technology
NSA	National Security Agency
OTC	Over-The-Counter
OTP	One-Time Pad

P2P	Peer-To-Peer
P2PK	Pay to PubKey
P2PKH	Pay to PubKey Hash
P2SH	Pay to Script Hash
PBFT	Practical Byzantine Fault Tolerance
PBKDF2	Password-Based Key Derivation Function 2
PDF	Probability Density Function
PKI	Public-Key Infrastructure
PLB	Physical Lower Bound
PM	Programmable Money
PoA	Proof-of-Authority
PoB	Proof-of-Burn
PoC	Proof-of-Capacity or Proof-of-Concept
PoS	Proof-of-Stake
PoW	Proof-of-Work
PSCM	Pharmaceutical Supply Chain Management
QE	Quantitative Easing
QTM	Quantity Theory of Money
RE	Real Estate
RMSE	Root Mean Square Error
ROHF	Random Oracle Hash Function
RSA	Rivest-Shamir-Adleman
RV	Random Variable
S&L	Savings and Loan
S2F	Stock-To-Flow (ratio)
SC	Smart Contract
SEC	Securities and Exchange Commission
SegWit	Segregated Witness (method)
SEIR	Susceptible-Exposed-Infected-Removed
SHA	Secure Hash Algorithm
SIC	Schwarz Information Criterion
SIR	Susceptible-Infected-Removed
SIS	Susceptible-Infected-Susceptible
SOV	Store of Value
SPV	Simplified Payment Verification
TCP	Transmission Control Protocol
TpS	Transactions per Second

TUSD	TrueUSD (coin)
TWh	TeraWatt-hours (of electricity)
TXID	Transaction ID
UNL	Unique Nodes List
UR	Universal Resolver
USC	Utility Settlement Coin
USDC	USD Coin
USDT	Tether
UTXO	Unspent Transaction Output
VAR	Value-at-Risk
VASP	Virtual Asset Service Provider
VC	Venture Capital
W3C	World Wide Web Consortium
WEF	Wallet Export Format
WIF	Wallet Import Format
WTO	World Trade Organization
ZK	Zero-Knowledge
ZKP	ZK Proof
ZKPoK	ZKP of Knowledge
zk-SNARK	ZK Succinct Non-Interactive Argument of Knowledge
ZLB	Zero Lower Bound

Preface

This book concentrates on distributed ledger technology (DLT) and its potential impact on society. This technology, which became extremely popular over the last decade, allows solving many complicated problems arising in economics, banking, and finance, industry, trade, and many other fields. DLT develops new mechanisms for distributed consensus, using advanced tools from cryptography, game theory, economics, finance, scientific computing, etc. It offers an optimal and elegant solution in many situations, provided that it can overcome some of its inherent limitations and is used appropriately. However, its recent applications have been shrouded in misconceptions because DLT is often equated with cryptocurrencies such as Bitcoin, Ethereum, Ripple, and their numerous extensions.

A recent cryptocurrency bubble and its subsequent burst created a charged atmosphere, which is not particularly conducive to a balanced and well-reasoned discourse. Our book provides a much needed impartial assessment of the situation rooted in scientific reasoning rather than speculation, enthusiastic naivete, and personal attacks on one's opponents. As J.J. Rousseau put it in his Confessions: "I, therefore, have in the number of facts and in their species all what it takes to make my narrations interesting. Perhaps despite this, they will not be, but it will not be the fault of the subject, it will be the writer."

We cover several important topics: (a) the modern payment and banking system and reasons for its substantial overhaul; (b) theoretical foundations of decentralized cryptocurrencies, including game theory aspects and Byzantine fault-tolerant consensus; (c) the foundations of cryptography and its applications to cryptocurrencies; (d) the cryptocurrency universe, including Bitcoin, Ethereum, Ripple, and some other protocols and coins; (e) the role of decentralized finance and stablecoins in the financial ecosystem of the future; (f) potential financial and non-financial applications of decentralized ledgers, including their pros and cons.

We dedicate much attention to the operational features of the Bitcoin trading platform, including storage, mining, wallets, and similar concepts. In addition to Bitcoin, we cover alternative platforms, including Ethereum, Ripple, Stellar, Zcash, and others, and smart contracts. This book aims to offer a detailed and self-contained introduction to the founding principles behind DLT accessible to a well-educated but not necessarily mathematically-oriented audience. The intended readership includes master's degree and Ph.D. students from quantitatively-oriented fields such as mathematics, computer science, finance, economics, banking, supply chain management, and the likes; as well as academics and industrial researchers, working in the fields of computer science, finance, banking, and others; as well as anyone with a suitable background wishing to understand the subject for the first time.

The book seeks to balance the theory with practice and liberally use computational approaches. Each chapter includes the presentation of a particular topic, including its economic and mathematical foundations. It discusses the theoretical and practical aspects (some theoretical results are presented in detail, others are mentioned or referenced) and contains numerical results, a list of exercises, and references to further reading and notes. Starting from the basic ideas, we guide the reader by developing and building up a distributed ledger suitable for their area of interest. The book aims to equip the reader with the ability to participate in the crypto economy meaningfully.

While strong mathematical skills are not required, readers should learn necessary background materials from the book itself. We clearly and accessibly explain and articulate rather sophisticated ideas underpinning DLT, so that it is accessible to anyone with a modicum of understanding of computer science, mathematics, and economics. The book is self-contained and provides all the necessary theoretical background for the reader to understand how DLT operates in both theory and practice and, if the need occurs, build a simple distributed ledger from scratch. It can serve as a primary textbook for a course on DLT and crypto-economics and a supplementary text for courses on economics, finance, cryptography, and others.

As mentioned above, DLT is currently an intense area of research by representatives of various disciplines and a hot topic of popular discourse (mostly neither balanced nor impartial). Many practitioners and academics are investigating and advancing it further in a variety of directions. However, the number of books that are solely devoted to DLT and cryptocurrencies is relatively limited. Moreover, several of these are partisan pamphlets rather than real books, let alone textbooks, and omit the requisite technical details altogether. Simultaneously, few technically oriented books provide no necessary economic and applied background, leaving readers without an adequate understanding of DLT in its proper context and its practical usage.

Some of the existing books' strongly biased nature is also likely to lead to unbalanced opinions on the subject by making the readers either distributed ledger zealots or their archenemies. Our goal is to offer a generally accessible book that is balanced and fair in its approach, while, at the same time, technical enough not to be superficial. We provide an in-depth discussion of several important topics, such as the pros and cons of distributed consensus, crypto monetary policies, stablecoins, tokenization, and regulatory aspects of crypto-economics, which may appeal to a broad audience. We include exercises and pseudocode to help the reader to master the subject properly. One of our goals is to dispel numerous misconceptions surrounding DLT in general and cryptocurrencies in particular by providing numerous historical examples and allowing one to put the subject in the proper context. Using real online data, we enable readers to understand the fundamental mechanisms behind DLT successes, failures, and future promises.

In the process of writing this book, the authors enjoyed the advice and support of various colleagues, including Marsha Lipton, Alex (Sandy) Pentland, Thomas Hardjono, Irving Wladawsky-Berger, Yaniv Altshuler, David Gershon, Zvi Wiener, Rachel Lipton, Ofira Eliav, Darrell Duffie, Matheus Grasselli, Damir Filipovic, Marcos Lopez de Prado, David Shrier, Shamir Karkal, Angela Angelowska-Wilson, Isaac Hines, Paolo Tasca, Christopher Brummer, Paul Glasserman, Richar Olsen, Richard Senner, Didier Sornette, Tim Swanson, Robert Sams, Ferdinando Maria Ametrano, Johannes Hoehener, Aetienne Sardon,

Christian Schuepbach, Fabian Schär, Mathias Imbach, Manuel Krieger, Gerald Goh, Max Kantelia, Juzar Motiwala, Amrit Kumar, Alexander Eydeland, Nicolas Dorier, Giulia Traverso, and many others. We gratefully acknowledge WSPC's Max Phua, Zvi Ruder, Yubing Zhai, and Yulin Jiang for all their help.

This book contains sample addresses and secret keys. Never send bitcoins to or import any sample keys. If you do so, your money might disappear forever.

The views expressed in this book are the authors' personal views. They do not represent the views of their respective companies, Sila and Metaco. Authors' opinions should not form the basis for making any investment decisions or engaging in investment transactions. Data used in the book is deemed reliable but is not guaranteed accurate by the authors. Readers are responsible for verifying the accuracy of all information they wish to use.

Alexander Lipton, Chicago
Adrien Treccani, Lausanne

Background

It is because every individual knows little and, in particular, because we rarely know which of us knows best that we trust the independent and competitive efforts of many to induce the emergence of what we shall want when we see it.

Friedrich August v. Hayek, The Constitution of Liberty

1.1 Introduction

The Global Financial Crisis (GFC) has demonstrated that the existing banking and payment system, while still working, is outdated and struggling to support the continually changing requirements of the modern world; see Lipton et al. (2019). It would be an understatement to say that the GFC turned into a wasted opportunity to reorganize the world financial ecosystem. Too-big-to-fail banks have dramatically increased (rather than decreased!) in size, disproportionally amplifying their banking business share, while the number of banking institutions has significantly reduced. For instance, the size of JPMorgan's balance sheet is presently nearly twice as large as it was at the end of 2006, at the onset of the crisis; similarly, the balance sheets of China's four systemically important banks have more than tripled over the same period. The clearing of many over-the-counter derivatives was mandatorily transferred to central clearing counterparties (CCPs), which have become potential points of failure for the system as a whole. This situation is further exacerbated by the high level of interconnectedness of CCPs, due to numerous general clearing members they have in common.

Central banks engaged, *nolens volens*, in the massive Quantitative Easing (QE) efforts and embraced the fractional reserve modus operandi. At the same time, commercial banks reassessed their priorities and are switching to the Narrow Bank (NB) model, partly by choice (to satisfy their risk preferences), and partly by necessity (not being able to lend enough to the real economy). As central banks employed short-term interest rates policy tools for achieving their macroeconomic goals, the short-term interest rates moved close to zero for much of the past decade, reflecting the effects of QE, low headline inflation, and low productivity growth; see, e.g., Rudebusch (2018).

Due to the COVID-19 pandemic and associated lockdowns, the current worldwide economic recession forces central banks to push the short interest rates further into extremely low or outright negative territory. Given the unprecedented level of unemployment, the economic downturn is likely to pave the way for further use of the central banks' unconventional monetary policy, resulting in low or negative short-term rates for the foreseeable future; see Lipton (2020).

The frustration of the general public with the status quo is palpable. This frustration manifests itself in various aspects of social and economic life.

This book argues that the seemingly squandered opportunities to reshape the financial system are not all lost. More specifically, we show that, if used deliberately, new technologies, including blockchains and distributed ledgers, can create new business models. New technologies will put pressure on the incumbents. More importantly, they will allow newly formed fintech companies to enter the market in earnest, thus providing considerable benefits to the general public.

One can expect a great deal of innovation from DLT. Assuming that newcomers understand banking and its role in society, and regulators allow competition between various business banking models, we hope to see several hotly contested races. Specifically, we anticipate contests between fractional reserve banks vs. narrow banks, digital cash vs. physical cash, fiat currencies vs. asset-backed cryptocurrencies, and, most importantly, centralized payment systems vs. distributed payment systems. The outcome of these races is going to reshape the entire future financial ecosystem. In a few years, it might change beyond recognition. Potentially, DLT can have critical geopolitical implications and be a significant weapon in the inevitable struggle for supremacy between different central bank digital currencies (CBDCs).

Besides, DLT will reshape other, less financially oriented aspects of the economy, such as accounting, biomedical research, supply chain management, and many others.

1.2 Distributed ledgers in a nutshell

Simply put, blockchain is a shared, distributed ledger designed to record transactions in a business network and track resulting changes in assets ownership; see, e.g., Gupta (2017). The corresponding assets can be tangible (money, shares, real estate) and intangible (intellectual property, including patents, trademarks, copyrights, goodwill, and brand recognition). In theory, ownership of anything of value can be tracked and traded on a blockchain network, reducing risks by increasing business interoperability, and cutting costs for all involved by removing intermediaries. Besides, it might be possible to use blockchain to establish individuals' and businesses' digital identities and radically reorganize business *modus operandi*.

As shown in Figure 1.1, we can think about three complementary ways of organizing information: (a) centralized, (b) decentralized, (c) distributed. At present, a centralized system, also known as the hub and spoke model, is underpinning most industries, for example, banking in a single country, with the hub being the central bank, and spokes being individual commercial banks. A decentralized system with several hubs and spokes is typical for multi-country arrangements, such as cross-border banking. Finally, a distributed system characterizes direct or peer-to-peer (P2P) business organization (currently a rare instance) and potentially can be viewed as the most robust of the three.

Given the recent developments in DLT, we can think about alternative ways of organizing business activities in general. In Figure 1.2, we show two possibilities: (a) a current

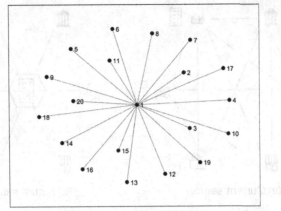

(a) Centralized network.

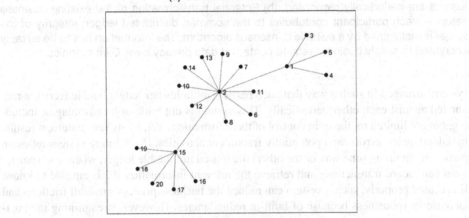

(b) Decentralized network.

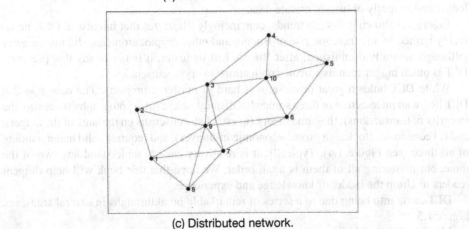

(c) Distributed network.

Figure 1.1 (a) Centralized; (b) decentralized; and (c) distributed networks. Own graphics.

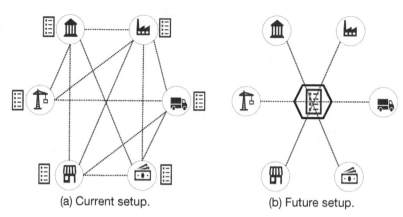

(a) Current setup.   (b) Future setup.

Figure 1.2 (a) Current business setup — each participant maintains their own ledger; ledgers are periodically reconciled; (b) Potential transformation of the existing business setup — each participant contributes to the common distributed ledger; integrity of the ledger is maintained by a suitable consensus algorithm. The information has to be suitably encrypted to satisfy business requirements and data privacy laws. Own graphics.

system, arranged in such a way that each participant holds her ledger, and ledgers are reconciled against each other periodically. This system is not without its advantages, including, but not limited to, the tight control of the information. Yet, by its very nature, it results in redundancies, errors, and potentially fraudulent activities. (b) A future system relies on parties maintaining (one way or the other) the shared immutable ledger, where each participant can record transactions and retrieve the relevant information she is entitled to know. If executed properly, such a system can reduce the business process's overall frictions and increase its robustness because of built-in redundancies. However, streamlining the business process comes with substantial costs, required to maintain consensus on the shared ledger and properly obfuscate private data.

Figure 1.3 showing Google trends, convincingly illustrates that interest in DLT, undeniably ignited by the meteoric rise of Bitcoin and other cryptocurrencies, did not go away (although naturally diminished) after Bitcoin lost its luster. It is fair to say that presently DLT is much bigger than its narrow applications to cryptocurrencies.

While DLT holds a great promise, it is hard to master it properly. The reason is that DLT lies at an intersection of three somewhat disjoint fields: (a) cryptography (to ensure the integrity of transactions); (b) game theory (to establish consensus on the state of the ledger); and (c) economics (to design proper economic initiatives), and requires solid understanding of all three (see Figure 1.4). Typically, it is relatively easy to understand any two of the three, but mastering all of them is a tall order. We hope that this book will help diligent readers to climb the ladder of knowledge and experience.

DLT came into being due to a series of remarkable breakthroughs in several areas, see Figure 1.5.

Many authors, including the present ones, argue that DLT is a fundamental innovation, which can be viewed on par with some other fundamental discoveries closing conceptual

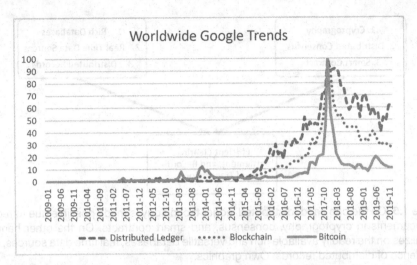

Figure 1.3 Google trends reflecting interest in distributed ledgers, blockchains, and Bitcoin. Own graphics. Source: google.com.

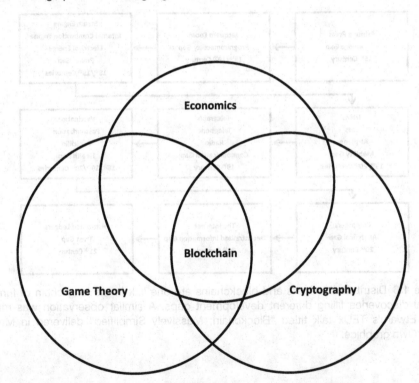

Figure 1.4 It is hard to master distributed ledgers and blockchains because the underlying concepts come from such disparate fields as economics, cryptography, and game theory. Own graphics.

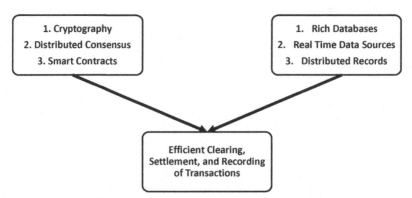

Figure 1.5 On the one hand, distributed ledger technology is made possible due to recent advancements in cryptography, consensus, and smart contracts. On the other hand, it capitalizes on the readily available rich and versatile databases, real time data sources, and prevalence of distributed records. Own graphics.

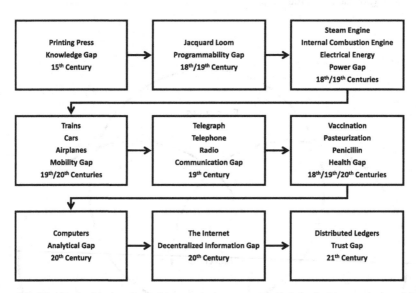

Figure 1.6 Distributed ledgers and blockchains are one link in the long chain of fundamental discoveries filling different development gaps. A similar observation was made in R. Etwaru's TEDx talk titled "Blockchain; Massively Simplified" delivered in March 2017. Own graphics.

gaps in human progress. Figure 1.6 shows that DLT closes one of such gaps, specifically, the trust gap in the modern Internet-oriented economy.

The development of cryptocurrencies is one of the most critical areas where DLT plays a central role. Such currencies can be viewed as emerging competitors of the existing fiat currencies. Many leading authorities on the subject, including Hayek, viewed the developments in the area of monetary innovation as highly beneficial; see Hayek (1943).

1.3 A retrospective view

The idea of transferring valuable objects and property rights via chains using some form of consensus is nothing new. For example, genealogical trees of royal and aristocratic families can be thought of as (block)chains, since they transfer birth rights according to a more or less strict set of rules from one ruler to the next; see Lipton (2018). Figure 1.7 shows two representative family trees for the houses of Capet-Valois and Hapsburg.

One needs to have an excellent command of genealogy and heraldry to correctly interpret the unbelievably detailed information encoded in these family trees. However, one feature is beyond dispute — much like with Bitcoin transactions (see below) — the ownership is transferred from one owner to the next according to the set of predetermined rules. These transfers have to be accepted by a group of peers who enforce the rules and validated by distributed parties, including the country's authorities, where the transfer takes place, and other royal courts.

What happens when things go wrong? If parties cannot agree upon a transaction, a hard fork might occur, much like it does in blockchains, when various interested parties cannot agree on what to do as a group (see below). The Hundred Years' War (1337–1453) is a case in point. It was a conflict between the House of Plantagenet, rulers of the Kingdom of England, and the House of Valois, rulers of the Kingdom of France. The prize was control of the Kingdom of France. It is a typical example of the hard fork — in France, the House of Valois was considered legitimate, but in England (and subsequently Britain) — the House of Plantagenet and its offshoots.[1] Another interesting example is the War of the Austrian Succession (1740–1748) — a conflict that involved most of Europe's powers to settle the question of Pragmatic Sanction and decide Maria Theresa's succession to the realms of the House of Hapsburg. As before, the consensus was achieved by the force of arms.

The transition of power in a democracy is another example of a chain-like arrangement, with consensus achieved by tallying the votes. Unfortunately, it is not always smooth; failure to reach consensus results in social upheavals.

To move from politics to economics, we present another example of blockchains — land registry title deeds in England, which have been in continuous use since the Middle Ages. According to the UK government: "Title deeds are paper documents showing the chain of ownership for land and property. They can include: conveyances, contracts for sale, wills, mortgages and leases."

In many instances, these chains are very long, since titles can be traced to medieval times. It is clear that titles are blockchains; instead of miners, succession is verified by notaries; and held in the central repository. Titles are meaningful candidates for being treated on a distributed ledger; we discuss further details in Chapter 11.

Of course, one can argue that we are talking about remote historical events, which are not relevant to our narrative. Here is a much more recent and more appropriate example.

[1]George III stopped using the title of "King of France" on December 31, 1800. An excellent way to start a new century!

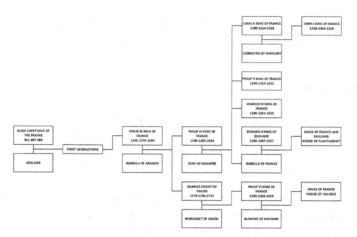

(a) Capet-Valous family tree.

(b) Hapsburg family tree.

Figure 1.7 Historical examples of blockchains. (a) Capet-Valous family tree. (b) Partly imaginary Hapsburg family tree depicted by Albrecht Durer as a Triumphal Arch. Not only this tree is shown as a chain of blocks, but it is also distributed! Sources: (a) own calculations, much helped by reading Druon (2014); (b) Wikipedia.

On July 26, 2016, in his post "Onward from the Hard Fork," Vitalik Buterin promulgated (much like various kings and emperors of the old in their edicts) that theft of funds initially allocated to a distributed autonomous organization (DAO) cannot be tolerated.[2] Accordingly, the recent history of transactions on the Ethereum blockchain ought to be rewritten: "The foundation has committed to support the community consensus on the admittedly difficult hard fork decision. Seeing the results of various metrics, including carbonvote, dapp and ecosystem infrastructure adoption, this means that we will focus our resources and attention on the chain which is now called ETH (i.e. the fork chain). That said, we recognize that the Ethereum code can be used to instantiate other blockchains with the same consensus rules, including testnets, consortium and private chains, clones and spinoffs, and have never been opposed to such instantiations."

As a result, a hard fork occurred, and two versions of the original Ethereum blockchain emerged — Ethereum and Ethereum Classic. So much for blockchain immutability!

1.4 A perspective view

As mentioned earlier, a distributed ledger, shared among participants, is used to record transactions and track assets in a highly secure manner. DLT reduces transaction costs, including, but not limited to, the cost of transaction verification, and enhances networking opportunities of a business ecosystem; see Catalini and Gans (2016). Its main benefits are as follows: increase of transactional trustworthiness, transparency, and reliability; improvement of data quality and accuracy; reduction of fraud and cybercrime.

The introduction of natively digital assets, achieved by asset tokenization on a distributed ledger, simplifies and automates business processes by allowing participants to access the history of past transactions and establish current ownership of the corresponding tokens. The execution of a smart contract can enact a change of ownership. In theory, it will enable to eliminate a centralized intermediary and operate a marketplace or a supply chain in a decentralized fashion. However, in practice, it is easier said than done. Profitable application of DLT requires a phased approach. First, it is necessary to standardize and digitize the internal business process by issuing digital tokens and reorganizing settlement and reporting. Second, businesses can start sharing digital tokens across the entire ecosystem.

Financial services constitute one of the most prominent areas where DLT can be a game-changer. Janet Yellen, who served as the Federal Reserve Chair from 2014 to 2018, articulated her thoughts on the subject in Yellen (2017) as follows: "[Blockchain] is a very important, new technology that could have implications for the way in which transactions are handled throughout the financial system. We're looking at it in terms of its promise in some of the technologies we use ourselves and many financial institutions are looking at it. It could make a big difference to the way in which transactions are cleared and settled in the global economy."

---

[2]Strictly speaking, the very possibility of the so-called theft was caused by a glitch in the code, so that the removal of funds from the DAO should not be viewed as a theft in the first place.

1.5 Book's structure

The remainder of the book proceeds as follows. In Chapter 2, we discuss the modern financial system and identify its numerous pain points. We argue that using DLT can address some of these issues. Chapter 3 covers a nontechnical introduction to distributed ledgers and cryptocurrencies. In Chapter 4, which is technical and rather mathematically involved, we equip the reader with a handy cryptographic toolkit used throughout the rest of the book. Among other topics, we discuss elliptic curve cryptography and hash functions. In the short but crowded annals of cryptocurrency history, the so-called Big Three crypto protocols — Bitcoin, Ethereum, and Ripple — have been in the leading position right from their inception. Hence, we dedicate Chapters 5, 6, 7 to their detailed analysis. We compare and contrast the three and articulate their strengths and weaknesses. In Chapter 5, we describe Bitcoin protocol in considerable detail by covering transaction verification, block formation, proof-of-work consensus, and other relevant topics. In Chapter 6, we discuss Ethereum protocol, which can be viewed as a consensus as a service provider and a self-styled distributed world computer. In Chapter 7, we look at non-crypto Ripple protocol — a SWIFT-based system for executing cross-border money transfers. Chapter 8 covers very important practical applications — CBDCs, digital trade coins (DTCs) and stablecoins. In Chapter 9, we describe wallets and industrial-strength key management. Chapter 10 covers cryptocurrencies from the mathematical finance standpoint. We discuss the Big Three price formation, their temporal dynamics, Bitcoin options, and other related topics. Chapter 11 is dedicated to potential applications of DLT in such diverse areas as audit and accounting, biomedical research and healthcare, data and identity management, distributed finance and programmable money, global payments, government services, supply chain management, tokenization of real assets, and, finally, trade execution, clearance, and settlement. In Chapter 12, we cover some of the hottest topics in DLT research, numerous problems that prevent its wide adoption, and discuss its future.

1.6 Summary

While blockchains and distributed ledgers are not new concepts, modern technology gives them a new lease of life. In addition to financial applications, DLT can have numerous applications in the real economy. To start with, DLT opens new possibilities for making conventional banking and trading activities less expensive and more efficient by removing unnecessary frictions. Among many financial applications of DLT, it is not entirely clear which ones should be handled first. Exchanges, payments, trade finance, rehypothecation, syndicated loans, and other similar areas, where frictions are relatively high, are attractive candidates. CBDCs, DTCs, and stablecoins are another auspicious venue. Moreover, if used with skill, knowledge, and ambition, it can restructure the whole financial system on new principles. We emphasize that achieving this goal requires overcoming not only technical but also political obstacles.

There are currently numerous DLT proof-of-concept pilot projects in such diverse areas as accounting, healthcare, supply chain management, mobility, and smart cities, to mention but a few. Yet, many applications of DLT appear to be misguided. On occasion, they are driven by a desire to apply these tools for their own sake rather than because of the significant added value. In other cases, they are caused by a failure to appreciate that the current systems may not be in their current state because of technological reasons, but rather because of business and other considerations.

The future of DLT technology will be bright, provided that it is deployed in the areas where it is genuinely needed, and its implementation is done correctly. Reducing the hype and increasing the useful output is the way to improve this exciting technology's overall prospects.

1.7 Further reading

Satoshi Nakamoto' seminal paper, Nakamoto (2008), ignited the world's fascination with blockchains and, by extension, distributed ledgers. This paper expands ideas of Wei (1998), and uses electronic document time-stamping techniques developed by Haber and Stornetta (1990, 1997), and proof-of-work ideas proposed by Back (2002).

Several blockchain books, covering different aspects of this fascinating field (some better than others) might be of interest to the reader:

(1) Ammous (2018) analyzes Bitcoin's history, describes the economic properties that have allowed its rapid expansion, and discusses Bitcoin's likely economic, political, and social implications.

(2) Antonopoulos (2017) presents a broad introduction to the Bitcoin protocol and its underlying blockchain.

(3) Antonopoulos and Wood (2018) provide useful information about building smart contracts and DApps on Ethereum and other virtual-machine blockchains.

(4) Bheemaiah (2017) critically surveys fractional-reserve banking, examines the emerging blockchain technologies, and shows how they can challenge the financial sector's status quo.

(5) Berentsen and Schär (2017); Schär and Berentsen (2020) provide an introduction to cryptocurrencies and blockchain technology and a guide for practitioners and students.

(6) Burniske and Tatar (2017) produce an innovative investor's guide to an entirely new cryptoasset class.

(7) De Filippi and Wright (2018) argue that blockchain technology cannot be harnessed productively without new rules and new legal thinking approaches.

(8) Girasa (2018) identifies the key actors in digital technology and articulates its benefits and risks.

(9) Gupta (2017) explains what blockchain is, how it works, how it can enhance business networks, and how to get started building on blockchain today.

(10) Lewis (2018) presents a non-technical guide covering the history and basics of crypto-currencies and blockchains.

(11) Narayanan et al. (2016) provide a useful introduction to the new technologies, which underpin digital currencies, with an emphasis on Bitcoin.

(12) Pentland et al. (2020) lay out a vision of how to reforge our societies' social contract and how institutions, systems, infrastructure, and the law should change in support of this new order.

(13) Popper (2015) offers a non-technical but entertaining account of the Bitcoin protocol and the corresponding cryptocurrency, BTC.

(14) Shrier (2020) provides a non-technical explanation of what blockchain is and shows how it will transform the way people work and live.

(15) Swan et al. (2019) produce a practical introduction to the economic models based on DLT; these models are characterized by three factors: open platform business models, cryptotoken money supplies, and Initial Coin Offerings.

(16) Tapscott and Tapscott (2016) present a non-technical introduction to blockchain technology, and convincingly argue that it will be powering our future.

(17) Vigna and Casey (2016, 2019) provide a current snapshot of blockchain technology developments and review its challenges, risks, and opportunities.

Blockchain surveys (of variable quality) are currently a popular genre as well; here is a representative selection of useful surveys:

(1) Alharby et al. (2018) perform a systematic mapping study of all peer-reviewed technology-oriented research in smart contracts; their objectives are twofold: to provide a survey of the scientific literature; to identify academic research trends and uptake.

(2) Ali et al. (2019) cover the evolution of blockchain-based systems, bringing a renaissance in the existing, mostly centralized, space of network applications; they re-imagine these applications with blockchain and highlight various common challenges, pitfalls, and shortcomings that can occur.

(3) Al-Jaroodi and Mohamed (2019) review several industrial application domains where the usage of blockchain technologies have been proposed and explore the opportunities, benefits, and challenges of incorporating blockchain in various industrial applications.

(4) Belotti et al. (2019) provide the community with a general presentation of blockchain, going beyond its usage in Bitcoin and surveying a literature selection that emerged in the last few years.

(5) Blandin et al. (2019) examine the global cryptoasset regulatory landscape and expose the reader to technology-enabled financial innovation's operationalization.

(6) Bodó and Giannopoulou (2019) study the concept of decentralization, intending to understand the social, legal, economic forces that produce more or less decentralized techno-social systems.

(7) Cachin and Vukolić (2017) discuss the process of assessing and gaining confidence in the resilience of consensus protocols exposed to faulty and adversarial nodes and advocate following the established practice in cryptography and computer security, relying on public reviews, detailed models, and formal proofs.

(8) Casino et al. (2019) present a systematic literature review of blockchain-based applications, including their current status, classification, and open research topics.

(9) Conti et al. (2018) present a systematic survey covering Bitcoin's security and privacy aspects and give an overview of the Bitcoin system and its major components, along with their functionality and interactions within the system.

(10) de Leon et al. (2017) clarify widespread misconceptions about blockchain technologies' properties and describe challenges and avenues for correct and trustworthy design and implementation of a distributed ledger system (DLS) or technology (DLT).

(11) Garay and Kiayias (2020) systematize knowledge in the landscape of consensus research in the Byzantine failure model, starting with the original formulation in the early 1980s up to the present blockchain-based new class of consensus protocols.

(12) Glaser et al. (2019) show that the platform approach is the dominant strategy for large companies to operate an extensible, digital medium of exchange for products, information, and services.

(13) Härdle et al. (2020) provide insights into the mechanics of cryptocurrencies, describe summary statistics, and focus on potential future research avenues in financial economics.

(14) Iansiti and Lakhani (2017) argue that blockchain, being an open, distributed ledger that records transactions safely, permanently, and very efficiently, could slash the cost of transactions and eliminate intermediaries like lawyers and bankers, transforming the economy.

(15) Khezr et al. (2019) review emerging blockchain-based healthcare technologies and related applications, describe the open research matters in this fast-growing field, and show blockchain technology's potential in revolutionizing the healthcare industry.

(16) Liu (2018) analyzes the basic features and categories of blockchain and delineates their practical applications, aiming to find blockchains' development prospects through analysis of existing applications and technologies.

(17) Morhaim (2019) provides a synthetic sketch of issues raised by blockchains and cryptocurrencies development by presenting them through the link between the technological aspects involving technologies and network structures and the issues raised from applications to implications.

(18) Mulligan et al. (2018) develop a toolkit based on real-world experience of blockchain in various projects across a variety of industries.

(19) Shahaab et al. (2019) reviewed 66 known consensus protocols, classified them into philosophical and architectural categories, and provided their visual representation.

(20) Tasca and Tessone (2017) perform a comparative study across the most widely known blockchain technologies with a bottom-up approach by deconstructing blockchains into their building blocks and building a taxonomy tree to provide a navigation tool across different blockchain architectural configurations.

(21) Tschorsch and Scheuermann (2016) introduce the Bitcoin protocol and its building blocks and explore the design space by discussing existing contributions and results, some of which are applicable far beyond Bitcoin itself.

(22) Wang et al. (2019a) give comprehensive explorations on the cryptographic primitives, including their functionalities and usages in blockchain and their developments.

(23) Wang et al. (2019b) provide a comprehensive survey of the emerging applications of blockchain networks in telecommunications and explain how the consensus mechanisms impact these applications.

(24) Yaga et al. (2019) provide a high-level technical overview of blockchain technology to help readers understand how blockchain technology works.

(25) Yang et al. (2019) address blockchain integration to secure Internet services and identify the critical requirements of developing a decentralized, trustworthy Internet service.

(26) Zheng et al. (2017) provide an overview of blockchain architecture and compare some typical consensus algorithms used in different blockchains, articulate technical challenges, and lay out possible future blockchain trends.

Since cryptocurrencies exist mostly online, a lot of up-to-date information can be found there. In this regard, the following websites are particularly useful:

- https://blockchain.com;
- https://en.bitcoin.it;
- https://coin.dance;
- https://coinmarketcap.com;
- https://etherchain.org;
- https://etherscan.io;
- https://learnmeabitcoin.com;
- https://xrpl.org.

The Global Financial System and its Pain Points

"If you look at mainstream economics there are three things you will not find in a mainstream economic model — Banks, Debt, and Money. How anybody can think they can analyse capital while leaving out Banks, Debt, and Money is a bit to me like an ornithologist trying to work out how a bird flies whilst ignoring that the bird has wings."

Steve Keen

2.1 Introduction

In this chapter, we briefly describe the global financial system and identify its numerous pain points. Subsequently, we show that some of the issues it is struggling with can be alleviated by employing DLT.

The chapter is organized as follows. In Section 2.2, we start our discussion with a very brief discussion of money, since understanding money's evolution is necessary to put cryptocurrencies into a proper context. In Section 2.3, we present a short exposé of banks, which, until very recently, have been the leading creators of money in the modern economy. Since the GFC, money creation has shifted from commercial banks to central banks, whose balance sheets increased dramatically. In Sections 2.4 and 2.5, we cover domestic and cross-border payments and explain that their hierarchical structure causes many frictions, which negatively affect the process of value transfer, and make it both slow and expensive. In Section 2.6, we posit that some of the thornier issues afflicting the global financial system can be solved, at least in theory, by skillfully and deliberately applying DLT. A brief summary is given in Section 2.7.

2.2 Money in retrospective and perspective

2.2.1 The role of money

Money in modern society is multi-faceted. Money is very concrete and abstract at the same time. It is one of the greatest inventions of humankind, on par with writing.[1] Several main attributes of money are beyond dispute:

(1) money is a medium of exchange;

---

[1] As we shall see shortly, initially, the writing was invented by the Sumerians to express economic concepts.

(2) money is a means of payments;

(3) money is a store of value;

(4) money is a unit of account;

(5) money is a perpetual call option on goods and services.

Historically, anything used to discharge tax obligations eventually becomes money. The following opinion of P.H. Wicksteed, see Wicksteed (1910), summarizes the situation best: "Inconvertible paper money had a positive value squarely on its being made acceptable by the government for the payment of taxes."

Given this fact, in a modern, legally compliant economy, money has to be linked to identity one way or the other. Besides, Graziani (2003) and Keen (2001) argue that in a monetary (as opposite to barter) economy:

(1) money has to be represented by a token;

(2) money has to be accepted as a means of final settlement of all transactions terminating all credit and debt relationships between the parties;

(3) money should not grant privileges of seigniorage to any agent making a payment, thus requiring the presence of a bank as a third party to any non-cash transaction.

Okamoto and Ohta (1991) succinctly articulate requirements for electronic money as follows:

(1) Online payment — can be securely used online;

(2) Offline payment — can be securely utilized offline;

(3) Non-reproducibility — cannot be copied and reused;

(4) Anonymity (or pseudonymity?) can be transacted without revealing parties' identities;

(5) Transferability — can be transferred to others;

(6) Divisibility — can be subdivided as needed.

In the next few subsections, we discuss how money was represented over time by various means.

2.2.2 Money as objects

Initially, money was represented by tokens of value such as cowry, electrum, gold, and silver coins; see, e.g., Davies (2010). The beauty of the idea of a physical object representing money is that one can freely exchange it for goods and services without any third party directly involved in the exchange. Thus, the "money object" becomes a material expression of a very abstract idea. The money represented by objects has all the attributes of the abstract money. Of course, the actual material used is hard to obtain, and its content is not diluted. Various objects representing money tokens are shown in Figure 2.1. We briefly discuss them below and try to emphasize their aspects, which can be extended to cryptocurrencies.

In this figure, we show:

(a) A Chinese cowry shell. Shells are an ancient form of commodity money used in various parts of the world. Shells are very durable. They are easy to recognize and handle. Naturally, shells should be impossible (or very hard) to find in the area where they are used as money, although they can be abundant elsewhere. For example, cowries were collected in the Maldives and other East Indian islands, where they are plentiful, and

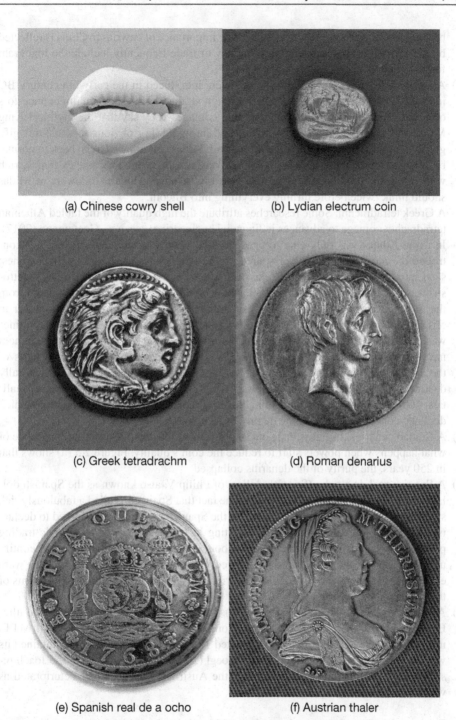

(a) Chinese cowry shell

(b) Lydian electrum coin

(c) Greek tetradrachm

(d) Roman denarius

(e) Spanish real de a ocho

(f) Austrian thaler

Figure 2.1 Money as objects: (a) a Chinese cowry shell; (b) a Lydian electrum coin; (c) a Greek tetradrachm; (d) a Roman denarius; (e) a Spanish real de a ocho; (f) an Austrian thaler. Sources: The Met.

brought to China, where they are scarce. The importance of cowries in China is reflected by the fact the logograms relating to money or trade frequently include the logogram for cowry shell; see, e.g., Goetzmann (2017).

(b) A Lydian electrum coin. Electrum coins were introduced in the early 6th century BC in Lydia. From there, they propagated throughout the world. They are direct ancestors of Greek, Roman, Byzantium, and Islamic coinage. They say that the mythical King Midas was able to turn anything he touched into gold. For obvious reasons, this gift, given to him by Dionysus, turned to be a curse. Midas was allowed to go back to normal by washing the Dionysus gift away in the river Pactolus, which became extremely rich with electrum, subsequently used for the early coinage. Modern followers of Midas should think twice before turning everything into Bitcoin.

(c) A Greek tetradrachm. Some researches attribute the high quality of the fabled Athenian tetradrachm coinage to advances in financial engineering; see, e.g., Goetzmann (2017). In truth, Athens was only able to produce high-quality tetradrachms because it controlled the most abundant silver source on the Peloponnese at Laurium silver mines; see Lipton (2017). In this regard, classical Athens was not dissimilar to today's petrostates, with silver being the ancient oil analog. While the first silver was extracted from the mines as early as 3200 BC, it wasn't until Peisistratus, who came into power in 561 BC, that Athens started to exploit its potential in earnest. Revenue from the mines was used by Themistocles to build a fleet of triremes and make Athens the pre-eminent naval power in Greece; see Ardaillon (1897). However, as Sparta captured parts of the area and the mines became depleted, the tetradrachm coinage's quality naturally declined. No amount of financial engineering could fix these simple physical realities. Tetradrachm was the foundational layer of the Classical Greek civilization, which declined concomitantly with the quality of its coinage.

(d) A Roman denarius. Initially made out of pure silver, the denarius is a cautionary tale of what happens when powers start to reduce the coin's quality. Figure 2.5 (a) shows that in 250 years, the purity of the denarius collapsed.

(e) A Spanish real de a ocho (Spanish dollar) of Philip V, also known as the Spanish dollar. The high quality of the coin is due to the fact that Spain controlled a fabulously rich silver source in Potosí, Peru. Despite that, the Spanish monarchy was forced to declare nine sovereign defaults in 110 years, starting in 1557, because of its myopic trading policies, the expulsion of the Jews and Moors in 1492, and, last but not least, continuous wars in Europe and elsewhere. Moreover, when the Potosí mines in Peru were exhausted, the Spanish dollar quality deteriorated — a story familiar to the students of the Athenian tetradrachm.

(f) An Austrian Maria Theresa thaler (MTT). MTT is a silver bullion coin named after Empress Maria Theresa, shown on the coin. It was first minted in 1741; all MTTs minted after Maria Theresa's death are dated 1780. For centuries MTTs maintained its worldwide status as a trade coin; see Tschoegl (2001). Once again, when the Joachimsthal mines in Bohemia were exhausted, the Austrian thaler's quality deteriorated as well.

Coins made of precious metals, such as gold or silver, rely on their respective metal content and only partially (if at all) on the government enforcement. However, government markings are significant to signal their quality and simplify transactions. Nicholas Oresme associated money with information and certification as in Oresme (1956): "When men first began to trade, or to purchase goods with money, the money had no stamp or image, but a quantity of silver or bronze was exchanged for meat and drink and was measured by weight. And since it was tiresome constantly to resort to the scales and difficult to determine the exact equivalent by weighing, and since the seller could not be certain of the metal offered or of its degree of purity, it was wisely ordained by the sages of that time that pieces of money should be made of a given metal and of definite weight and that they should be stamped with a design, known to everybody, to indicate the quality and the true weight of the coin, so that suspicion should be averted and the value readily recognized."

The so-called trade coins have been known since antiquity and flourished since the 16th century as Spanish pieces of eight and Austrian thalers. They have been traded outside of their respective countries' borders, based on their gold or silver content alone.[2] A modern version of a trade coin is, potentially, central bank digital currency, such as the digital Yuan currently being developed in China.

2.2.3 Paper money

Eventually, money started to be represented by objects without discernable intrinsic value, such as paper money. In Figure 2.2 we show various forms of paper money and banknotes.
In this figure, we show:

(a) The first known paper money was developed in China in the 11th century, after earlier attempts to use iron and copper coins.[3]

(b) The first European paper money was issued by the Stockholms Banco in 1666.[4]

(c) A continental dollar, also known as "Continental" was issued by the Continental Congress after the American Revolutionary War began in 1775. Continental currency depreciated rapidly during the Revolutionary War, not least because the British counterfeited them on a large scale. Benjamin Franklin tried to counter their efforts by developing nature prints using leaves' patterns, making counterfeiting the notes difficult. Regardless, by May of 1781, Continentals became worthless and ceased circulation; see Wright (2008).

(d) A very early British pound issued in 1805 during the Napoleonic wars due to silver and gold shortages.

(e) A Soviet Chervonets. Being fully convertible into gold (at least in theory), the Chervonets kept its value well. It can be viewed as a precursor of stablecoins, discussed in Chapter 8.

(f) A 100 trillion Zimbabwean dollar banknote — a sad outcome of a failed monetary policy. Cryptocurrency designers can learn a lot from futile attempts to issue money

---

[2] In some parts of the world, the Maria Theresa thaler continued to be used as currency long after being demonetized in Austria proper. For instance, during the Second World War, the British minted 18 million thalers in Bombay and used them during their campaign against the Italians in Ethiopia.

[3] For a short period, the Chinese tried to use white deer hide as money, but this effort proved unsuccessful, not least due to lack of white deerskin.

[4] We see that Sweden has a long tradition of financial innovation extending to its current experiments with digital money.

(a) Chinese paper money.   (b) Swedish paper money.

(c) Continental dollar.   (d) Early British pound.

(e) Soviet chervonets.   (f) Zimbabwean dollar.

Figure 2.2 Paper money: (a) The first known paper money developed in China starting in the 7th century; (b) the first European paper money, issued by the Stockholms Banco in 1666; (c) a Continental dollar; (d) a very early British pound; (e) a Soviet chervonets; (f) the 100 trillion Zimbabwean dollar banknote (10^{14} dollars). Sources: Wikipedia.

without any restrains. For example, algorithmically stabilized coins, described in Chapter 8, are not possible in practice.

Smith (1977) summarized the advantages of paper money with his usual flair: "The substitution of paper in the room of gold and silver money replaces a very expensive instrument of commerce with one much less costly, and sometimes equally convenient. Circulation comes to be carried on by a new wheel, which it costs less both to erect and to maintain than the old one."

The value of paper money, or lack thereof, has been a subject to much debate. While very difficult to replicate by a third party, paper money can be issued more or less at will by its respective government. Most states tend not to print excessive amounts of money all at once. However, episodes of unlimited printing of money do occur from time to time.

2.2.4 Anti-counterfeiting measures

Counterfeiting is as old as money itself. It has always been viewed as a significant threat to sovereign power and, more often than not, was punishable by death.[5]

Since the 7th century BC, when the first electrum coins were minted in Lydia from dies engraved with a primitive reversed design, counterfeiting efforts started in earnest. In many cases, these efforts were simple — shaving, or clipping, of coin's edges, to get the precious metal and mixed it with the base metal, thus producing counterfeit coins. To fight against this practice, the Romans started to make serrated edges on the denarius in the 2nd century BC. However, coin sweating, i.e., putting coins in a bag, shaking the bag, and collecting the metal dust, wears coins naturally (provided that it is not overdone) and is very hard to detect, and hence prevent.

In Europe, paper money appeared in the 17th century and became a juicy target for counterfeiters from the start. The manufactures of money put a lot of effort into making their product hard to replicate. They started to use specially prepared paper, flatbed printing plates, and complicated designs, including the so-called nature prints reproducing actual patterns of real leaves. The invention of the geometric lathe capable of engraving complex patterns proved to be very useful. Polymer notes, first introduced in Australia in 1988, proved to be incredibly hard to counterfeit.

However, another level of protection is seldom mentioned in conjunction with counterfeiting, namely an ingenious way they number banknotes. We use as an example the Deutsche Marks and explain in detail how their alphanumeric codes were chosen; see Tlustý and Šulista (2017). We emphasize that the only reason we can reveal this information is that the Deutsche Mark is no longer a legal tender.[6] Equally elaborate rules for banknotes, which are legal tenders at present, such as the USD or EURO, are better left unexplained.

---

[5] It is not a coincidence the "To counterfeit is Death" was printed on the reverse of many Continentals. Regardless, the British produced so many counterfeit Continentals that eventually they became worthless.

[6] It is an amusing and very revealing fact that even now the Italian Mafia reputedly continue to use Italian Lira as means of settlement of their internal transactions.

(a) Ten DM note.

(b) Hundred DM note.

Figure 2.3 (a) A ten DM note; (b) a hundred DM note. What is the story behind their alphanumeric codes? Sources: Wikipedia.

Consider typical ten and hundred Deutsche Mark notes shown in Figures 2.3 (a), (b).[7] To start with, the 11-characters alphanumeric code of the ten DM banknote, GG6414493L3, is transformed into the 11-digit number according to the following mapping rule:

$$A, D, G, K, L, N, S, U, Y, Z \Rightarrow 0, 1, 2, 3, 4, 5, 6, 7, 8, 9. \tag{2.1}$$

As a result, the note gets its 11-digit number, namely, 22641449343. Numbers $0, \dots, 9$ are mapped into numbers $0, \dots, 9$ according to the following permutation rule:

$$0, 1, 2, 3, 4, 5, 6, 7, 8, 9 \Rightarrow 1, 5, 7, 6, 2, 8, 3, 0, 9, 4. \tag{2.2}$$

Thus, $\pi(0) = 1, \pi(1) = 5$, etc. A group operation $\bullet$ is defined for m, n, $0 \le m, n \le 9$ in Table 2.1.

---

[7]It is very telling that Germany was using her most outstanding scientists and artists' portraits on banknotes. The ten Mark note depicts Karl Friedrich Gauss (1777–1855) and his most significant discovery — the ubiquitous bell curve, also known as the Gaussian distribution. Despite its ubiquity, in Chapter 10, we show that the Gaussian distribution is not suitable for describing Bitcoin and other cryptocurrencies' returns. The hundred Mark note depicts Clara Schumann (1819–1896) — one of the most distinguished pianists of the Romantic era.

Table 2.1 The multiplication table.

·	0	1	2	3	4	5	6	7	8	9
0	0	1	2	3	4	5	6	7	8	9
1	1	2	3	4	0	6	7	8	9	5
2	2	3	4	0	1	7	8	9	5	6
3	3	4	0	1	2	8	9	5	6	7
4	4	0	1	2	3	9	5	6	7	8
5	5	9	8	7	6	0	4	3	2	1
6	6	5	9	8	7	1	0	4	3	2
7	7	6	5	9	8	2	1	0	4	3
8	8	7	6	5	9	3	2	1	0	4
9	9	8	7	6	5	4	3	2	1	0

Thus, we have $3 \cdot 6 = 9$, $6 \cdot 3 = 8$, etc. Checksum rule reads

$$\pi\left(a_1\right) \cdot \pi^2\left(a_2\right) \cdot \ldots \cdot \pi^{10}\left(a_{10}\right) \cdot \pi^0\left(a_{11}\right) = 0, \tag{2.3}$$

where $\pi^0(a) = a$. Thus, the 11th digit is used to check that the first 10 digits form a proper number. Symbolically, we can write the sum above as follows

$$\sum_{n=1}^{11} \pi^{n\,\mathrm{mod}\,11}\left(a_n\right) = 0. \tag{2.4}$$

We can easily check that this is the case for the banknote in question, as well as for the hundred DM note; see Table 2.2.

Table 2.2 Checking the validity of banknotes' numbers. Here $n' = n \bmod 11$.

n	Code	a_n	$\pi^{n'}\left(a_n\right)$	Sum	Code	a_n	$\pi^{n'}\left(a_n\right)$	Sum
1	G	2	7	7	A	0	1	1
2	G	2	0	7	D	1	8	9
3	6	6	3	9	5	5	4	5
4	4	4	1	8	2	2	5	0
5	1	1	2	6	0	0	4	4
6	4	4	8	3	3	3	3	2
7	4	4	9	7	4	4	9	6
8	9	9	9	3	1	1	1	5
9	3	3	6	9	6	6	3	7
10	L	4	7	2	U	7	1	6
11	3	3	3	0	6	6	6	0

2.2.5 Money as records

Since the late Medieval times, money has gradually assumed the form of records in various ledgers. This aspect of money is all-important in the modern world, especially in the context of DLT. At present, the vast majority of money in circulation is nothing more than a sequence of transactions, organized in ledgers. Various private banks maintain these ledgers, while central banks provide the means (central bank cash) and tools (different money transfer systems), which are used to reconcile these ledgers.

Different mechanisms for representing money as records are shown in Figure 2.4. This figure illustrates the progression from clay tablets to modern electronic bank balance sheets. In Figure 2.4 we show:

(a) A Sumerian cuneiform clay tablet. Accadians and Sumerians have been true pioneers in numerous domains, including writing and money; see, e.g., Goetzmann (2017). They came up with an idea of debt and interest and developed writing to record the corresponding transactions using inexpensive and extremely durable clay tablets. Surprisingly, most of the existing cryptocurrency systems do not incorporate the idea of debt.

(b) An Egyptian papyrus record. The evolution from tablets to papyrus illustrates that, sooner or later, a hard-to-use recording medium, such as clay tablets, is replaced by a more convenient one.

(c) Split tally sticks used in medieval Europe, to record bilateral exchange and debts; see, e.g., Clanchy (1979). A stick is marked with a system of notches and then split lengthwise. The details of the transaction had to be written on the stick to make it a proper record. This way, each transacting party received half of the marked stick as proof. Typically, the two halves of the stick were of different lengths. The longer part of the stick, called the stock, was given to the lender of money or goods. The shorter portion of the stick, called the foil, was given to the borrower. Stock and foil provided both parties with an immutable record of the transaction. Due to natural irregularities, only the correct halves would perfectly fit when stuck together at the time debt was repaid. The split tally stick served as a form of currency in medieval Europe, chronically short of gold and silver coins, from the 12th century till the 19th century. An essential element shared by blockchains and tally sticks in the immutability of records.

(d) A large disk carved from calcite, an example of famous Rai, or Fei stone money used on the Micronesian island of Yap; see, e.g., Gillilland (1975). These disks vary in size between 4 cm and 4 m in diameter. Presently, there are about 6,000 large disks on the island. They were quarried on Palau and other Micronesian islands and transported to Yap by boat to use the currency for trading and ceremonial purposes. Because the largest of these stones are too heavy to move, a transaction invokes changing the ownership of either the whole stone or possibly, a fraction of it. Oral history records the deal, so there is no need to move the stone physically. Yapanese money is particularly interesting to us since it is the first cryptocurrency and contains all the Bitcoin protocol's essential elements. The value of Yapanese money is based on a proof-of-work (just try to get a colossal stone disk from a remote island on a canoe). The money itself is represented as a chain of transactions. Yapanese money also carries a cautionary tale for those who think that Bitcoin (or their favorite substitute for it) would exist forever based on the

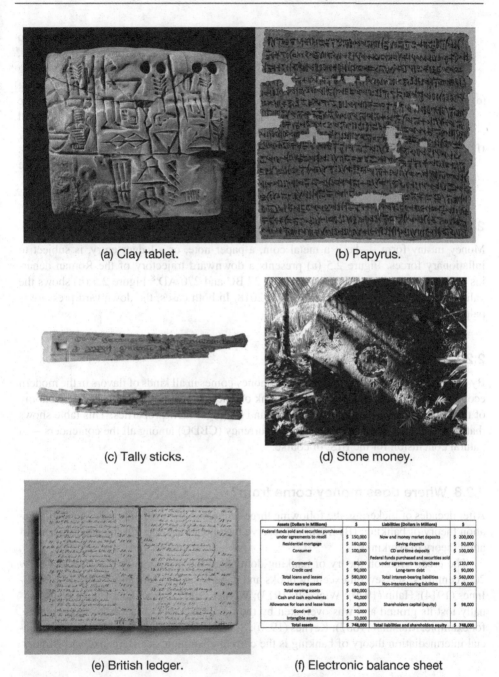

(a) Clay tablet.

(b) Papyrus.

(c) Tally sticks.

(d) Stone money.

(e) British ledger.

(f) Electronic balance sheet

Figure 2.4 Money as records. From clay tablets to modern electronic bank balance sheets: (a) Accadian and Sumerian cuneiform clay tablets; (b) Egyptian papyrus records; (c) Medieval English tally sticks; (d) Yapanese stone money; (e) early 19th-century British ledger; (f) modern simplified electronic bank balance sheet. Sources: (a–e) The Met and Wikipedia, (f) Own graphics.

fact that it has been around for a long time. When the Europeans arrived and started to use heavy machinery to cut and transport stones, rapid inflation followed. Eventually, Yap money became extinct, and some of the stones were used by the Japanese during the Second World War to build airstrips.

(e) An early 19th-century British ledger. While nothing special, it illustrates that the art and science of double-entry bookkeeping were well understood even by small enterprises.

(f) A modern electronic bank balance sheet, which is greatly simplified for readability. Apart from the fact that the balance sheet is recorded in a new medium, it differs little from the old banks' balance sheet.

2.2.6 Money and inflation

Money in any form — be it a metal coin, a paper note, or a ledger entry, is subject to inflationary forces. Figure 2.5 (a) presents a downward trajectory of the Roman denarius' silver content over 300 years between 27 BC and 270 AD.[8] Figure 2.5 (b) shows the value of the US dollar between 1799 and 2018. In both cases, the downward pressure is palpable.

2.2.7 BIS taxonomy of money

By now, the reader should understand that money comes in all kinds of flavors in the modern economy. In Table 2.3, inspired by the Bank of International Settlements (BIS) taxonomy of money, we list various types of money and highlight their properties. This table shows that the "best one" is central bank digital currency (CBDC) among all the contenders — a natural conclusion for the BIS, of course.

2.2.8 Where does money come from?

After decades of bickering, the following three schools of thought emerged: (A) credit creation theory of banking; (B) fractional reserve theory of banking; (C) financial intermediation theory of banking.

The credit creation theory of banking dominated the discourse in the 19th and early 20th centuries; see several excellent books and papers, such as Macleod (1905); Mitchell-Innes (1914); Hahn (1920); Wicksell (1913); Werner (2005, 2014). Unfortunately, it gradually lost its ground and was overtaken by the fractional reserve theory of banking; see, for example, Marshall (1887); Keynes (1930); Samuelson and Nordhaus (1995). The financial intermediation theory of banking is the current champion; see Bernanke and Blinder

---

[8]Depreciation of the denarius had a lot of unintended consequences. In the later Roman Empire, legionaries received their regular pay in worthless denarii (stipendium), a ration allowance paid in kind (annonae), and a bonus paid in gold coins and silver ingots. The bonus was paid on the accession of a new Emperor and every five years of his reign (donatives); see Elton (1998). In particular, the accession bonus was five gold solidi and a pound of silver. As a result, the fabled Roman legions became a pale shadow of their former selves, and the Empire collapsed. A cautionary tale for those who advocate bankers' bonuses or huge hedge fund managers payouts.

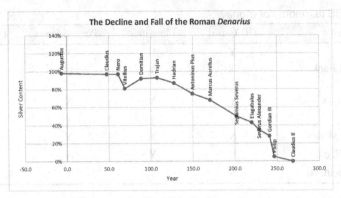

(a) Silver content of the Roman denarius.

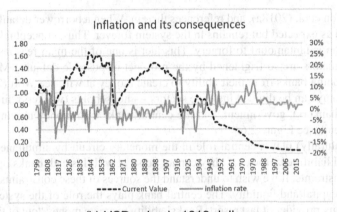

(b) USD value in 1913 dollars.

Figure 2.5 Money and inflation come hand-in-hand, regardless of the material used to represent money: (a) decline and fall of the silver content in the Roman denarius over a period of 300 years; (b) decline of the value of the US dollar over a period of 200 years. Own graphics. Sources: (a) https://pisoproject.wordpress.com/the-reduction-of-silver-content-in-the-roman-denarius/, (b) The Fed and others.

(1988); Keynes (1936); Tobin (1969); Tobin and Golub (1998), which give a representative description of this theory. In our opinion, this theory puts a grossly insufficient emphasis on the unique and special role of the banking sector in the process of money creation. It hence fails in its stated objective of explaining the function of the banking system.

The so-called modern monetary circuit theory provides the most convincing explanation of the process of money creation and annihilation. The main conclusion of this theory is that commercial banks create money when they lent to their clients. Banks destroy money upon repayment. Interest, which, in effect, comes from the next round of borrowing, stays in the system for good. Monetary circuit theory has a long and illustrious history; see, Godley and Lavoie (2007); Kalecki (2007); Keen (2013, 2015); Petty (1899); Quesnay (1991); Zezza and Dos Santos (2004). For the modern take on this theory; see McLeay et al. (2014); Lipton

Table 2.3 BIS taxonomy of money.

Type \ Property	CB Issued	Electronic	Peer-to-Peer	Universal
Cash	Y	N	Y	Y
Commercial Bank Deposits	N	Y	N	Y
Central Bank Deposits	Y	Y	N	N
CBDC	Y	Y	Y	Y
Commodity Money	N	N	Y	Y
Crypto (USC, etc.)	N	Y	N	N
Crypto (Bitcoin, etc.)	N	Y	Y	Y

(2016a); Lipton et al. (2018a), and references therein. When a borrower defaults, money is not destroyed as expected but remains in the system forever. Thus, conceptually speaking, default on loan is tantamount to forgery. This fact is one of the main reasons why banks and their regulators are so frightened by the prospect of possible defaults. Many authors advocate stripping banks from their ability to create money at will; see Wolf (2014).

The economy can be divided into five sectors — households (workers), firms (capitalists), banks (bankers), government, and the central bank, with money moving a gigantic monetary circuit, see Figure 2.6.

Banks (naturally) play a central role in the monetary circuit by simultaneously creating assets and liabilities. However, they cannot do so at will. Instead, banks operate under numerous constraints. If we wish to understand the sources of these constraints, we need to look at bank capital and liquidity. The central bank plays the role of the system regulator and the liquidity provider of last resort. Banks' ability to create money "out of thin air" by a simple act of lending is somewhat exaggerated because, in reality, they lend money against collateral. The monetary circuit framework helps to think about such important concepts as digital currencies, distributed ledgers, and similar constructs. Besides, by understanding the circuit, we can envision the impact of negative interest rates on the system as a whole. We note in passing that negative interest rates are not a new phenomenon; they have been known for a long time, for example, in the form of medieval demurrage and, more recently, in the form of so-called stamp scrip money *a la* Fisher (1933); Gesell (1958); see below.[9]

2.3 How the financial system operates at the moment

2.3.1 Current situation

In modern society, nontrivial quantities of money can only be held at banks in various deposit accounts. Thus, banking institutions play a unique societal role as record keepers.

---

[9]The crucial difference between inflation and negative rate regimes is that cash is highly undesirable under the former, and very valuable under the latter.

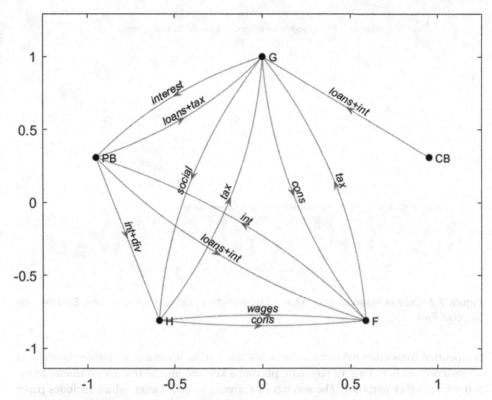

Figure 2.6 Sketch of the monetary circuit. G — government, CB — central bank, PB — private banks, F — rms, H — households. Own graphics.

In general, in developed economies, the proportion of cash versus bank deposits is relatively small. However, when large-denomination notes are available, they are occasionally used instead of bank accounts. Despite all the verbal acrobatics, depositors *nolens volens* become unsecured junior creditors of banks and therefore have the lowest priority in the event of a bank liquidation; see Devaynes (1816). If a bank were to default, it would generally cause partial destruction of deposits. For small deposits, the bank's default risk is alleviated by insurance provided by the Federal Deposit Insurance Corporation (FDIC); however, this is not an option for large deposits. For instance, when IndyMac Bancorp filed for bankruptcy in the summer of 2008, its large depositors lost a significant portion of their money. Hence, banks are required to keep sufficient capital cushions, as well as ample liquidity.

At present, banks do not have sufficiently attractive opportunities to lend money. Hence they pile up reserves with the central bank; see Figure 2.7.

As private banks are unable or unwilling to find suitable borrowers, since the inception of the GFC and, significantly, the onset of the COVID-19-induced recession, central banks massively increase their lending operations, and, as a result, their balance sheets; see Figure 2.8.

In general, money is an obligation of private banks. However, there is an important exception — cash, which is an obligation of the central bank. While cash is facing

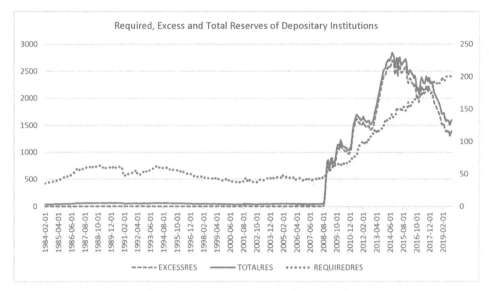

Figure 2.7 Excess reserves and failure of quantitative easing. Own graphics. Source: The St Louis Fed.

competition from other payment instruments, the results of numerous studies suggest that it remains a resilient form of payment, playing a key and unique role for consumer transactions and other purposes. The amount of currency in circulation, which includes paper currency and coin held both by the public and in the vaults of depository institutions, grows at a healthy pace. While the quantity of currency increases steadily, its velocity currently decreases, meaning that money is being used less efficiently.

The velocity of money is the number of times a unit of money is spent on purchasing goods and services per unit of time, say quarterly. The increasing velocity of money implies that more transactions happen between economic agents, while decreasing velocity indicates the opposite. Simply put, the quantity of cash in circulation continuously grows; however, its velocity slows down; see Figure 2.9.

The velocity of stablecoins such as Tether is discussed in Chapter 8.

2.3.2 Challenges faced by the current financial system

The current financial system emerged as a result of centuries-long evolution. Its roots can be traced back to the fabled city-states of Sumer, which was the site of the earliest known civilization. Located between the Tigris and Euphrates rivers, Sumer was the birthplace of numerous financial ideas, including debt, interest, and others, which are the cornerstone of modern finance. Subsequently, Italy's city-states, especially Florence, Venice, and Genoa, became hotbeds of financial innovation. Later, Northern Europe's great trading cities, such as Antwerp and Amsterdam, evolve into true finance leaders. Eventually, London and New York replaced them. Presently, the main high-finance centers are New York, London,

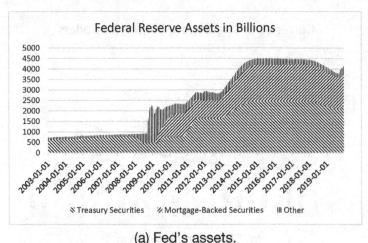

(a) Fed's assets.

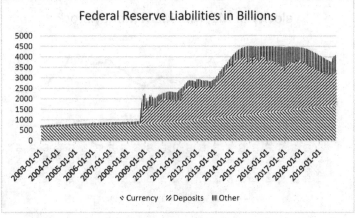

(b) Fed's liabilities.

Figure 2.8 Transformation of central banks into fractional reserve banks. While commercial banks become more and more conservative and narrow, central banks become more adventurous. (a) Fed's assets have increased dramatically since the GFC, particularly during the COVID-19 pandemic. (b) And so have Fed's liabilities. Own graphics. Sources: The St Louis Fed.

Tokyo, Hong Kong, Shanghai, Singapore, and a few regional ones. Which region will reign supreme in the future remains to be seen.

An objective observer has to conclude that today's financial system cannot continue in its present form much longer. The reasons are easy to fathom (but hard to fix): finance is too complicated for its own good, involving lending, payments, clearing, and settlement across multiple systems and geographies. Most banks and other financial institutions rely on legacy information technology (IT) systems, which are past their due date. The financial

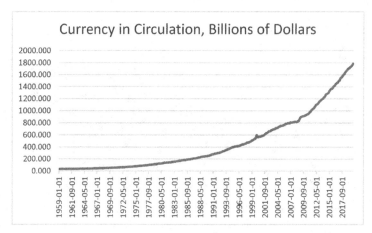

(a) Money in circulation.

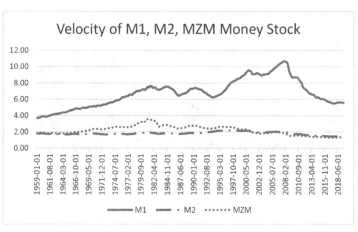

(b) Velocity of money.

Figure 2.9 (a) Money in circulation. (b) Velocity of money. Own graphics. Sources: The St Louis Fed.

system depends on numerous intermediaries and supporting actors and is chronically unable to put its own house in order. It lacks transparency and cannot be understood not only by the outsiders but, much more alarmingly, by the players themselves. While ostensibly better capitalized, banking institutions have increased in complexity to such a degree that their stability and creditworthiness can be established with certainty, neither by regulators nor by depositors, investors, and, somewhat surprisingly, by their management. The balance sheets of Tier 1 banks have become so opaque that their complexity is beyond quantitative analysis. As a result of this complexity, many banks and other financial institutions have become too-big-to-manage. Tier 1 banks have to spend billions of dollars annually (and their smaller competitors tens to hundreds of millions) developing, validating, and maintaining complex models and IT systems. The *raison d'être* of these systems is to demonstrate to the regulators that banks are compliant with capital and liquidity requirements. The Comprehensive Capital Analysis and Review (CCAR) is an annual rite of passage that every systemically important bank has to pass *nolens volens*.

Moreover, information asymmetry built into the system results in an inevitable heavy concentration of wealth (with the top 1% controlling about 50% of worldwide wealth). Due to its heavily concentrated nature, the system is intrinsically unstable. It cannot be satisfactorily controlled by major centralized institutions, such as governments, central banks, and various national and supranational regulatory bodies. It creates numerous geopolitical and economic risks in under-developed nations, which suffer from hyperinflation, slow economic growth, and more than two billion people are financially unserved or underserved. The economic consequences of the ongoing COVID-19 pandemic exacerbate the overall situation further.

2.3.3 Detailed analysis of money creation

Our goal is to understand the potential applications of DLT in finance. In this regard, it is advantageous to have a schematic pictorial representation for the banking system's inner working.

We start with a simple case of a single bank, or, equivalently, the banking system as a whole. We assume that the bank does not operate at full capacity capital and liquidity wise and can lend money. When a new creditworthy borrower approaches the bank and asks for a reasonably-sized loan, the bank issues an on-demand loan. The bank simultaneously creates on its books a deposit (the borrower's asset), and a matching liability for the borrower (the bank's asset). Figuratively speaking, the bank has created money "out of thin air." However, strictly speaking, this is not true, because more often than not, a borrower has to possess suitable collateral against the loan she wishes to obtain. Of course, when the borrower pays the loan back, the process is carried in reverse, and the money is "destroyed." Assuming that the interest charged on loans is higher than the interest paid on deposits, bank's capital increases. The interest comes from the next borrowing round, of course. A little reflection shows that the bank cannot destroy the money if the borrower defaults instead of repaying the loan. It this case, the bank's capital naturally decreases to compensate for the default. However, since the capital is an abstract accounting construct, rather than the actual cash or central bank money, the amount of money in circulation increases disproportionally — to put it differently, the money borrowed but not repaid is forged.

The money creation process is graphically illustrated in Figure 2.10.

The process, initiated when the bank lends some amount to a new borrower, results in the following changes in the bank's balance sheet shown in Table 2.4.

We now consider a more complicated case of two (or, possibly, more) banks. In this case, liquidity is of paramount importance. To account for liquidity, we must incorporate a central bank into the financial ecosystem. We assume that banks keep part of their assets in reserves (cash), representing a central bank's liability. Here cash is understood as an electronic record in the central bank ledger, but some can be in the form of physical banknotes. The money

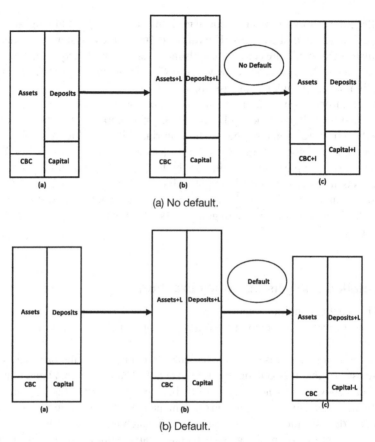

Figure 2.10 Money creation by a single bank. (a) Borrower does not default. (b) Borrower defaults. We see that the balance sheet goes through a series of state transitions. This idea is borrowed by Bitcoin and all other cryptocurrency protocols. Own graphics.

creation process consists of three stages. First, a creditworthy borrower asks the first bank for a loan. The bank obliges by simultaneously issuing the loan (an asset for the bank and liability for the borrower) and transferring the funds electronically to the second bank — a transaction accompanied by a matching transfer of reserves from the first bank to the second. In either case, the liquid assets (reserves + cash) of the first bank go down, possibly decreasing its liquidity below the desired level, whereas the liquid assets (reserves + cash) and those of the second bank go up, perhaps increasing its liquidity above its desired level. Third, the first bank approaches the second bank to borrow its excess reserves. If the second bank deems the first bank creditworthy, it lends its surplus reserves, creating a link between itself and the first bank. Alternatively, if the second bank refuses to lend its excess reserves to the first bank, it must borrow from the central bank. To obtain a loan from the central bank, the first bank uses its performing assets as collateral, or outright acquires reserves by selling assets, such as government bonds in open market operations. Thus, the central

Table 2.4 Money creation by one bank. State 1: Balance sheet of Bank I before the lending round commences. State 2: Changes in the balance sheet caused by a client borrowing 4mm. Bank I's assets and liabilities simultaneously increase by 4mm. State 3a: No default. Bank I's assets and liabilities simultaneously decrease by 4mm, while its cash and capital increase by 0.5mm (interest); 4mm worth of money is destroyed. State 3b: Default. Bank I's assets and capital simultaneously decrease by 4mm; 4mm worth of capital is destroyed instead of money.

	State 1	State 2	State 3a	State 3b
External Assets	50	54	50	50
Interbank Assets	20	20	20	20
Cash	10	10	10.5	10
External Liabilities	60	64	60	64
Interbank Liabilities	13	13	13	13
Capital	7	7	7.5	3
Assets minus Liabilities	0	0	0	0

bank lubricates the wheels of commerce by providing liquidity to creditworthy borrowers. Its willingness to provide reserves to commercial banks determines, in turn, their desire to issue loans to firms and households. When the borrower repays the loan, the process plays in reverse.[10] The money creation process, initiated when Bank I lends some amount to a new borrower, depositing money at Bank II, results in the following changes in two banks' balance sheets; see Table 2.5.

In summary, in contrast to a non-banking firm, whose balance sheet can be adequately described by simply reflecting its assets, A, liabilities, L, and equity, K, $A = L + K$, the balance sheet of a typical commercial bank must, in addition to real economy assets and liabilities, incorporate additional items, such as interbank assets $\tilde{A}$ and liabilities $\tilde{L}$, as well as central bank cash C, representing simultaneously bank's assets and central bank's liabilities,

$$A + \tilde{A} + C = L + \tilde{L} + K, \tag{2.5}$$

see Figure 2.11.

In a supply and demand-driven economic system, money is treated on par with other goods. The dynamics of demand for loans and lending activity are understood in the supply–demand equilibrium framework. Increasing demand for loans from firms and households leads banks to lend more. We should emphasize that the ability of banks to generate new loans is not infinite. We can draw a parallel with physical goods, whose overall production at full employment is limited by the physical plant capacity. The money (loan) creation is determined by the banking system's capital and liquidity capacity. Besides, as was mentioned earlier, typically, money is lent against collateral, which is another limitation on how much money banks can create at any given moment. Typically, during boom times, the value of collateral, such as real estate, is greatly overestimated, so that money creation

---

[10]The authors are grateful to Matheus Grasselli for his valuable comments on the money creation process.

Table 2.5 Money creation, the case of two banks. State 1: Balance sheets of Banks I and II before the lending round commences. State 2: Changes in the balance sheet caused by a client borrowing 4mm from Bank I. Bank I's assets increase by 4mm while its central bank cash position simultaneously decreases by 4mm. Bank II increases its cash position and its external liabilities by 4mm. State 3: Bank I borrows 4mm of cash from Bank II. State 4a: The borrower repays the loan to Bank I. Bank I's assets decrease by 4mm, its cash position increases by 4.5mm (including interest), its capital increases by 0.5mm. Bank II's external liabilities and central bank cash decrease by 0.5mm. State 5a: Bank I's repays 4.25mm (including interest) to Bank II. Bank I's cash position and capital decrease by 4.25mm, and 0.25mm, respectively. Bank II's cash position and capital increase by 4mm and 0.25mm, respectively. State 4b: The borrower defaults on loan to Bank I. Bank I's assets decrease by 4mm, its capital decreases by 4mm. Bank II's external liabilities and central bank cash decrease by 4mm. State 5a: Bank I's repays 4.25mm (including interest) to Bank II. Bank I's cash position and capital decrease by 4.25mm, and 0.25mm, respectively. Bank II's cash position and capital increase by 4.25mm and 0.25mm, respectively.

	State 1	State 2	State 3	State 4a	State 5a	State 4b	State 5b
	Bank I	Bank I	Bank I	Bank I	Bank I	Bank I	Bank I
External Assets	50	54	54	50	50	50	50
Interbank Assets	20	20	20	20	20	20	20
Cash	10	6	10	14.5	10.25	10	5.75
External Liabilities	60	60	60	60	60	60	60
Interbank Liabilities	13	13	17	17	13	17	13
Capital	7	7	7	7.5	7.25	3	2.75
Assets minus Liabilities	0	0	0	0	0	0	0
	Bank II	Bank II	Bank II	Bank II	Bank II	Bank II	Bank II
External Assets	80	80	80	80	80	80	80
Interbank Assets	15	15	19	19	15	19	15
Cash	15	19	15	11	15.25	11	15.25
External Liabilities	70	74	74	70	70	70	70
Interbank Liabilities	30	30	30	30	30	30	30
Capital	10	10	10	10	10.25	10	10.25
Assets minus Liabilities	0	0	0	0	0	0	0

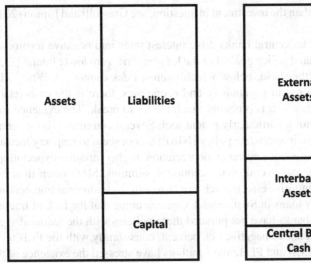

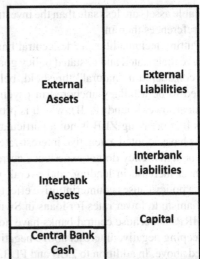

(a) Firm's balance sheet.   **(b) Bank's balance sheet.**

Figure 2.11 (a) Firm's balance sheet. (b) Bank's balance sheet. It is clear that banks do not exist in a vacuum and become naturally interconnected as part of doing business. Own graphics.

is getting out of control. On the contrary, during bust times, collateral loses its value, with all the negative implications.

Once we have embedded the flow of money in the supply–demand framework, we can extend the model to several interconnected banks that issue loans in the economy. These banks compete with each other for business, while, at the same time, help each other to balance their cash holdings, thus creating interbank linkages. These linkages are posing risks because of the potential propagation of defaults in the system. Therefore, a fractional reserve banking system is inherently unstable, which becomes painfully evident during periodic financial crises.

2.3.4 Negative interest rates

Negative interest rates are a disturbing sign of our times. Due to their ubiquity, we tend to think of negative interest rates as a new normal, while, deep down, we know that the very idea of negative interest is wrong. While not exactly unheard of, negative interest rates were last prevalent in the Middle Ages in the form of demurrage. There were some sharply localized episodes of negative rates here and there, for instance, in the 1970s in Switzerland. In principle, there is nothing technically abnormal about negative interest rates as such. Instead of receiving a positive return on a bank deposit, a bond, or a similar investment, it is entirely possible to be charged for holding an asset. This situation makes sense if all other available assets are less safe than the investment in question; see Grasselli and Lipton (2019) and references therein.

Putting technicalities aside, central banks drive interest rates into negative territory to achieve their stated and unstated policy goals. For the longest time, zero lower bound (ZLB) was considered a "natural" threshold, below which interest rates cannot go. While ZLB proved to be a fallacy anchored in psychology and economics, there is the cash-related physical lower bound (PLB), which is probably much harder to break. The evidence suggests that breaking ZLB is not a particularly potent tool. Several countries whose central banks have adopted a negative interest-rate policy (NIRP) do not seem to reap any measurable benefits. They do not experience currency depreciation, higher inflation expectations, a robust increase in lending, and overall macroeconomic stimulus. More often than not, such a policy causes counterintuitive effects, such as a failure of the traditional transmission mechanism to lower rates for loans in Switzerland. One can argue that the lack of traction of NIRPs is because central banks have not pursued these policies with the required vigor by keeping negative interests above negative one percent, consistently with the PLB mentioned above. In addition to ZLB and PLB, many authors have stressed the existence of the economic lower bound (ELB) for interest rates, below which the adverse effects of lower bank profitability become prevalent. We show below that central bank digital currencies can help make the NIRP a potent tool because it eliminates the PLB and pushes banks into a more efficient *modus operandi*.

2.3.5 Banking regulations and monetary stability

In addition to their lending businesses, private banks perform ledger-maintaining and transactional functions for their clients. As part of these efforts, private banks play two critical roles, which central banks cannot achieve. They are the system gatekeepers, who provide Know Your Customer (KYC) services, and system police, who provide Anti-Money Laundering (AML) services. In addition to the more prominent areas for DLT, such as in digital currencies, including CBDCs, DLT can solve such complex issues as trust and identity, with an emphasis on the KYC and AML aspects. Further, given that all banking activities boil down to maintaining a ledger, judicious applications of DLT can facilitate trading, clearing and settlement, payments, trade finance, and so on. However, if applied without a clear understanding of the underlying business, DLT can make the situation even less satisfactory than it is at the moment.

2.3.6 The pros and cons of the current system

The biggest issue afflicting the existing banking system is that it is overly complicated due to commingling three distinct activities: (a) creation and annihilation of credit money through lending; (b) record keeping; (c) execution of transactions. Separating these activities is vital for making banking more agile and efficient. We shall concentrate below on transactional aspects of banking and argue that judicious use of DLT can help bring transactional banking into the 21st century.

Though the chance of a significant bank collapsing is not high, it is non-negligible either. By lending money, banks risk both their capital and depositors' money. FDIC has deposit insurance funds of only US$85 billion, constituting a little more than 1% of all existing deposits in the US. FDIC funds are not sufficient to cover the losses of even one of the largest 23 banks in the US. Because of the above, one has to conclude that depositors implicitly subsidize banks by hundreds of basis points annually by not being paid interest covering their credit risks.

During crises, bankers become acutely aware of banks' riskiness and stop unsecured lending to each other, so that London Interbank Offered Rate (Libor) breaks. If banks, who understand their brethren's inner workings, are reluctant to lend money to each other, why should depositors lend money to them for free?

2.4 Domestic payments

2.4.1 The protagonists

When dealing with payments, be they conventional or crypto-based, in this book, we shall call the payer (buyer) Alice, and payee (seller) Bob. This terminology is traditional in cryptography that proved its usefulness over many decades. On occasion, we shall deal with other parties — Ben, the banker, Charlie, the impostor, etc.

2.4.2 Settlement

The current financial system's hallmark is a long chain of middlemen engaged in moving money between a buyer and a seller, not least in the form of correspondent banks. Not surprisingly, intermediaries thrive in most situations. Various fees can easily reach 3% or more, which is a significant amount for most participants.

The domestic payment and settlement system is shown in Figure 2.12.

Given its complexity, this system is prone to error, such as a recent blunder by Citi, who incorrectly paid US$900 million due to a human oversight; see, e.g., Doherty and Surane (2020).

Moving money between two bank accounts is illustrated in Figure 2.13.

We have to distinguish two cases: (a) Alice and Bob have accounts in the same bank; (b) Alice and Bob have accounts at two different banks. Algorithmically, we represent the process when Alice and Bob have accounts in the same bank in Table 2.6. and the process when Alice and Bob have accounts at two different banks in Table 2.7.

2.4.3 Credit card payments

While for the consumer, who has a valid credit card, using it for payment is magically simple, the existing architecture is very complicated. Figure 2.14, adapted from FFIEC (2016),

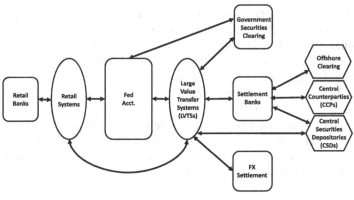

(a) Clearing and settlement network.

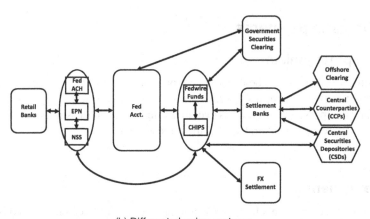

(b) Different clearing systems.

Figure 2.12 (a) USD clearing and settlement network topology; (b) Specific clearing systems. Own graphics. Sources: PRC (2016).

illustrates the payment and information flows for a typical credit card transaction between Alice and Bob percolating through the clearing and settlement systems.

The actual steps can be summarized as follows: (1) Alice initiates credit card payment; (2, 3) Bob transmits the data via the bankcard network to Bank A; (4, 5, 6) if the transaction is approved, the information flows in the opposite direction and Alice signs the credit card slip; (7, 8) Bob submits a bunch of transactions to his Bank B and receives payment; (9, 10) Bank B sends the sales draft data to the bankcard company, which forwards it to Bank A; (11) the bankcard company determines the net positions of Banks A and B by the end of the day and coordinates the settlement between these banks, which takes place via the Fedwire R; (12) Bank A sends Alice her monthly statement; (13) Alice makes payments. Further, even more arcane details can be found in FFIEC (2016).

It shows that this system is very inefficient — to move money along just four arrows, multiple flows of information are needed. The actual settlement process occurs using a

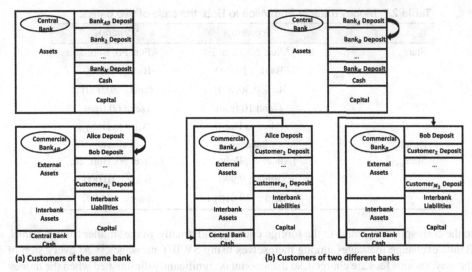

Figure 2.13 Domestic money transfers: (a) Customers of the same bank. (b) Customers of two different banks. Own graphics.

Table 2.6 Money transfer from Alice to Bob, the case of one bank.

	Scenario$_1$	Scenario$_2$
State$_n$	{Alice: 50, Bob: 20, Bank$_1$: 10,000}	{Alice: 50, Bob: 20, Bank$_1$: 10,000}
TX	{send 10 from Alice to Bob}	{send 60 from Alice to Bob}
Result	Success	Failure
State$_{n+1}$	{Alice: 40, Bob: 30, Bank$_1$: 10,000}	{Alice: 50, Bob: 20, Bank$_1$: 10,000}

separate payment network such as Fedwire R, shown in Figure 2.12. It is not surprising that card payments are costly, particularly compared to very slim profit margins in the retail sector; see Rigby (2014); Ross (2018).

2.5 Cross-border payments

2.5.1 Cross-border transactions and correspondent banking

Cross-border payments from the USA to Europe and other destinations are structured in a highly hierarchical fashion, first flowing from smaller to bigger banks within a country, then

Table 2.7 Money transfer from Alice to Bob, the case of two banks.

	Scenario$_1$	Scenario$_2$
State$_n$	{Alice: 50, Bob: 20,	{Alice: 50, Bob: 20,
	Bank$_1$: 10,000	Bank$_1$: 10,000
	Bank$_2$: 30,000}	Bank$_2$: 30,000}
TX	{send 10 from	{send 60 from
	Alice to Bob}	Alice to Bob}
Result	Success	Failure
State$_{n+1}$	{Alice: 40, Bob: 30,	{Alice: 50, Bob: 20,
	Bank$_1$: 9,990	Bank$_1$: 10,000
	Bank$_2$: 30,010}	Bank$_2$: 30,000}

to the correspondent banks in the foreign country, and, finally, to the smaller foreign banks. Banks exchange messages among themselves using SWIFT messages.[11] At every step of the way, various fees are charged, so the amount is significantly diminished when the money reaches its recipient. Foreign exchange payments result in a real middlemen bonanza.

Figure 2.15 shows a typical payment pyramid.

According to this figure, money goes from Alice to Bob via a complex network of inter-mediaries. Various fintech companies try to intermediate the banks and cut the payment pyramid at different levels. The reality of the situation is that, under the hood, they have

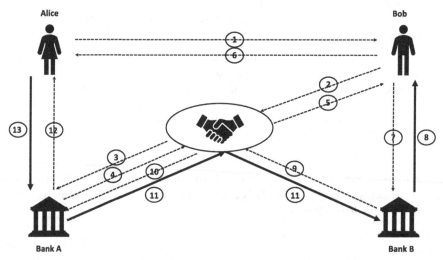

Figure 2.14 Origination, clearance, and settlement of a credit card transaction. Own graphics. Source: FFIEC (2016).

---

[11] SWIFT stands for the Society for Worldwide Interbank Financial Telecommunication. It was founded in Brussels in 1973. Currently, it connects more than 11,000 financial institutions located in more than 200 countries and territories.

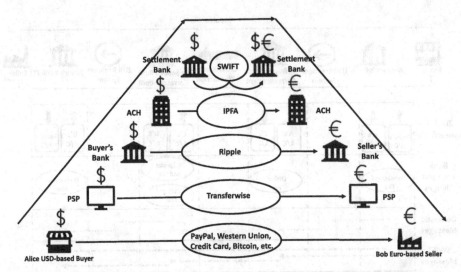

Figure 2.15 The payment pyramid and its disintermediation. Own graphics. Source: SWIFT.

to rely on banks to move funds. Hypothetically, Bitcoin and other cryptocurrencies can be deployed at the foundational, peer-to-peer level. However, unless Bitcoin becomes a means of payment, these transactions still have to use banking rails, as explained in the following chapters. Besides, to get some Bitcoin or other cryptocurrency such as Ripple, Alice has to become a centralized exchange customer, thus negating distributed aspects of the protocol.

Figure 2.16 illustrates that chain of transactions when the banking system is used throughout. The funds flow from Alice's bank to its corresponding bank to Bob's corresponding bank, and, finally, to Bob's bank. Given the number of intermediaries, transaction costs accumulate, and the transaction's timing becomes uncertain.

2.5.2 Forex trading

For Alice to move her money across the border, someone has to be willing to take Alice's dollars and exchange them for foreign currency of Bob's choice. These exchanges are the job of forex traders, who operate the forex market, which is currently moving about US$6.6 trillion daily. Typically, they charge 2–5 basis points, called pips in forex trading, in the interbank market. Banks themselves charge significantly more from their customers, 1–2% is not uncommon. An alternative to the route discussed in the previous subsection is shown in Figure 2.17.

For readability, we cut the intermediaries' chain to just four: Alice's USD Bank, Bob's EURO Bank, Market Maker's USD, and EURO banks. Besides, we show the respective central banks — the Federal Reserve Bank and the European Central Bank, since the only way the four banks in the picture can communicate is through their central bank deposits.

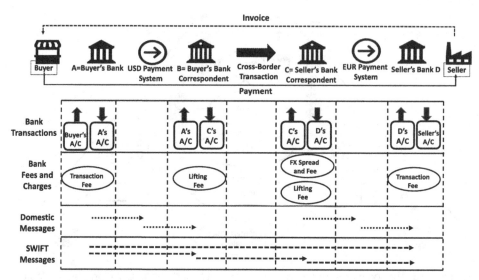

Figure 2.16 Correspondent banking example — movement of funds from the USA to Europe. Own graphics.

2.5.3 Informal funds transfer systems

It is worth noting that, in addition to the established banking system, there are several informal funds transfer systems. The best known of these is the Hawala payment system, illustrated in Figure 2.18. It can be viewed as an example of a very efficient cash transfer mechanism. Its main idea is that money stays in place but changes ownership; see, e.g., El-Qorchi (2002).

Of course, this system relies on trust much more than conventional transfer systems and is afflicted by additional issues, such as lax KYC and AML practices. One of Hawala's main problems is that typically money flows in one direction, for instance, in remittances from the Gulf to India and Pakistan or from Hong Kong to the Philippines. Thus, to balance accounts, some reverse Hawala route has to be executed. It can have different forms. The most common one is for the Hawalador in the developed country to ship goods to the Hawalador in the developing country. Thus, one of the Hawaladors becomes a natural exporter, and the other becomes an importer.

2.6 Blockchain payments

Compared to the current situation, the potential blockchain solution, shown in Figure 2.19, looks decisively attractive. Putting technicalities aside, and assuming that Alice owns funds, all she needs to do is to broadcast her intention to pay Bob to the blockchain ecosystem. The relevant blockchain participants would automatically handle the rest and ensure that

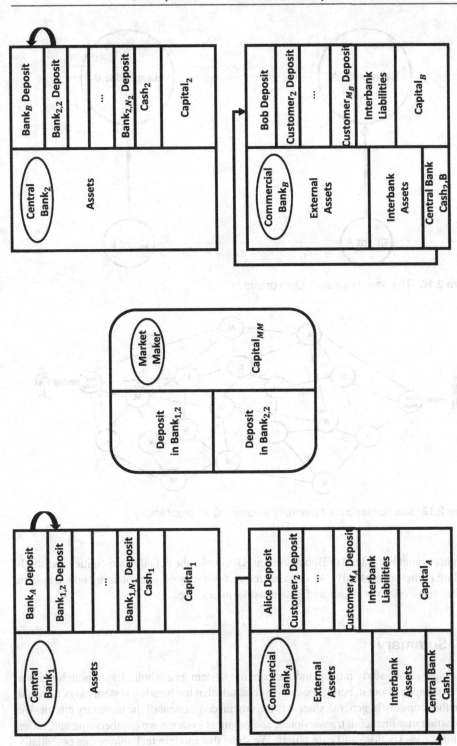

Figure 2.17 Cross-border money transfers: the unbalanced flow case. Own graphics.

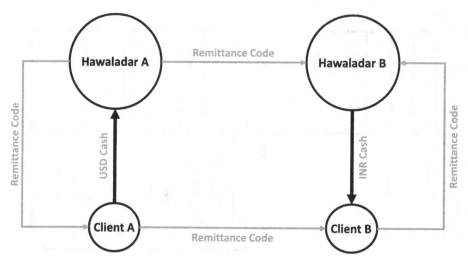

Figure 2.18: The hawala system. Own graphics.

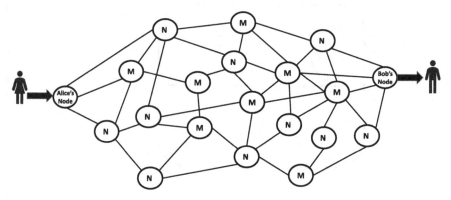

Figure 2.19 Blockchain-based payment solution. Own graphics.

the right amount is credited to Bob's account. Of course, the actual technicalities are mind-boggling in their complexity. We dedicate the rest of the book to explaining how a system like that can work in principle and how it works in practice.

2.7 Summary

In this chapter, we study money and the banking system as a whole. Even though many a party is unhappy about it, being honest, we conclude that the banking system plays a pivotal role in the economy in general, since it is the crucial component of the monetary circuit. We demonstrate that financial, transactional, and payment systems are cumbersome and suffer from numerous frictions and pain points. We show that modern technology can potentially simplify and streamline the existing system. In particular, we argue that many of the transactional frictions can be reduced by using DLT. Details of how record keeping, regulatory requirements, domestic and cross-border payments can be dramatically improved using distributed ledgers are discussed in the rest of the book.

Additional references: Begenau and Stafford (2019); Collomb and Sok (2016); Kavuri and Milne (2019).

2.8 Exercises

1. Describe the process of money creation in the modern economy. Who is the leading party in this process? Why do commercial banks have to keep reserves with the central bank? Why do they need capital? Where does the interest on loans come from?
2. Compare and contrast the existing financial system and a potential financial system based on distributed ledgers. What are the pros and cons of each? How could money transfer mechanisms be improved by using DLT?

3 A Primer on Cryptocurrencies and Distributed Ledgers

"The Times 03/Jan/2009 Chancellor on brink of second bailout for banks."

Metadata in Bitcoin's genesis block

3.1 Introduction

In 2008, an anonymous researcher working under the name of Satoshi Nakamoto publishes the white paper *Bitcoin: A Peer-to-Peer Electronic Cash System*; see Nakamoto (2008). He describes the functioning of a decentralized and *peer-to-peer* (P2P) technology capable of maintaining a monetary ledger. Such platform allows, he claims, a new form of money exempt of central governance or intermediaries. It also guarantees high availability and location independence, and is inherently censorship resistent. Satoshi Nakamoto releases the first implementation of Bitcoin in 2009. Eleven years later, the value of all bitcoins (BTC) combined exceeds US$200 billion and has inspired thousands of other similar projects.

Satoshi Nakamoto's main contribution is to specify the architecture of a distributed ledger maintained endogenously by its own users, free of any preliminary permission requirement. This distributed ledger, also called *blockchain* due to its data structure, maintains a complete audit trail of every transaction in the system and offers a freely accessible view over the complete accounting. Because the blockchain is distributed amongst many parties, it is highly resilient to most sorts of attacks and can enforce the core Bitcoin rules — such as the BTC monetary policy — with no single point of trust. Satoshi designs it such that it incentivizes collaborative behaviors and punishes disruptive actors despite their anonymity.

A series of challenges may be apparent to the reader. As new peers may join or leave the network arbitrarily and anonymously, how can one establish a trust relation whose standards are high enough that they can support a financial system? How to open an account in absence of a bank? How to demonstrate ownership of an account in an anonymous system void of any legal identities? Where to store the accounting ledger to maintain balances and transactions in absence of a trusted data store provider? How to avoid data tampering of the accounting in presence of dishonest network participants and without a trusted auditor reviewing the accounts and committing to their integrity? How to regulate the monetary policy in absence of a central bank?

Despite the magnitude of these challenges, all building blocks of Bitcoin, including digital signatures, Merkle trees, chaining *a la* Haber and Stornetta, proof-of-work based on cryptographic hash functions, and proof-of-work *a la* hashcash invented by Back, have been known for some time. Packaging these building blocks so as to create a secure system with appropriate incentives has been the impressive achievement of Satoshi Nakamoto. However, BTC is not the first digital currency by a long shot and very likely not the last major one (see, e.g., Chaum (1983); Chaum et al. (1988); Damgård (1988); D'Amiano and Di Crescenzo (1994); Di Crescenzo (1994); Brands (1993); Kim and Oh (2001); Fan et al. (2013); Narayanan and Clark (2017); Sherman et al. (2019)). Its precursor include Ecash and Digicash invented by David Chaum in 1983 and 1990. Other, and closer, precursors are bit gold invented by Nick Szabo in 1998 and presented on his blog in 2005[1], and Wei Dai's b-money.

In this chapter, we review the genesis of Bitcoin, the main principles and challenges of distributed ledger technology (DLT), and the trends initiated by this technological innovation. The chapter serves as an introduction to a more formal treatment of cryptography in Chapter 4 and an in-depth analysis of the Bitcoin protocol in Chapter 5 and Ethereum protocol in Chapter 6. We introduce the concepts of public key cryptography, blockchain, double spending and proof-of-work in an attempt to provide a horizontal review of all the building blocks of a cryptocurrency network. Section 3.2 presents the notion of digital identity. Section 3.3 covers the mains principles of a peer-to-peer networking and introduces the main pillar of DLT, the blockchain data structure. Section 3.4 explains the challenge of double spending in a decentralized setup. Section 3.5 reviews the core principles of consensus protocols for permissionless DLTs. Section 3.6 lays down the key milestones in the evolution of the digital asset ecosystem over the 2010–2020 decade. Section 3.7 describes a path to the after-Bitcoin. Finally, a summary is given in Section 3.8.

3.2 Distributed identity

3.2.1 Legal identification

In today's economy, government is trusted to certify identities and to provide identity verification material such as ID cards, passports, and driving licenses. One is who he is because some government says so, and because he has the ability to demonstrate it with a high enough credibility that a counterpart will trust it.

Facing an identification document, a border control officer assesses two critical requirements: First, that the document is an original one — that it was authentically issued by a trusted government and that it is still valid; for that, a set of anti-counterfeiting measures are in place including security ink, special paper and material, hard-to-copy designs and printing techniques, special identifiers, and generally micro-chips including some form of cryptography technology. Second, the officer verifies that the passport belongs to its bearer: does the portrait photo match, does the color of the eyes match, the height, the fingerprints,

---

[1]https://unenumerated.blogspot.com/2005/12/bit-gold.html.

the age? The combination of these several traits being unlikely to match multiple individuals, the ownership of the passport can be confirmed. After correlating the document with both its issuer, the government, and with its bearer, trust in the government is translated into trust in the bearer's legal identity.

Satoshi Nakamoto's ambition to fully distribute Bitcoin clashed with this principle. One cannot have a centrally-managed identity issuer in a system whose *raison d'être* is the absence of a central party. A decentralized system must rely on an identification protocol guaranteeing independence from central authority and allowing the participation of any user with no preliminary permission requirement. The progress made with public key cryptography in the previous decades allowed Satoshi Nakamoto to define a simple model based on so-called public key cryptography.

3.2.2 Digital identification and wallets

In a traditional legal environment, one authenticates a paper copy of a contract with a handwritten signature. The underlying assumption is that only the legitimate contractor knows how to design his own signature — in other words, that it is hard to replicate — but that any external observer may easily compare it to some identification documentation to assess its validity. In practice, it has become clear that this assumption is unreasonable: counterfeiting a signature is a relatively straightforward process whether it is achieved manually or with computer graphics tools, and assessing the authenticity of the signature is a difficult task that may lead to a high risk of false positives or false negatives.

A *digital signature* process aims to rely on mathematical constructs to formally demonstrate that information was authenticated by a specific party, with no risk of mischaracterization. It relies on two related primitives: a *secret key* (also called private key), which must be held confidential by the signer, and a *public key* derived from the secret key, which acts as a public identifier. A digital signature can only be computed from the secret key and can be verified by comparing it to the public key. The foundational assumption of signature algorithms is that it is computationally infeasible to derive the signature of a message corresponding to the public key without knowing the secret key. Consequently a message authenticated with a signature verifiable with public key is assumed to originate from an entity with knowledge of the corresponding secret key.

Satoshi Nakamoto defines a digital *wallet* as a secret and public keypair[2]. A wallet is a form of decentralized bank account, which can receive funds, execute transfers out, and be attached to a balance and a transaction history. The public key of a wallet acts similarly to a bank account number: it may be shared with other network participants to allow transfers in. The secret key, assumed to be known only to the wallet owner, is used to authorize transfers out through the computation of a digital signature authenticating the transfer. An external observer may assess the legitimacy of a transfer by verifying the validity of the digital signature as compared to the transfer data and the public key of the sender. A digital, yet physical wallet is shown in Figure 3.1.

---

[2] A wallet may include a *collection* of multiple keypairs in practice. In particular, it is a good practice for a BTC wallet to create new keypairs for every incoming and outgoing transaction.

Figure 3.1 A digital, yet physical wallet of the form of a banknote, as issued by Swiss company BitcoinSuisse. The banknote exposes a public key and hides the corresponding secret key. It can store an arbitrary amount of BTC.

This paradigm allows the generation of digital wallets on the fly and at will, with no pre-authorization requirement, in a privacy-preserving manner. The secret key of a wallet is the only proof of ownership: if lost, destroyed, or stolen, so is the asset: it represents a form of distributed, anonymous identifier — a poor man's digital identity. We refer to Chapter 9 for a more extensive review of the main challenges and practices related to key management and how related risks can be mitigated.

3.3 Decentralized network

3.3.1 Client-server model

Historically, our interaction with other users and services on the Internet has been predominantly based on a client-server paradigm. Servers are centralized computers that allow connections in and act as information broadcasters, relays, data stores, and processors. Rather than favoring direct connection between users, a client-server model relies on an intermediary server controlling the information flow, as demonstrated in Figure 3.2.

This creates an asymmetric relation where the service consumer acts as a client only and delegates the servicing role to specialized third-party computers. This multi tier architecture has shown to be easier to work with in the past given the simplicity of its topology: knowing the IP address of a server and communicating with a unique endpoint removes the need for network discovery, data broadcasting, and synchronization between multiple nodes, the management of intermittent connections, the management of persistent storage, etc.

In a decentralized system, one aims to remove the dependency on specific centralized servers. Such servers may fail, may corrupt business processes, or may misrepresent information with an ill-intent, because of an attack, or by mistake: they are antithetic to the

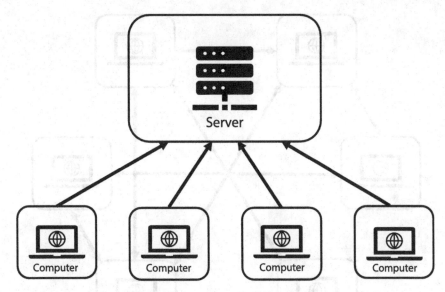

Figure 3.2 A centralized client-server communication model. Own graphics.

decentralization objective and represent a single point of failure and trust which has been repeatedly exploited by attackers to destabilize online services. Data corruption, data thefts, denial of service, or phishing attacks are just some examples that servers experience and that clients suffer from. In addition, control over central servers can easily be abused by the service provider, or by powerful third-parties such as governments and regulators; surveillance or censorship are the typical low hanging fruits that such parties aim to achieve.

3.3.2 Peer-to-peer model

Bitcoin builds on a so-called *peer-to-peer network*, graphically presented in Figure 3.3, where every *peer* (or *node* or *user*), may take the role of both a client and a server at the same time. Peers consume information from other peers but also can serve information to other peers by maintaining bilateral connections with each other.

This model favors a symmetrical network in which every node has the capacity to consume data from other nodes but also to become an active intermediary and data server. Because of the duplication of servers and the possibility of a dense interconnectivity, well developed peer-to-peer networks are inherently resilient to many forms of network failures, are highly available and persistent. Because data can be massively replicated over geographically distributed and heterogeneous servers, interrupting the network purposedly or because of a failure is often unrealistic. Some of the mainstream applications of peer-to-peer networks have included files sharing, streaming, distributed computing or telecommunication systems.

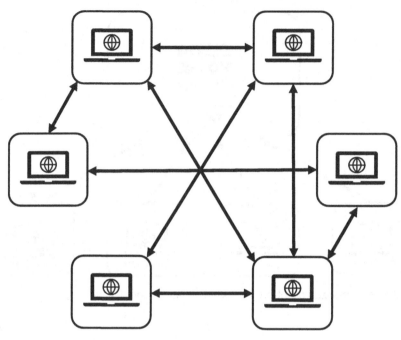

Figure 3.3 A decentralized, peer-to-peer communication network. Own graphics.

3.3.3 Network nodes

Any Internet user may deploy a Bitcoin node acting both as a client to the network and as a relay server. Each node interacts with other nodes to maintain a dynamic view over the underlying Bitcoin ledger, called *blockchain*, and to audit its content and integrity. Some nodes, called *archive nodes*, may decide to store and offer a complete copy of the blockchain history since its inception in 2009 until the latest contemporaneous updates; others may decide to prune the data and rely on archive nodes when necessary. *Seed nodes* advertise their IP addresses and serve as entry points for new network participants aiming to download their own copy of the blockchain. In 2020, it was estimated that the total number of Bitcoin nodes exceeds 10,000, with a wide distribution around the globe, which guarantees an extreme resilience to catastrophic risks, jurisdiction risks, technology failure, and cyber attacks. Top 10 countries involved in BTC mining as of October 9, 2020 are listed in Table 3.1.

Bitcoin nodes do more than store and serve the blockchain data. They also act as protocol enforcers: they actively monitor the traffic, verify the validity of transactions, the validity of the blockchain, and the good application of protocol rules in general to police any dissident behavior. Invalid transactions, incorrect accounting operations, and other forms of illicit actions are intercepted by cooperative nodes and purposely ignored so that they do not pollute the network. Because nodes are built from open-source software and may be audited, compiled, and executed by any Internet user for free, a user may always have a high degree of certainty that his node strictly enforces the protocol rules and locally verifies that the

Table 3.1 Top 10 countries with their respective number of reachable nodes on October 9, 2020.

Rank	Country	Nodes
1	Germany	1842 (17.28%)
2	United States	1834 (17.20%)
3	France	556 (5.21%)
4	Netherlands	426 (4.00%)
5	Canada	310 (2.91%)
6	United Kingdom	291 (2.73%)
7	Singapore	259 (2.43%)
8	Russian Federation	235 (2.20%)
9	Japan	220 (2.06%)
10	(others)	2629 (24.66%)

Source: bitnodes.io.

actions taken by the network do not infringe on any rule agreed on at the inception of the network, as originally defined by Satoshi Nakamoto. One says that nodes enforce the *consensus rules*.

3.3.4 Bootstrapping of a node

In practice, a node follows the below lifecycle:

(1) bootstrap phase: a new node connects to one or multiple seed nodes whose IP addresses are publicly available; seed nodes may be any cooperative nodes already connected to the network.

(2) peer discovery: the node asks seed nodes for the IP addresses of other nodes known to them in the network, possibly recursively in order to build a local map of the available nodes and their respective IP endpoints.

(3) state recovery: the node downloads the ledger history, in part or fully, from any combination of cooperative nodes discovered.

(4) network servicing: the node optionally becomes a cooperative node itself and starts relaying information received from peers to its own peers and to allow other peers to request copies of the locally stored ledger. While doing so, the node validates consensus rules and blocks any information that diverges from them.

Once the node is set up and up-to-date with the rest of the network, it starts listening to all network traffic in order to learn about new instructions and transactions being broadcast and processed by the decentralized system. It also acts as a local gateway to the network, from which all sorts of protocol-related instructions can be broadcast: this is how new cryptocurrency transactions are communicated to the whole set of network participants: like a sort of exponential chain reaction, a node broadcasts the instruction to a set of connected peers that immediately share it with their own respective peers, and so on until most of the nodes have been informed of the new request.

3.4 Distributed ledger

A distributed ledger is an accounting database jointly maintained by multiple loosely coupled and autonomous parties communicating over a peer-to-peer network. The principal use case of this technology is to share a unique reference accounting between different, possibly distrusting parties, and guarantee that none may take a sole control over the network, manipulate its good execution, access private insider data, selectively censor the network, or hold other parties hostage of a particular business model. Although there are many different architectures and protocols capable to support a distributed ledger, Satoshi Nakamoto's model, including a blockchain data structure, has been the first successful attempt at building a fully decentralized, permissionless, and secure ledger.

3.4.1 Permissioning framework

Although a blockchain may be deployed as a centralized or semi-centralized data store, it shines when openly shared on an open peer-to-peer network and distributed amongst multiple parties following an appropriate consensus algorithm. Who these parties are depends on the nature of the ledger itself with regards to its permissioning framework as defined in its protocol: in the case of cryptocurrency networks such as Bitcoin or Ethereum, one speaks of *permissionless* ledgers because any user may join them, download the full blockchain, and assume an active maintenance role with no preliminary human authorization process. One refers to a *permissioned* ledger when restrictions on whom can read it, update it, or access it for any purpose, may apply (e.g., R3 Corda, Hyperledger Fabric)[3]. Whichever the framework and specific rules, distribution is the best assurance against risks of data loss, data corruption, and tampering: by having many redundant copies of the same information — copies ideally geographically distributed and under the control of diverse, independent entities — such networks become prohibitive to attack as each and every participant fully contributes to its full shape and form.

3.4.2 Blockchain data structure

A blockchain is a cryptographic chain of data blocks. Each block contains a batch of transactions that chronologically succeeds transactions of previous, parent blocks and precedes transactions of possible future, children blocks. With the exception of the original block, also called the *genesis block*, each subsequent block maintains an explicit cryptographic

---

[3] See https://www.r3.com/corda-platform/ and https://www.hyperledger.org/use/fabric.

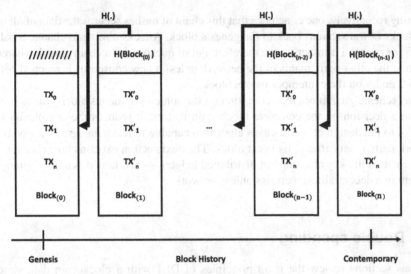

Figure 3.4 A simple blockchain data model with each block storing a list of transaction and a hash pointer to the previous block. Own graphics.

pointer to its parent such that the chain may be naturally traversed and the complete transactions graph reconstructed. The blockchain therefore maintains the exhaustive list of validated transactions from genesis to contemporaneous updates, as exhibited in Figure 3.4.

Formally, a simplistic blockchain data structure may be specified as follows:

(1) Let $hash(message)$ be a cryptographic hash function. By construction, $hash$ is a one-way in that one practically cannot generate a message that yields a given output value nor find two different messages with the same output value (more details in Chapter 4). Hash can be used to compute the digital equivalent to a fingerprint, such that any change to the original message, as minor may it be, invalidates its hash.

(2) Let $block_0 : (transactions_0)$ be the genesis block of a blockchain, where $transactions_0$ is a list of transactions included in the block.

(3) Let $block_i : (hash(block_{i-1}), transactions_i)$ be the data structure of any descendent of the genesis block, where $i > 0$ and $transactions_i$ is the list of transactions included in $block_i$.

(4) A blockchain is a sequence of blocks $(block_i)_{i=0}^N$, where N is the index of the contemporaneous block.

By constructing the ledger with the above structure, one is able to create trust in the integrity of data despite the untrusted environment within which the ledger is often exposed and maintained. As long as the most recent block $block_N$ is known and agreed on, the complete history is strictly defined and cannot be subject to any tampering without breaking the chain validity. The reason is evident in its recursive definition: $block_N$ contains the hash of $block_{N-1}$, therefore $block_{N-1}$ cannot be tampered with undetected. $block_{N-1}$ contains the hash of $block_{N-2}$, therefore the same data integrity guarantee applies. Applying a similar reasoning recursively, one concludes that this chain of hashes secures the data of all historical blocks of transactions back to the genesis block. Attacks aiming to change a balance, modify, or cancel a transaction are therefore out of question on a chain already shared and known to the other participants of the network unless a new consensus is reached between network nodes on the contemporaneous block.

This scheme guarantees that a decision on the contemporaneous block $block_N$ is equivalent to a decision on the complete blockchain history. It reduces the complexity of the protocol to the design of a consensus algorithm capable to reach an agreement with other network participants around its latest block. The next section explains how reaching such a consensus is the key challenge of distributed ledgers — and how it is a surprisingly hard problem in a decentralized, permissionless network.

3.5 Double spending

Previous sections review the main principles of DLT with a blockchain data structure. Communication between nodes is free of intermediary, or peer-to-peer; the ledger is persisted by any voluntary node such that many redundant copies exist; data is authenticated with appropriate hashing schemes such that the whole history is tamper-evident up to the agreed contemporaneous block; user accounts are pseudonymous secret-public key pairs, where the public key is a shareable account identifier and the secret key is used for transactions authorization with a digital signature.

The above model is powerful but in itself does not naturally enforce basic accounting principles. A payment network must provide high guarantees that a user cannot spend more than he owns. This may sound like a simple problem to solve — in fact, it generally amounts to simply verifying that the sum of outgoing transfers is no larger than the total value of incoming transfers — but distributed ledgers come with their own complexities that make the problem extremely complex. We illustrate the main issue, commonly called double spending, below.

3.5.1 Illustration in a centralized setup

Consider the example of a traditional bank account with a balance of US$1,000. Let us also assume that the account owner would like to buy a red bike and a blue bike, each worth US$1,000. Intuitively, the owner can only purchase either the red bike, or the blue bike given that the total value of both goods is US$2,000, which exceeds his total wealth of US$1,000. Let the owner try to cheat the system by creating two payment requests in a short instant, the first one paying the US$1,000 invoice for the red bike and the second one paying the US$1,000 invoice for the blue bike. The bank receives a notification for both payment requests, sequences them (for instance based on a timestamp or a priority), and initiates the processing of the initial payment request. After a simple check, the bank can verify that the balance is sufficient for the first payment, process it, and update the balance to US$0.00. The processing of the second payment request is then initiated but fails before execution because of the insufficient balance. The owner is notified that his transaction was not executed.

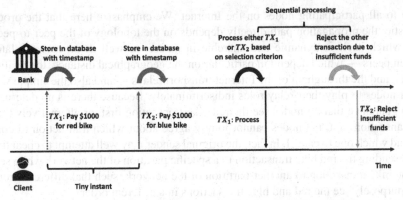

Figure 3.5 An attempt of double spend in a centralized setup. Own graphics.

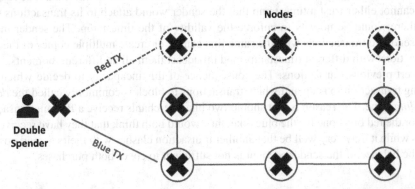

Figure 3.6 An attempt of double spend in a decentralized, peer-to-peer setup. Two conflicting transactions are propagated through the network, reaching peers at different instants. Own graphics.

This process looks trivial in a centralized context, where the bank is the only party to process payments because it is, see Figure 3.5.

3.5.2 Illustration in a decentralized setup

Let us now consider a similar situation in a decentralized setup such as Bitcoin. The account owner now holds 1 BTC on the Bitcoin ledger and he would like to buy both a red bike and a blue bike, each worth 1 BTC. Intuition suggests that the owner can only purchase either the red bike, or the blue bike, as the total value of both goods is 2 BTC, which is larger than his total wealth. The owner aims to cheat the system and he creates two transactions almost at the same time: the first transaction sending 1 BTC to pay for the red bike and the second transaction sending 1 BTC to pay for the blue bike. Details are shown in Figure 3.6.

The peer-to-peer broadcast of both transaction requests starts, and each network node notified of either transaction immediately broadcasts it to its own set of neighbor nodes. Following a cascade effect, one, or the other, or generally both transactions quickly become known to all participating nodes on the Internet. We emphasize here that the process is stochastic: the propagation path directly depends on the topology of the peer-to-peer network, which itself is dynamic and anarchic in nature, as well as on network latencies between peers — which depend in particular on the geographical distance, the quality, the quantity, and the throughput of the Internet transport relays — and also the simple willingness of nodes to play their relay roles indiscriminately. Because there is no deterministic way to guarantee that all nodes see the red bike transaction first, or alternatively the blue bike transaction first, the nodes cannot jointly agree about which transaction to consider valid and which one to cancel. In fact, the original sender may well attempt to cheat the system by sending the red bike transaction to a specific partition of the network while sending the blue bike transaction to another partition of the network, such that different groups of nodes purposely see the red and blue transactions in a different order.

As there is no natural ordering of transactions, there is no network-wide agreement on which of the two transaction should be considered valid and which one should be rejected. One cannot either trust a timestamp that the sender would attach to its transactions given that there would be no way to enforce the validity of the timestamp. The sender may as well send two transactions with the same timestamp, or create multiple copies of the same transactions with different timestamps and broadcast them all at different moments, trying to revert previous transactions. The consequence of this incapacity to decide which conflicting transaction to keep and which transactions to block is commonly called the *double spending* issue. The reason is that should two bike merchants receive a Bitcoin transaction, one for the red bike, one for the blue bike, they would both think that they have received the funds while it may very well be that another transaction clashes with theirs — and possibly that the balance of the sender account is not sufficient to cover both purchases.

3.6 Network consensus

It is obvious that a distributed ledger needs a robust solution against double spending, or in other words, it must embed a *consensus algorithm* allowing each network participant to reach high certainty that their view on which transaction to keep is shared by the rest of the network. A consensus algorithm is precisely about agreeing on a decision when there is more that one node involved, and making sure that the agreement is robust to different kinds of failures or attacks. In the case of double spending, the consensus aims to promote a decision about which transaction to keep and which to disregard. The consensus need not guarantee that the earliest transaction be kept; it merely drives the general agreement around one that the "network" agrees with.

Many different consensus algorithms may be considered for the problem at hand, although the most adapted algorithm will depend on the topology of the network (e.g., how many nodes) and the degree of trust between participants. A permissioned ledger, for which the nodes are generally pre-approved, well-known, and in a small number, many consensus algorithms have existed for years; see the review and comparaison of such algorithms in Chaudhry and Yousaf (2018). For a permissionless ledger such as the Bitcoin blockchain

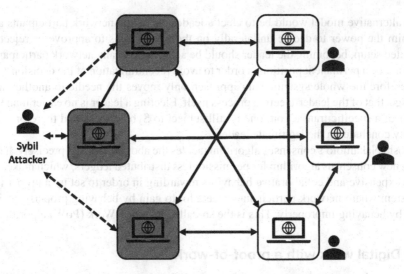

Figure 3.7 A Sybil attack illustrated with a single user controlling three different nodes and gaining 50% of the network representation. Own graphics.

where nodes are unknown, can join and leave the network at any time and for low cost, the problem is complex. Further details are given in Chapters 5, 6, and 7.

3.6.1 Sybil attacks

Consider a simple model where a majority of votes by the network participants is required to approve a transaction. Under this model, one waits that a majority is reached before concluding that a transaction is final. In practice, the bike merchant only hands in the bike after verifying a 51% vote in favour of the transaction paying for his good. Would such a model suffice?

A careful analysis quickly reveals that it would not. In fact, one cannot enforce a fair allocation of the voting rights in a permissionless setup. How can the system guarantee that a network participant only votes once when it has the technical capability to multiply its Internet connections and artificially inflate its presence at will, undetected? This is the so called *Sybil attack*: a network participant attacks the system by creating an arbitrary number of pseudonymous identities and uses them to increase his influence on the outcome of an agreement, as demonstrated in Figure 3.7. Sybil attacks are particularly effective in that the creation of identities can be achieved at low cost and therefore massively scaled so as to destroy any trust in the voting process. Related to this, one also concludes that nothing prevents a voter from voting in favor of two incompatible outcomes. Why would voters be trusted to only vote on a single possible scenario? The voter may decide to vote in favor of the red bike transaction and then additionally vote for the blue bike transaction, possibly after the red bike transaction has been agreed on by a majority already. Again, because there is no cost to voting, the assumption should always be that some malevolent network participants will exploit the weakness and create distrust in the system.

An alternative model would be to elect a leader out of the network participants and to grant him the power to decide unilaterally on the transactions to approve or reject. In a distributed setup, becoming the leader should be achievable to any network participant and should not be a permanent position in order to avoid recentralization of the decision-taking, and therefore the whole system. This approach only moves the needle to another similar challenge, that of the leader-election process itself. Electing a leader is no easier than voting in favor of a specific transaction: one is still subject to Sybil attacks and therefore cannot naturally conclude on the legitimate leader.

Satoshi Nakamoto's consensus algorithm tackles the above challenge precisely. It establishes a new consensus algorithm for permissionless distributed ledgers, which makes Sybil attacks expensive and collaborative behaviors rewarding in order to set up a proper incentive system where network participants have a lot to gain by behaving properly, and a lot to lose by behaving improperly. This is the so-called Proof-of-Work (PoW) algorithm.

3.6.2 Digital work with a proof-of-work

In a context where every network participant can inflate its perceived presence at will — for instance by artificially multiplying its Internet connections — and where participants have no reason to trust each other, how can one set up a process under which an network-wide agreement may be reached? Should there be a leader taking decisions, and if so, how can the participants agree on whom this leader should be? Should there be a leaderless, direct democracy model, and if so, how to assure that every participant only votes once?

It seems clear that Sybil attacks prevent any form of naive democratic model because of the inherent distrust between anonymous participants and their capacity to abuse a voting scheme at low cost and with no penalty. To work, Bitcoin must demonstrably limit the ability of participants to inflate their voting power arbitrarily. Bitcoin must set up a fair framework where each and any participant may contribute, but voting is a costly enterprise. It must create an incentive system encouraging users to behave well and penalizing dissidents for misbehaving. Bitcoin must create an identity-less democratic system that does not allow fraud.

Satoshi Nakamoto sets the Bitcoin protocol so that the validity of a transaction block is conditioned on the preliminary resolution of a computationally expensive problem — or *work*. The protocol guarantees that only blocks with a valid proof of resolution to the problem — or proof-of-work — are accepted as valid blocks; blocks without a valid proof-of-work are rejected by the network nodes. Satoshi purposely designs the problem such that finding a solution is a computationally expensive, random challenge, while verifying the solution is a trivial and deterministic process. Also, verifying the solution reveals the estimated amount of work invested by the solver of the problem. This approach still allows any network participant to submit a new block of transactions to the network as long as this block contains a valid proof-of-work and therefore demonstrates to any observer that the work was achieved by the submitter — work which directly translates into an investment both in time and money. The specifics of this work are left to Chapter 5.

This requirement has two main consequences. First, it reduces the rate of blocks. Because any network user must first solve a complex problem, submitting a valid block can take a long time. The lower frequency and inherent stochasticity of the solving process decreases the probability that multiple valid blocks appear at the same time, in particular in presence of network latencies. Second, and most importantly, it makes block submission (i.e., vote on the next block) costly. Generating two blocks is twice as expensive as generating one block because it requires twice as much work, in average. We insist here that this work is not just a theoretical construct: it corresponds to a concrete operational cost related to the purchase and maintenance of computers and the financing of electricity. Performant hardware processors will produce a solution faster on average than less performant hardware; horizontal scaling of processors will, too, accelerate the resolution time. By relating the production of valid blocks to a liability, Satoshi alleviates the incentive to trick the network by submitting multiple conflicting blocks.

3.6.3 Building the chain

The construction of the blockchain works according to the following process.

(1) Download the existing chain of blocks up to the contemporaneous block from other peers on the network.
(2) Assess the validity of the chain, of its accounting, of its transactions, and its proofs of work in respective blocks.
(3) Select a set of pending transactions waiting to be included in the blockchain. These transactions, which have been exchanged between the peers of the network after the broadcast of their respective senders, are stored in a *memory pool* of respective network nodes, waiting to be included in a new block.
(4) Build a block with the selected transactions, set the parent hash within the block, and start computing the proof of work.
(5) If and when a valid proof is identified, include it within the block and broadcast the block to the other nodes.
(6) Let other nodes verify the integrity of the block, including its proof-of-work, and decide accordingly on whether to include it to their local copy of the blockchain.

Interestingly, other network participants do the same exact process to submit their own opinion on the next block. Many questions remain however. Given the cost associated with block production, why would anybody willingly engage into this activity? How should the complexity of the work be defined, by whom, and should it be time varying? Which transactions would be selected for inclusion within a block? Which block to choose in cases where multiple conflicting, yet valid, blocks are broadcast to the network with the same parent?

3.6.4 Block rewards and miners

The production of blocks, which is conditioned on obtaining a valid proof-of-work, is an expensive enterprise. Satoshi Nakamoto knew too well that network users would not engage into this activity without a proper incentivization scheme. He set up as part of the protocol a *reward*, paid in BTC, that the creator of a block receives when successfully getting his block approved by the rest of the network.

Due to the loose comparison to the mining of gold, where extracting gold nuggets requires infrastructure, physical effort, and time, block production was metaphorically named *mining* — and users performing this work called *miners*. Some participants of the Bitcoin network would decide to become active maintainers, or miners, and would start tracking the network to gather pending transactions from other users, group them into blocks, start solving the proof-of-work, and contingent on its resolution, broadcast the block to receive the reward. The reward would need to cover running costs (e.g., electricity, infrastructure maintenance, depreciation) to allow a sustainable business model.

The Bitcoin protocol includes two components as part of the reward.

(1) *block reward*: freshly minted BTC adding up to the total number of BTC in the network. Importantly, the block reward schedule — or monetary policy — is part of the core consensus protocol of Bitcoin, which guarantees the immutability of the scheme even against influential participants of the network. Satoshi Nakamoto set the block reward such that the inflation rate is originally high and then predictably converges to zero. One can loosely compare this model to gold mining, which was relatively easier to extract in the 19th century given the abundance of unexploited mines, and is getting rarer and more expensive to extract as mines get emptier.

(2) *transaction fees*: the sum of all transaction fees of transactions included in the block, as paid by the respective senders. Fees, although optional, may be defined by the transaction sender to incentivize inclusion in a block. By setting a higher fee, the sender makes his transaction more attractive to the miner, which may process it with higher priority and therefore include it in an earlier block.

The sum of both rewards constitutes the total BTC reward paid to the miner of a block.

3.6.5 A computational power race

The definition of the difficulty of the proof-of-work problem remains. We start with what it should *not* be: a problem of constant complexity. Let's assume that at the inception of the Bitcoin network, only Satoshi Nakamoto himself is mining the network, and that he is doing this with his personal computer. Setting the mining complexity can be calibrated such that solving the problem takes an arbitrary amount of time, for instance, 10 minutes, in average. Satoshi mines block after block and confirms that on average, it takes 10 minutes to solve the problem and therefore produce a new contemporaneous block. Assume now that a new network participant, say, Hal Finney, starts mining blocks. The processing power allocated to mining is now twice as large as when Satoshi operated alone. Consequently, the average block production rate is now 5 minutes. Half of the time, Satoshi finds the solution first; half of the time Hal finds it first. Now let's assume that Hal wishes to get more control over the network. Hal invests in eight new equivalent computers and gets them running side by side for mining. As a result, the block production rate is now 10 times as fast as originally; also, Hal produces a block 90% of the time while Satoshi solves the problem first only 10% of the time; he therefore gets 90% of BTC rewards but also has significantly higher operational costs.

Considering the openness, or permissionlessness, of the network, allowing any user to connect powerful computers and compete for block rewards, one quickly perceives the obvious outcome: that of a computational power race leading to a growing block production

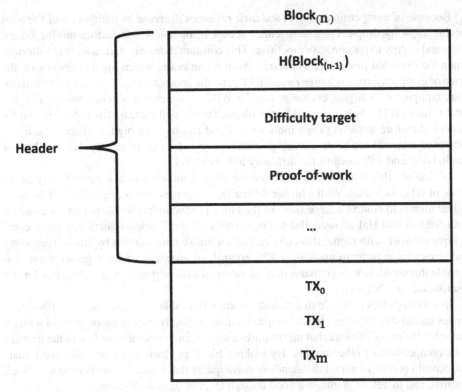

Figure 3.8 A block of transactions with additional data in its header to track the current difficulty and the proof-of-work. Own graphics.

rate. Eventually, block production may be so fast that the whole proof-of-work mechanism becomes useless. One concludes that mining cannot be a constant complexity problem: it must be adaptive to the amount of computing invested by miners such that for instance, the resolution time remains constant on average.

3.6.6 Difficulty under competition

The *difficulty* target defines the hardness of the proof-of-work problem. Difficulty is endogenously defined and time-varying such that the expected block production rate is stable, or equivalently, that the time between each blocks is constant on average. The specific duration has no clear justification except that it is slow enough that two valid blocks should rarely be submitted at the same time by miners, and fast enough that waiting for a blockchain update is still tolerable; Bitcoin sets it at 10 minutes. This approach ensures that as the aggregate computational power of miners increases, the difficulty target rises; alternatively, as the computational power of miners decreases, the difficulty target drops. The outcome is a constant target of block production rate. Because of its importance, the difficulty level target is included in a block's header, as demonstrated in Figure 3.8.

Because of competition, miners see their revenues decrease as their share of the total mining capacity drops. They must either accept it, or invest into further mining equipment and energy to remain as competitive. This constant race only finds a local equilibrium when the expected mining gains balance the running costs, which highly depends on the price of electricity, the exchange rate of BTC, and the computational efficiency of the hardware equipment. A higher exchange rate for BTC — which equivalently means a higher total value of BTCs being secured by the blockchain — will increase the potential gain for miners, therefore attracting even more miners and leading to a higher difficulty; a lower exchange rate will see some miners go bankrupt or reduce their activity due to a lack of profitability and will readjust the difficulty to lower levels.

Ultimately, this scheme sees an increasing pool of miners and a rising difficulty as the value of BTC increases. With a higher difficulty, it becomes increasingly difficult for individual miners to control a large stake of the block production: whereas the above example with Satoshi and Hal allowed Hal to control 90% of the block production, a more competitive network with higher difficulty requires a massive investment by Hal in computers and electricity in order to maintain a 90% control; an investment which, passed some reasonable threshold, is so high that it may be assumed to be virtually impossible. For further discussion, see Chapter 5.

It is through this principle that Satoshi Nakamoto is able to create a decentralized consensus model that regulates the block production. Anybody may attempt to build a block and solve the proof-of-work, but this requires a significant investment and faces the tremendous competition of other miners. By making block production a free, capitalistic market, Satoshi is able to attract thousands of participants that naturally compete to win block rewards, and therefore maintain a good enough level of decentralization.

Is this enough though? Not quite. The above scheme reduces the pace of block production and defines a arguably fair, decentralized block production algorithm. However, we have not described the behavior of the network when multiple blocks may clash; in other words, when the blockchain forks.

3.6.7 Forks and consensus chain

A *fork* happens when two or more valid blocks refer to the same parent. In that case an ambiguity exists over which block to consider, and which block to build on for further blocks. Such forks may happen with no ill intent, for instance when multiple miners happen to solve the proof-of-work almost at the same time, broadcast their freshly produced block, and create confusion with the other network participants who cannot decide which to follow, as illustrated in Figure 3.9. A fork may also be the result of an attack to the system where miners aim to revert or modify some part of the accounting by creating an alternative view of the blockchain where transactions may differ.

A selection rule is required. Such a rule that unambiguously allows any network participant to decide on which subchain to trust and build on. But because all blocks involved in the fork may be technically valid, a rule based on the block itself may not be possible. Here different kinds of distributed ledger may have different answers. Bitcoin defines a simple policy based on the so-called *longest chain* rule: in case of a fork (which may involve two or more branches), the branch with the longest cumulative proof-of-work difficulty is the

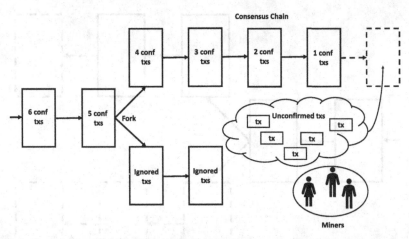

Figure 3.9 A general overview of the blockchain with the example of a fork, and with unconfirmed transactions pending insertion by a miner. Own graphics.

one that is chosen as the consensus chain. The longest cumulative difficulty generally coincides with the longest branch in terms of its number of blocks, but this is not guaranteed. Whenever a fork occurs, network participants wait until additional blocks are created in order to conclude that one of the branches is longer (difficulty-wise) than the other; ambiguity is therefore resolved by only considering the subchain that seems to be the chosen one by miners. We say that this is the *consensus chain*, and this defines the proof-of-work consensus algorithm.

3.6.8 The 51% attack

Because the block reward is paid within the respective produced block, the miner only gets paid if its block ends up in the consensus chain; the reward is otherwise entirely ignored. Miners must target the eventually longest chain or, when they don't, must hope that their shorter chain grows faster than the consensus chain so it catches up with the longest chain and becomes the new consensus chain. Growing a chain faster, for obvious computational reasons, implies that the shorter chain must gather more than 50% of the total mining capacity of the network — so that new blocks are statistically created at a higher frequency than on the competing fork and can catch up with the consensus chain.

How would the creator of a block included in a shorter chain convince half of the network to grow it despite the risk of losing their own rewards for the blocks yet to create, while other miners could simply commit to the already longest chain without the uncertainty? In practice, rational miners never do it. The danger that a shorter chain never catches up the consensus chain makes it suboptimal for all miners to diverge from the natural equilibrium, that is, mining the consensus chain.

However, the possible existence of a colluding group of miners surpassing 50% of the mining capacity of the network changes the dynamics: such miners may produce blocks at a

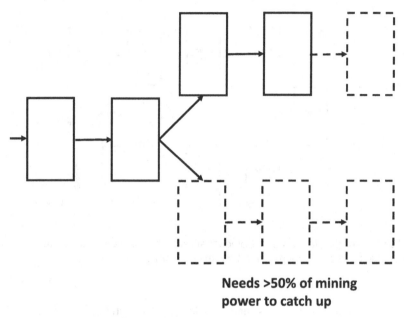

**Needs >50% of mining
power to catch up**

Figure 3.10 A fork with two branches, one longer than the other. The shorter chain needs more than 50% of the total mining power persistently to catch up with the longest chain. Own graphics.

higher rate than the sum of any other network participants, and therefore unilaterally define the longest chain, as exhibited in Figure 3.10. This is the so-called *51% attack*. The attacker is in control of the longest chain and may prevent any other miner from contributing to the consensus chain by forking them away, or may even create an alternative fork starting in a past and slowing catching up with the consensus chain to replace it entirely — reverting all accounting. For the consensus algorithm to be trusted, the risk of such an event happening must be low enough that it can be ignored. For this reason, the security of the Bitcoin network is loosely defined by its total mining power (or its difficulty target): the higher the difficulty, the more expensive and unlikely it is for an attacker to surpass 50% and take ownership of the block production.

Interestingly, Satoshi Nakamoto's consensus algorithm does not assume that miners are honest whatsoever. As long as no colluding group of miners — honest or not — reaches a large enough control of the total mining power, the system is safe against most forms of attacks. We note however that the above explanation forgets about many exogenous factors which may derail the natural incentives. For instance, a wealthy attacker may attempt to harm the system with no immediate need to breakeven (e.g., governments); or may attempt to manipulate the BTC market by hurting the trust in the system and profiting from independent trading strategies, e.g. short selling; or may implement advanced behaviors dangerously shifting the incentives and game theoretic equilibrium. In practice, and despite many sorts of theoretical attacks, Bitcoin has remained resilient — anti-fragile — and is still perceived as the most secure, permissionless distributed ledger.

3.6.9 Confirmations and finality

One important consequence of the consensus algorithm is the lack of deterministic transaction finality. Once broadcast, a transaction is quickly seen by all network nodes but it is not guaranteed to ever be included in the blockchain. We say that a transaction pending block inclusion is *unconfirmed*. Once the transaction is included in the chain, we say that it is *confirmed*. There is still no guarantee that the block within which it lies will be part of the consensus chain later on however; there exists a tiny chance that a fork may appear which may invalidate the block within which the transaction was included.

Because long forks shall rarely occur in absence of a 51% attack, a confirmed transaction included deep enough in the history of the blockchain is considered more secure than a transaction being part of the contemporaneous block of the consensus chain. Therefore one speaks of *number of confirmations* to refer to the depth of the transaction within the chain. A *1-confirmation* transaction implies that the transaction is within the latest, contemporaneous block; a *2-confirmations* transaction implies that the block within which the transaction is included has been built on by a single new contemporaneous block. The higher the number of confirmations, the lower the probability of a reversal due to a fork catching up with the consensus chain.

In theory, there is never an absolute guarantee that any transaction, as deep as it is within the blockchain, is absolutely final; in practice, one considers that a few confirmations may be sufficient to reach a great level of certainty. The number of confirmations is a rule-of-the-thumb, which highly depends on the specifics of the consensus algorithm and its parametrization. It varies from a cryptocurrency to another — and a reasonable number of confirmations on Bitcoin may be very different from a reasonable number of confirmations on another platform, such as Litecoin or Ethereum.

3.6.10 The limits of proof-of-work

Although the proof-of-work algorithm of Bitcoin has been a revolution in providing a robust consensus within a permissionless, decentralized network, it does not come for free:

(1) it consumes an important amount of energy, which depends on the total processing power allocated to mining (and therefore indirectly varies with the market value of BTC);
(2) it produces blocks at a slow rate (which can be reduced, but not without tradeoffs);
(3) it has inherent scalability limits, which prevent high frequency, high throughput applications on chain;
(4) it is subject to a risk of recentralization due to economies of scale benefitting large miners;
(5) it never guarantees the finality of a transaction;
(6) it requires high performant purpose-built hardware and low electricity prices for mining to remain competitive.

There has been a considerable amount of research to alleviate the above drawbacks, that is, to decrease the electricity footprint, accelerate the production of blocks, increase the maximum transaction throughput, mitigate the risk of centralization, reach deterministic transaction finality, and make the consensus process more accessible without specific hardware requirements (e.g., Chaudhry and Yousaf (2018); Sompolinsky and Zohar (2013); Kokoris-Kogias et al. (2018); Saleh (2020); Baird (2016)).

3.6.11 Staking as an alternative to working

Proof-of-stake (PoS) is an alternative class of consensus algorithms for distributed ledgers. As compared to proof-of-work, proof-of-stake does not require the resolution of a computationally expensive problem; rather, it allows its users to directly vote on the next valid block with a digital signature. In order to prevent Sybil attacks, the voting influence is proportional to the amount of *staked coins* — larger financial commitments to the voting process leading to a larger influence. Because a right to vote relates to a staked coin (i.e., a financial commitment on the ledger) rather than to a number of connections, Sybil attacks are naturally prevented. Notably, Peercoin, Tezos and Ethereum 2.0 rely on some variant of a proof-of-stake consensus algorithm.

Proof-of-stake has numerous advantages in principle:

(1) it consumes a negligible amount of energy;
(2) it supports a higher rate of blocks;
(3) it potentially facilitates scaling techniques such as sharding.

However, proof-of-stake also comes with its complexities and weaknesses. Whereas proof-of-work inherently punishes miners for mining on a wrong fork (i.e., they pay electricity for their work but do not get a reward if the block is not in the consensus chain), proof-of-stake makes staking costless. Therefore, a naïve implementation of the algorithm may allow users to vote for not just one block, but possibly multiple incompatible forks, with no penalty. In such a setting, proper incentives and an equilibrium may not exist; it may become a best response for users to vote in favour of any possible block, be they incompatible or should they include double spends. This is the so-called *nothing at stake with proof-of-stake* paradox: one votes with its money, but has nothing to lose in case of incorrect or malicious vote.

Proof-of-stake also may be less resilient than proof-of-work in case of extreme events, such as a long-term network split. Consider the case of a complete network split between two regions of the world. Each partition of the network is only aware of its own fork of the chain. When reconnected, the network must decide which fork to keep. Bitcoin achieves consensus trivially by selecting either fork based on the longest chain rule (therefore reverting the dominated chain entirely). Under a proof-of-stake model, both splits are convinced that their fork is valid and there may not be a universally agreeable rule for chain selection.

These only cover some examples of challenges associated to proof-of-stake, which is an active field of research in academia and in the cryptocurrency community.

3.7 Path to the after-Bitcoin

3.7.1 Pizza day

After its 2009 launch, Bitcoin slowly captures the interest of cryptographers and engineers who start perceiving its massive technological potential — but potentially do not fully anticipate its soon-to-be commercial and political impact. The first documented use of BTC as

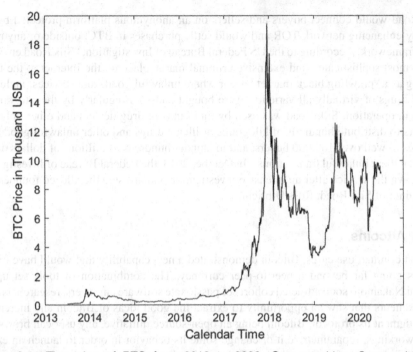

Figure 3.11 The price of BTC from 2013 to 2020. Own graphics. Source: https://coinmarketcap.com.

a currency was tagged on May 22, 2010, as software developer Laszlo Hanyecz ordered two pizzas for the price of 10,000 BTC, implicitly setting the BTC value at approximately US$0.004 per unit. The famous event became so significant in establishing the feasibility of BTC as a means of payment that it became widely known as the "Bitcoin Pizza day" and continues to be celebrated every year. In 2020, Laszlo's 10,000 BTC would be worth about US$100 million.

Two months later, exchange Mt Gox launches its BTC trading venue and starts gaining massive traction. For the first time, the value of BTC is tracked and the market liquidity is aggregated on a single dominant platform. The impressive traction of the asset quickly brings the cryptocurrency to parity with the dollar, in the first quarter of 2011. The growth continues with BTC surging to US$31 in summer that same year before being subject to a significant crash. The high growth and high risk properties of BTC would remain preponderent during the whole decade as it alternates between high growth and depression periods yet managed to grow by orders of magnitude to value of approximately US$10,000 per coin in 2020. The winding trajectory of BTC price with spectacular surges and crashes since 2013 is presented in Figure 3.11.

3.7.2 Silk Road

In parallel emerged the dark web market Silk Road. Launched in 2011, Silk Road aimed at facilitating the exchange of illicit products and services with a focus on drugs and weapons.

Silk Road would connect buyers and sellers on an anonymous platform protected by the privacy-enhancing network TOR and would settle purchases in BTC, outside of any regulated framework. According to the US Federal Bureau of Investigation, "Silk Road emerged as the most sophisticated and extensive criminal marketplace on the Internet at the time, serving as a sprawling black market bazaar where unlawful goods and services, including illegal drugs of virtually all varieties, were bought and sold regularly by the site's users. While in operation, Silk Road was used by thousands of drug dealers and other unlawful vendors to distribute hundreds of kilograms of illegal drugs and other unlawful goods and services to well over 100,000 buyers, and to launder hundreds of millions of dollars deriving from these unlawful transactions." In October 2013, the Federal Bureau of Investigation shut down the service after an intensive investigation and arrested the alleged founder and maintainer of Silk Road, Ross Ulbricht.

3.7.3 Altcoins

Beyond criminal use cases, Bitcoin demonstrated a new capability that would have implications going far beyond a peer-to-peer currency. The combination of tools set up by Satoshi Nakamoto soon attracted cohorts of passionate software engineers, researchers, and entrepreneurs that saw an opportunity to extract the good ideas of Bitcoin and incrementally augment its protocol. Bitcoin being an open-source initiative, any user can browse its inner workings, reparametrize it or change some its behavior in order to launch an exact, or modified, copy of it. Once deployed, the replica bootstraps its own blockchain starting with a new genesis block and applies the consensus rules of the new network. In essence, the replica only shares design principles with Bitcoin, but it is not Bitcoin — the same way twin brothers share a similar DNA but are different instances of it.

In 2011, Litecoin was launched by Charlie Lee, an ex-employee of Google. Litecoin is similar to Bitcoin in many ways but it changes three main aspects of the protocol. First, it replaces the mining algorithm with an alternative scheme assumed to be more resilient to some recentralization risks associated with Bitcoin's primitives. Second, it decreased the block time down to 2 minutes, as compared to Bitcoin's 10 minutes. Third, Litecoin changes the inflation schedule while maintaining the ultimately capped monetary supply. Although minor, these changes were sufficient to attract the interest of investors, miners, and engineers. Litecoin saw its market value surge, as well as its reach of its ecosystem. In 2020, the market capitalization of LTC lies around US$3 billion.

Following the commercial success of Litecoin and its creator, hundreds of arguably purposeless copies of Bitcoin quickly appeared. Startups specialized in the deployment of reparameterized itcoin copycats even emerged, allowing their paying clients to deploy a new Bitcoin-like with, for instance, a custom name, a different inflation schedule, a custom average block time, within minutes. Naturally, most of these alternative coins, also called *altcoins*, quickly disappeared given the absence of differentiation, use cases, and ecosystem. Interestingly, some of them did manage to survive: amongst them dogecoin, a meme-inspired cryptocurrency whose only true innovation was to attach some funny dog picture to the coin (see Figure 3.12), managed to get enough media attention to bootstrap its ecosystem and grow both in market capitalization and in support. In 2020, dogecoin reaches a market capitalization of about US$400 million. Some seeing the irrational growth

Figure 3.12 The icon of DogeCoin inspired by a meme.

of Bitcoin replicas spoke of a massive bubble of bubbles; others just saw an interesting experiment of a free market of arbitrary, scarce tokens.

3.7.4 Ethereum and smart contracts

It is in 2013 that Vitalik Buterin, a cryptocurrency researcher, programmer, and young entrepreneur first announced his project to create a new cryptocurrency platform: Ethereum. Rather than being a mere copycat of Bitcoin, Ethereum aimed to reconsider multiple technical foundations of Bitcoin and build a new system capable of alleviating its big brother's weaknesses: Ethereum would be faster, more scalable, and more capable. For detailed discussion of Ethereum, see Chapter 6.

But more than anything, Ethereum aimed at introducing the notion of stateful and Turing complete *smart contracts*. Invented by Nick Szabo in 1993, a smart contract is a piece of software designed to automatically coordinate and enforce the execution of contractual terms. A smart contract helps disintermediate the contractual relationship between multiple

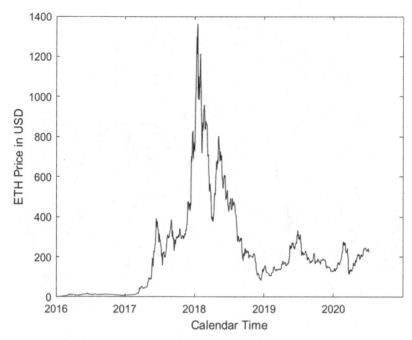

Figure 3.13 The price of ETH from 2016 to 2020. Own graphics. Source: https://coinmarketcap.com.

parties and decrease the risks of errors, frauds, and subjective interpretation, with the ultimate goal to decrease costs and uncertainties. The blockchain provided a powerful platform to consider the execution of smart contracts: with its strong guarantees of immutability and integrity, it could theoretically execute software in a fully decentralized, yet trusted way.

Satoshi Nakamoto designed Bitcoin to solve the use case of payment — not for the digitization of contracts. Although programmable to some extent, Bitcoin was, and still is, too limited to process contracts allowing complex behaviors and recurring interactions. This is what Ethereum promised: a platform that could not only process payments but also maintain and execute user-defined smart contracts capable of maintaining a state, receiving and triggering events, and enforcing terms over the flows of underlying assets. As compared to normal software, the terms of the smart contract, once created, become immutable and use the distribution of the blockchain to guarantee the integrity of the execution. The Ethereum foundation raised US$14 million in an online crowdsale in summer 2014 and the first version of the network was launched one year later. In 2020, ETH is the second largest cryptocurrency by market capitalization with a total value of US$50 billion, see Figure 3.13. In 2021, ETH capitalization has increased by an order of magnitude.

3.7.5 Tokenization

In parallel to the birth of Ethereum, a series of initiatives to digitize, or *tokenize*, traditional assets onto the blockchain emerged. The principle was simple: if the blockchain were so practical for native cryptocurrencies, why not to use it for other asset classes such as equities, real estate, commodities, fine art, or others? One would benefit from all qualities of distributed ledgers but for assets unrelated to cryptocurrencies. Bitcoin proponents started building so-called *colored coins* protocols with the aim to augment the native capabilities of Bitcoin to support the issuance and management of traceable tokens. In part due to the plurality of emerging colored coin protocols, the technology did not succeed at gaining enough momentum to become widespread. The lack of flexibility of the Bitcoin protocol also made it difficult to operate such protocols in practice or to build new functionality on top of the tokens. The increasing success of Ethereum and its flexible smart contract constructs quickly attracted most tokenization projects.

With its generic smart contract capabilities, Ethereum provided all the tools to create powerful token management contracts. Such contracts would need to maintain the balance of each token holder, to allow transfers, and depending on the kind of contract, allow the introduction of more advanced features related to the lifecycle management of the token, such as corporate actions for equity tokens, coupons payments for bond tokens, minting and burning for variable-supply tokens, etc. In 2015, a smart contract interface holding the name of Ethereum Request for Comments number 20 (ERC-20) is proposed to standardize the format of contracts managing fungible tokens and to define the common list of capabilities that an Ethereum token has to implement. This interface is not mandatory but it aims to standardize the interaction with token smart contracts such that the ecosystem can build new services to interact with tokens with the assumption that all tokens technically look the same. This push toward standardization, helped by the relative simplicity of the ERC-20 protocol, quickly becomes a massive success with hundreds of tokens building on this default interface. Colored coins on the Bitcoin ledger lose momentum and get eclipsed by the adoption of the new Ethereum standard.

3.7.6 Initial coin offerings

The Initial Coin Offering (ICO) craze started in 2016. The ICO terminology is coined after the more traditional Initial Public Offering (IPO) related to the listing of a company on an exchange. IPOs have historically been limited to large companies deciding to move away from a private shareholder structure to a public listing where their shares are traded freely at a market-defined value. Due to the complexity of existing exchange platforms, their rules, and the various regulatory and reporting requirements, IPOs have notoriously been inaccessible to earlier stage companies looking to raise capital or to improve the liquidity of their stocks — whether for cost reasons, or for lack of profitability. ICOs aimed to resolve that and offer a crowdfunding platform for early stage and later stage ventures also allowing the early listing of their shares, or tokens.

A series of early-stage companies prepared white papers and business plans to support the promotion of their crowdfunding based on an ICO. In practice, the issuing company distributes new tokens in exchange for established cryptocurrencies. The tokens may be denominated in a new cryptocurrency packaged with its own new distributed ledger, or it may leverage a token protocol on an existing ledger, such as ERC-20 contracts on Ethereum. This is not unlike the foundation event of a traditional company where initial investors get a share of the capital depending on the liquidity they bring into the company. However as compared to company shares, ICOs are generally open to any investor irrespective of their geographical location and the size of their investment; also, once issued, the tokens are freely exchangeable peer-to-peer and can often be immediately traded on secondary markets (through centralized exchanges or decentralized trading). As compared to publicly traded company shares, ICO tokens can be issued and held with no intermediary and they are highly interoperable.

After the fast-growing success of ICOs in 2016 and 2017 and their many failures — Ponzi schemes, exit scams, and other schemes defrauding their investors — regulators in most jurisdictions banned ICOs or at least formalized the regulatory processes around their implementation. Regulations being antithetical to the even principle of decentralized finance, the attractiveness of ICOs as they were implemented in 2016 and 2017 massively decreased. In January 2018, the total market capitalization of cryptocurrencies and tokens reached its all-time high at US$831 billion before collapsing by more than 85%.

3.7.7 Decentralized finance and autonomous organizations

In parallel to the growth of the adoption of cryptocurrency and tokens, the end of the decade was marked by the increasing importance of a broader ambition to decentralize finance and commerce as a whole. Rather than relying on centrally-regulated, "real-world" companies, a new notion of *decentralized autonomous organization*, or DAO, gained traction. DAOs are the decentralized equivalent to companies or associations: they rely on smart contracts and are tokenized on a distributed ledger such that "shareholders" are anonymous token holders. Rather than relying on articles of association, shareholders' agreements, the local law, or other complex legal documentation, the governance of DAOs, as well as its profits distribution, are enforced by its smart contracts directly; they can be arbitrarily defined within the contract with no relation to the laws and habits of usual regulated companies. Because of their transparency and immutability, DAOs are inherently robust to many forms of risks (e.g., legal, jurisdiction) and, by removing intermediaries such as commercial registers or banks, they provide a highly efficient vehicle for international, self-enforced, anonymous organizations.

DeFi, short for decentralized finance, has been the hottest development in 2020 with a complete financial system being replicated little by little in a fully decentralized way such that it does not depend on central banks, commercial banks, payment processors, governments, or regulators. From *stablecoins* providing a stable token indexed to some national currency or basket of national currencies (see Chapter 8), to peer-to-peer exchanges, derivative products, decentralized insurances, prediction markets and gambling contracts, decentralized oracles, and lending tools, DeFi is growing massively and opens the perspective of a fully disintermediated, alternative financial system where established financial institutions lose their dominant position. Whether DeFi is yet another massive bubble ready to pop, whether it is the future of finance or a short-term experiment, one thing is sure: blockchain technology is far from having reached its full potential and we are prone to discover exciting new use cases that today are still invisible to both skeptics and experts.

3.8 Summary

This chapter reviews the philosophical roots of Bitcoin and its growth alongside with the rest of the crypto economy. We introduce the fundamental building blocks of cryptocurrencies, such as the blockchain and the proof-of-work consensus, and explain how they solve the main challenge of decentralized currencies: double spending. We argue about the value of cryptocurrencies and introduce the latest trends emerging from decentralized finance.

For additional inputs, see Wüst and Gervais (2018).

3.9 Exercises

1. What is the main distinction between bank money creation and bitcoin mining, peer-to-peer lending, etc.?
2. Describe the main differences between unpermissioned public ledgers, permissioned public ledgers, permissioned private ledgers, and centralized ledgers. Come up with examples illustrating the usefulness of each of these.
3. We consider a 256-bit secret key. Express the number of possible keys in scientific, decimal notation. Assuming a powerful computer can generate 1 billion keys per second, how long would it take to iterate through every possible combination? What if one could generate 1 trillion keys per second? Why, in practice, it is not necessary to worry about key collisions if the key is truly random?

Essential Cryptographic Tools

"One of the most singular characteristics of the art of deciphering is the strong conviction possessed by every person, even moderately acquainted with it, that he is able to construct a cipher which nobody else can decipher. I have also observed that the cleverer the person, the more intimate is his conviction."

Charles Babbage, Passages from the Life of a Philosopher

4.1 Introduction

In Greek, the word Cryptography ($K\rho\upsilon\pi\tau\varpi\gamma\rho\alpha\phi\eta\sigma\eta$) means hidden writing. Cryptography has been used since times immemorial by various parties interested in hiding their messages from adversaries and preventing adversaries from forging a message purportedly coming from the original sender.

Until the middle of the 20th century, all cryptography relied on the symmetric key paradigm. To encrypt and decrypt a message, both the sender and the receiver must possess the *same* secret key. In the 1970s, several groups of researchers achieved a remarkable breakthrough and discovered the asymmetric key cryptography. Now, a sender can use the receiver's public key, readily available to all, to encrypt the message, while the receiver can use her secret (or private) key to decrypt it. Conversely, the sender can use her secret key to sign a message, which anyone can verify by applying her public key.

As we shall see below, in the original Bitcoin, Ethereum, and many other protocols, no information is encrypted per se. These protocols rely on the fact that asymmetric cryptography allows one to digitally sign messages, with a valid signature being an irrefutable proof that the signer owns the secret key associated with her public key. This feature allows one to establish the ownership of the bitcoins associated with a particular public address. The legitimate owner of the address can transfer this value to a different public address if she wishes to.

One more cryptographic construct is of paramount importance — a hash function. A useful hash function can map a message of an arbitrary length into a fixed-length message, say 256 bits. It is practically impossible to find another message with the same hash, even though, obviously, such messages do exist. Using hash functions, one can significantly reduce the computational burden associated with the requisite cryptographic operations, such as digitally signing a message, because it becomes possible to sign the hash of the message, rather than the message itself. In simple terms, one can think about a hash as a fingerprint of a person — it tells you nothing about who the owner is, but the owner can easily prove that the fingerprint is indeed hers.

Somewhat paradoxically, modern cryptography is simultaneously extraordinarily complex and relatively simple. On the one hand, the relevant cryptographic ideas are among the most complex achievements of humankind. On the other hand, at the end of the day, encrypting or signing messages relies on the fact that there are so many possible permutations that their straightforward analysis, called the brute force attack, is not possible given the abilities of modern computers. Besides, the so-called cryptographic breaks — operating faster than brute force attacks — should not be possible either.[1] One can think of the process of decrypting a message without knowing the key as searching for a needle in the stack of needles, not a needle in a proverbial stack of hay. Time will tell if bastions of modern cryptography begin to crumble with the advent of quantum computing.

A mathematical aside is in order — the entire edifice of modern cryptography is probabilistic in nature. A typical statement usually goes along the lines "if p is a huge prime number, then a particular algorithm is secure." However, if p is huge, as it should be in our day and age, it isn't easy to check that it is prime, other than probabilistically. Simultaneously, theorems in this field, such as the celebrated and extremely useful Hasse's theorem, described below, are, of course, absolute. One has to appreciate that these theorems are based on the assumption that p is indeed prime for their conclusions to be valid.

In this chapter, we provide a gentle introduction to cryptography. We cover both symmetric and asymmetric key cryptography and hash functions, digital signatures, and a few other concepts. Modern cryptocurrencies rely on the so-called elliptic curve cryptography (ECC). Since a frontal attack on the requisite ideas is hard, we start with the Finite Ring Cryptography (FRC) due to Rivest-Shamir-Adleman (RSA), which is easier to understand. Then we touch upon the Finite Field Cryptography (FFC) before proceeding to more advanced topics related to the ECC.

The chapter proceeds as follows. In Section 4.2, we start with a brief discussion of symmetric cryptography to put things in perspective. In Section 4.3, we introduce the all-important concept of asymmetric key cryptography based on the public-secret key pair. Section 4.4 contains a concise exposition of number theory, which are used in subsequent sections to build the set of distributed ledger technology-related cryptographic tools. In Section 4.5, we put these tools to good use by developing the so-called FRC, also known as RSA cryptography. We emphasize that no cryptocurrencies rely on the RSA digital signature algorithm directly; however, this framework is valuable as a starting point for more recent algorithms. In Section 4.6, we cover FFC including its most important derivatives — the celebrated ElGamal and Schnorr digital signature algorithms. These algorithms are particularly interesting since they are usable in the ECC framework, which we discuss in detail in Section 4.7. We compare three digital signature algorithms in Section 4.8 and conclude that at present, EC digital signature is best suited to DLT needs. However, in the long run, the EC-based version of the Schnorr signature algorithm will win. In Sections 4.9 and 4.10, we discuss the second cryptographic primitive needed for DLT — the so-called hash functions and secure hash algorithms. In Section 4.11, we cover Mekle trees and hash pointers.

---

[1]Of course, one can never be sure that this is true, hence, exhaustive adversarial analysis of any cryptographic algorithm is necessary before it is used in practice.

In Section 4.12, we outline potential threats to DLT emanating from the nascent field of quantum computing. We summarize the chapter in Section 4.13.

It is worth mentioning that this chapter is relatively technical. The reader should consume the material at her own pace. Taking the main conclusions for granted should not prevent the reader from mastering the rest of the book.

4.2 Symmetric key cryptography

4.2.1 Introduction

To start with, we introduce some terminology. A plaintext message contains the information in its standard form. A ciphertext or cryptogram is the transformed message. The key (keys) is (are) the secret parameter(s) for the encryption, which are known only to the sender and intended recipients. The secret key(s) are used to perform the transformation from the plain text to the cryptogram (encryption) and from the cryptogram to the plain text (decryption). Ciphers are based on a general algorithm employing a secret key to perform encryption and decryption. Cryptanalysis represents a method for transforming a cryptogram back to the original plaintext without previous knowledge of the key.

The general setup is shown in Figure 4.1; Alice uses a secret key to encrypt a message and broadcast it to Bob, who uses the same secret key to decrypt the cryptogram. As usual, Charlie, the adversary, tries to eavesdrop on Alice's to Bob's communications using cryptanalysis. If he succeeds, he can not only read the communications but, in principle, try to forge messages.

According to Kerckhoffs principle, also known as Shannon's maxim, a cryptographic scheme is secure when everything about it, except the key, is publicly known, and yet it cannot be broken. Shannon put it somewhat differently: "one ought to design systems under

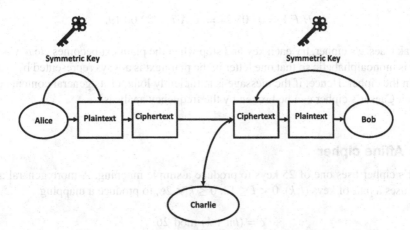

Figure 4.1 Conceptual framework for symmetric key cryptography, which relies on a shared key that is used to both encrypt or decrypt text/cyphertext. Own graphics.

the assumption that the enemy will immediately gain full familiarity with them." For details, see, e.g., Kerckhoffs (1883); Shannon (1949).

4.2.2 Caesar's cipher

As the name suggests, Caesar's cipher goes back to antiquity. Every plaintext letter is replaced by the letter located k positions further to the right in the alphabet, which is cyclically closed. For example, we can transform the letter A into F, say, so that:

$$
\begin{array}{c}
\text{ABCDEFGHIJKLMNOPQRSTUVWXYZ} \\
\Downarrow \\
\text{FGHIJKLMNOPQRSTUVWXYZABCDE}
\end{array}
\tag{4.1}
$$

We can replace every letter with a number, $A \to 0, B \to 1, ..., Z \to 25$, and express the key as an integer, for instance, $F \to k = 5$. Assuming that the i-th letter in the message is represented by an integer $m_i, 0 \le m_i < 26$, the i-th letter in the cryptogram is an integer c_i:

$$
c_i = \left(m_i + k \right) \bmod 26.
\tag{4.2}
$$

Thus,

$$
C\,A\,T = 02\ 00\ 19 \Rightarrow 07\ 05\ 24 = H\,F\,Y.
\tag{4.3}
$$

The decryption is performed as follows:

$$
m_i = \left(c_i - k + 26 \right) \bmod 26.
\tag{4.4}
$$

Thus,

$$
H\,F\,Y = 07\ 05\ 24 \Rightarrow C\,A\,T = 02\ 00\ 19.
\tag{4.5}
$$

To break Caesar's cipher, try each key and stop when the plaintext becomes clear. Caesar's cipher is monoalphabetic so that one letter in the plaintext is always represented by a single letter in the cipher. Hence, if the message is sufficiently long, or its general content can be guessed, Caesar's cipher can be broken by the frequency analysis.

4.2.3 Affine cipher

Caesar's cipher uses one of 25 keys to produce a simple mapping. A more general affine cipher uses a pair of keys $(l, k), 0 < l \le 12, 0 \le k < 26$, to produce a mapping

$$
c = (lm + k) \bmod 26.
\tag{4.6}
$$

The reader should verify that the number of different keys (l, k) is $12 \times 26 = 312$.

4.2.4 A simple substitution cipher

A simple substitution creates a table of plaintext characters and their corresponding crypto images. For example, this is a substitution table used in the first rotor of the celebrated Enigma machine (see below):

$$ABCDEFGHIJKLMNOPQRSTUVWXYZ$$
$$\Downarrow \qquad (4.7)$$
$$EKMFLGDQVZNTOWYHXUSPAIBRCJ$$

Similarly to Caesar's cipher, a substitution cipher is monoalphabetic, so that it can be broken by the frequency analysis, even though there is an enormous number of possible substitutions. More precisely, $26! = 4.03 \times e^{26}$. Thus, the number of possible substitutions alone does not make a cipher secure.

4.2.5 The Vigenère cipher

The cipher, mistakenly attributed to Blaise de Vigenère, but invented by Giovan Battista Bellaso in 1553, creates a stream of key letters, and uses them one after another for Caesar's encryption of the corresponding plaintext letter, thus using the polyalphabetic substitution; see Singh (2000). If the key is shorter than the message, which is usually the case, the key is repeated until it matches the message's length. For centuries, it was known as *le chiffre indéchiffrable* (the indecipherable cipher). Eventually, Friedrich Kasiski published a general method of deciphering Vigenère ciphers in 1863.

Assuming that the i-th letter in the message is represented by an integer is m_i, $0 \le m_i < 26$, the cryptogram is an integer c_i:

$$c_i = (m_i + k_i) \bmod 26, \qquad (4.8)$$

where k_i is the i-th letter of the key. For example, if we use a cipher with the key "KEY," the keys for each round are $10, 04, 24, \ldots$, or, equivalently, A–>K, A–>E, A–>Y, $\ldots$, then the word CAT becomes the word MER,

$$C\,A\,T = 02\ 00\ 19 \Rightarrow 12\ 04\ 17 = M\,E\,R. \qquad (4.9)$$

4.2.6 The Enigma machine

The Enigma machine, also called the Enigma, was invented by the German engineer Arthur Scherbius at the end of World War I. Initially, it was sold commercially mostly to private enterprises, who wished to keep their communications confidential. Eventually, the German Armed Forces and several other militaries started to use it in earnest. The Enigma relies on the same principle as the Vigenère cipher and produces a very long substitution sequence.

Not surprisingly, the secret source is in the automated process of the key generation. The electro-mechanical Enigma uses three rotors, a plugboard, and a reflector to achieve its goal. Rotors move after each keystroke, creating a very long key sequence, much longer than any of the messages sent. The number of possibilities is seemingly huge. To calculate their total number, one needs to know what rotors are used (three-out-of-five: 60 combinations),

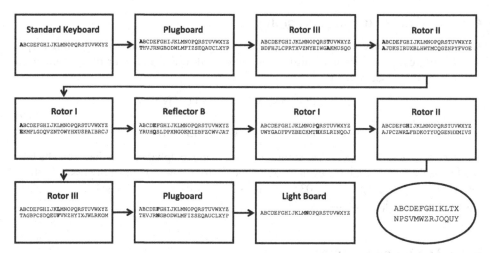

Figure 4.2 The inner workings of the Enigma Machine. Own graphics. Source: https://www.codesandciphers.org.uk/enigma/.

what are their starting positions ($26 \times 26 \times 26 = 17,576$ combinations), and what letters are exchanged (100,391,791,500 combinations). The Enigma exterior is well-known; its inner workings are shown in Figure 4.2.

In this figure, we show a particular implementation. All mappings are the actual mappings used during the war. Messages containing only letters (the sender has to spell the numbers out) are entered via a standard keyboard. A plugboard, reconfigured daily, is used for the initial interchange of letters. A set of three rotors further transforms transposed letters, then a reflector, and then the same rotors used in reverse. The plugboard performs the final interchange of letters. Finally, the enciphered letter is lighted on the light board, recorded by an operator, and sent via the Morse code by radio. Due to the presence of the reflector, the same machine can be used for decryption. However, there is a price to pay. The design is such that a letter will never be encoded into itself. At the time, this feature was viewed as a major strength, yet, turned to be a major weakness. A nice illustration of Babbage's quote used as an epigraph to this chapter. For the configuration in Figure 4.2, we get 13 pairs shown in an oval in the lower-left corner: (A,N),...,(X,Y). We highlight the path of the letter "A" through the machine in bold: A–>T–>A–>A –>E–>Q–>H–>L–>F– >N, so A–>N, as shown. We left it to the reader to show that N–>A. Of course, if this were all that Enigma did, it would be trivial to break. The trick is that Enigma performs a Caesar's shift after each letter is entered into the machine by shifting rotors against each other, starting with the right-most (Rotor III in the figure).

It is very well-known that some of the Enigma messages were broken by the British at Bletchley Park. What is somewhat less known is that most of the Royal Navy messages were also broken more or less in real-time by the *Beobachtungsdienst* (Observation Service) — the German analog of Bletchley Park.

4.2.7 A one-time pad

A one-time pad (OTP) is the only theoretically secure cipher. It was discovered by Frank Miller in 1882, and rediscovered by Gilbert Vernam and Joseph Mauborgne in 1918; see Kahn (1996); Miller (1882). In simple terms, the OTP generates a Vigenère sequence with a randomly chosen key as long as the message. The one-time pad has perfect security in the sense that

$$H(M|C) = H(M),\qquad(4.10)$$

where $H(.)$ is the so-called Shannon entropy and $H(.|.)$ is the conditional entropy. Thus, M and C are statistically independent. In other words, for every message M and corresponding ciphertext C, we can find at least one key K that binds them as a one-time pad.

The intuition behind OTP security is relatively simple. Assume that we have the complete statistics for every possible plaintext of the length L. For a given cryptogram and a plaintext M, there is a corresponding key K encrypting that plaintext M into the cryptogram C. Since every key is exactly as likely as another, one has no clue to which plaintext is the more likely one.

In binary notation, OTP operates with XOR (exclusive or) bitwise arithmetic, so that

$$0 \oplus 0 = 0, \;\; 1 \oplus 0 = 1, \;\; 0 \oplus 1 = 1, \;\; 1 \oplus 1 = 0.\qquad(4.11)$$

For instance, if

$$m = (1111100000), \;\; k = (0011000101),\qquad(4.12)$$

then

$$c = m + k = (1100100101), \;\; m = c + k = (1111100000).\qquad(4.13)$$

Despite its theoretical appeal, there is a problem with OTPs — to use the OTP properly, we must generate a truly random key sequence, which is as long as the message we want to encode. Thus, transferring the key is as hard as the message itself. For example, if you wish to encrypt a computer drive, you need another drive of the same length — obviously an issue! In practice, we can use a stream cipher, which we describe next.

4.2.8 A stream cipher

To succeed with an application of OTP, we need to have a perfectly random key. What if, instead, we try to generate a pseudo-random (but not truly random) "keystream" from a seed. We call the seed a "real key," the advantage being that it is much shorter than the full "keystream" added to the message. Still, the set of possible "real keys" or seeds has to be so sufficiently large, to make an exhaustive search of the seed space practically impossible. If this approach is used, pseudo-randomness (rather than full randomness) is a prerequisite, since both sender and receiver must generate the same keystream from the seed. Given the very nature of pseudo-random generators, their output is periodic, which is a problem in theory, but less so in practice, provided that the period is long enough. In this case, if the key is unknown, every output bit is unpredictable.

A simple way of generating a pseudo-random key is to use linear congruential generators (LCG); see, e.g., Knuth (1997). These are the oldest and the best-known pseudo-random generators. An LCG generates pseudo-random sequences as follows:

$$X_{n+1} = (aX_n + b) \bmod Y, \tag{4.14}$$

where the seed is the starting point of a sequence X_0, and, possibly, the integer coefficients (a, b). A popular choice for the modulus Y is $Y = 2^{32}$, but other values are used as well. Some properties of LCGs, such as speed, and small memory footprint, are useful, but some are not, since in the end of the day, the outputs are predictable and can be simulated in several independent streams. LCGs are not suitable for cryptographic applications.

Linear feedback shift registers (LFSR) generalize and improve LCGs; see, e.g. Press et al. (2007). An LFSR operates as follows

$$X_{n+k} = (a_0 X_n + a_1 X_{n+1} + ... + a_{k-1} X_{n+k-1}) \bmod Y. \tag{4.15}$$

The seed key consists of two sequences $(a_0, ..., a_{k-1})$ and $(X_0, ..., X_{k-1})$. The coefficients a_l are binary, i.e., a_l are 0s or 1s. LFSRs are easy to build by using simple electromechanical or electronic circuits. Provided that Y is sufficiently large, the output streams have almost uniform distribution and long periods. However, the linearity of LFSRs makes them vulnerable to brute force attacks.

4.2.9 Data encryption standards

In 1974 the National Bureau of Standards (NBS) called for proposals for a data encryption algorithm. An IBM team run by Horst Feistel submitted a successful algorithm currently known as the Data Encryption Standard (DES). At present, DES is obsolete. However, on occasion, DES is still used for non-crucial tasks.

DES uses both software and hardware to perform rather involved scrambling; see Diffie and Hellman (1977). The DES key is 64-bits long, but effectively it is 56-bits long since one bit in every byte is used for error checking. The DES algorithm uses the XOR operation and scrambling and works along the lines of the Enigma machine. A significant advantage of DES is that by reversing the key, you can decrypt the ciphertext using the same steps as before. Modern computers can break the DES ciphertext by brute force; hence, it is advisable not to use it in practice. Yet, it is historically significant and inspirational. Also, the so-called triple-DES (3DES) is still being used is some applications.

Advanced encryption standard (AES) is a modern method used to encrypt and decrypt electronic messages, which is a replacement for DES; for details, see Rijmen and Daemen (2001).

4.3 Asymmetric key cryptography

4.3.1 Introduction

It is natural to start with symmetric key cryptography from a pedagogical standpoint, assuming that the *same* key is used for encryption and decryption; see Figure 4.1. However, it is not the only way to proceed. In the 1970s, several researchers proposed to use *different* keys

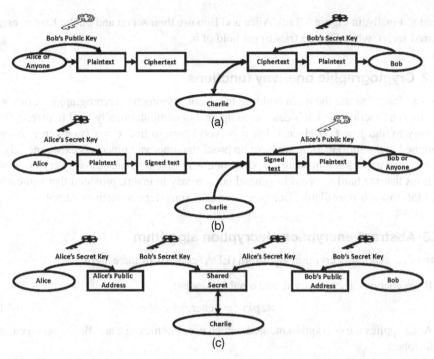

Figure 4.3 Conceptual framework for asymmetric key cryptography, which relies on two linked keys: (a) public key is used to encrypt text; secret key is used to decrypt cyphertext; (b) secret key is used to sign a message; public key is used to verify the signature; (c) both public and secret keys are used to create a common secret. Own graphics.

for encryption and decryption. The realization that this is possible and, in many instances, desirable, opens a new chapter in cryptography — the so-called public-key cryptography relying on public-key infrastructure (PKI). In Figure 4.3 (a), we graphically represent the corresponding encryption — decryption paradigm; in Figure 4.3 (b), we show the digital signature algorithm; while in Figure 4.3 (c), we show how two parties can create a common secret.

As always, we have two protagonists — Alice and Bob — and an antagonist, Charlie. In Figure 4.3 (a), Alice uses Bob's public key that is readily available to encrypt a message, which only Bob can decrypt using his secret key that is known only to her. Charlie tries to decrypt the message intended for Bob and possibly replace it with a forged one. In Figure 4.3 (b), Alice uses her secret key to sign a message or its hash. What is hashing? It is an algorithm for creating a *unique* digest of *fixed* length out of a message of an *arbitrary* length.[2] We shall discuss hashing algorithms later in this chapter since they are central to Bitcoin and all other blockchain protocols. Her signature can be checked by Bob (or anyone else for that matter). Charlie tries to create a forged message and pretends that Alice

---

[2]Typically, a hash function is 160–256 bit long and provides half as much, i.e., 80–128 "bits" of security due to the Birthday Paradox, which we discuss below.

signed it. Finally, in Figure 4.3 (c), Alice and Bob use their secret and public keys to create a shared secret, while Charlie tries to get hold of it.

4.3.2 Cryptographic one-way functions

One-way functions are the main building blocks of asymmetric cryptography. A one-way function is a function, which is easy to compute, but computationally hard to invert. Thus, it is easy to find $y = f(x)$ given x, but it is (very) hard to find $x = g(y)$ given y. As was mentioned in the introduction, we have no proof that one-way functions exist; and rely on evidence and experience instead. Trapdoor one-way functions are one-way cryptographic functions that are hard to invert in general but are easy to invert, provided that some additional information is available. Below we consider some representative examples.

4.3.3 Abstract encryption-decryption algorithm

Abstract encryption-decryption algorithm (EDA) consists of three steps:

(1) Bob generates a secret key sk and a public key pk:

$$(sk, pk) := \text{generatekeys}(); \tag{4.16}$$

(2) Alice applies the encryption method, which uses her message and Bob's pk to generate a cipher:

$$cipher := \text{encrypt}(pk, message); \tag{4.17}$$

(3) Bob applies the decryption method, which uses the cipher, and his sk to get the plaintext:

$$message := \text{decrypt}(sk, message). \tag{4.18}$$

In asymmetric cryptography, a recipient keeps his secret key secret, while making his public key readily available to everyone. By encrypting a message with her public key, the sender allows only the recipient to decrypt the cryptogram with the secret key. Figure 4.3 (a) shows the corresponding flowchart.

4.3.4 Abstract digital signature algorithm

Abstract digital signature algorithm (DSA) consists of three steps:

(1) Alice generates a secret key sk and a public key pk:

$$(sk, pk) := \text{generatekeys}(); \tag{4.19}$$

(2) Alice applies the sign method, which uses a message and her sk to generate a signature:

$$sig := \text{sign}(sk, message); \tag{4.20}$$

(3) Bob applies the verify method, which uses a message, a signature, and Alice's pk to verify if the signature is valid:

$$is\,Valid := \text{verify}(pk, message, sig). \tag{4.21}$$

In theory, Alice can sign a message; in practice, she signs its hash. Because a signature can be uniquely associated with a public key, these keys can be used as identities. Figure 4.3 (b) shows the corresponding flowchart. In asymmetric cryptography, a sender keeps her secret key secret, while making her public key readily available. The sender encrypts a message with her secret key and sends both the cryptogram and message to all interested parties. Anyone can decrypt the cryptogram with the public key and compare the result sent in the open with the decrypted message. If the original message and the one decrypted with the pubic key coincide, it becomes a verification of the sender's digital identity.

Thus, asymmetric cryptography is a perfect tool, which allows us to create a digital signature. Such a signature has many applications. In particular, it can prevent Charlie from forging messages and presenting them as if they were written either by Alice or Bob. Similarly to the physical signature, a digital signature is tied to both the signing user and the message. Encrypting with the secret key does the trick since only Alice can create a signature, which is valid only if the decryption and the plaintext coincide. In the most basic case, no attempts are made to hide the information. This approach is perfectly suitable for certain applications, such as the Bitcoin protocol. However, doing so might not always be desirable, since it might be necessary to keep the message private, so hashing (or digesting) might be more practical. Why do we need hashing? Two reasons: (a) by their very nature, all known signature algorithms (RSA DSA, FFDSA, ECDSA) are slow; (b) by construction, all known signature algorithms generate an output of the same size as the input, see below. In a different setting, we are facing the same issue as with a one-time pad! Thus, it is much more convenient to shorten (hash) the message first and sign its short hash. The required modification is obvious: the pair (*message*, sign(*message*)) morphs into (*message*, sign(hash(*message*))); the comparison is done between hash (*message*) and sign(hash(*message*))).

4.3.5 Abstract secret-sharing algorithm

The idea of a secret sharing algorithm is proposed in Diffie and Hellman (1976); Merkle (1978). Suppose that we have two one-way functions $f(.)$ and $g(.,.)$, which are symmetric in the following sense:

$$g(f(a), b) = g(f(b), a). \tag{4.22}$$

This setup cannot be used for encryption/signing purposes because we cannot necessarily recover a or b. However, it can be used for the key exchange. The corresponding steps are as follows:

(1) Alice chooses a secret random a and makes $f(a)$ public;
(2) Bob chooses a secret random b and makes $f(b)$ public;
(3) *Both* Alice and Bob create k of the form

$$k = g(f(b), a) = g(f(a), b). \tag{4.23}$$

Now, k is the shared secret, known to *both* Alice and Bob.

4.4 Elements of number theory

4.4.1 The set of integers modulo n

For a given n, we are interested in the set of integers $\{0, 1, ..., n-1\}$. Setting aside the trivial case of $n = 2$, we need to distinguish two cases: (a) n is an odd prime; (b) n is a composite number.

(a) Let n be an odd prime number p. We can define the finite field $\mathbb{F}_p$ as a finite field with elements represented by the set of integers: $\{0, 1, ..., p-1\}$. Recall that a field is a set on which addition, subtraction, multiplication, and division are defined and have properties similar to real numbers' corresponding operations.

We define addition and multiplication modulo p:

$$a + b = c \bmod p, \quad a \times b = c \bmod p. \tag{4.24}$$

It is easy to adapt standard algorithms for ordinary integer arithmetic for addition and multiplication in $\mathbb{F}_p$. The additive identity or zero element is the integer 0, and the multiplicative identity is the integer 1. Of course, we need to define subtraction and division of field elements. To this end, it is convenient to define the additive inverse (or negative) and multiplicative inverse of a field element. For $b \in \mathbb{F}_p$, the additive inverse $-b$ of b is $-b = p - b$. The multiplicative inverse b^{-1} of a in $\mathbb{F}_p$ is the unique solution x to the equation:

$$b \times x = 1 \bmod p, \tag{4.25}$$

which can be found via the extended Euclidean algorithm, discussed below. Division and subtraction are defined in terms of additive and multiplicative inverses:

$$a - b = (a + (-b)) \bmod p,$$

$$\tag{4.26}$$

$$a/b = a \times b^{-1} \bmod p.$$

(b) Let n be a composite number, say $n = pq$. Then the set $\mathbb{F}_n$ of integers $\{0, 1, ..., n-1\}$ is a ring, rather than a field. Recall that a ring is a set equipped with addition and multiplication operations with properties similar to the corresponding operations on real numbers. The reason why $\mathbb{F}_n$ is a ring and not a field is clear — while subtraction is always possible, so that $a - b$ is well defined for any $a, b \in \mathbb{F}_n$, division $a \times b^{-1}$ is only possible when b and n are co-prime so that the greatest common divisor of b and n is 1, $\gcd(b, n) = 1$.

It is important to understand that for asymmetric cryptography purposes, the plaintext and cyphertext spaces are either $\mathbb{F}_p$ or $\mathbb{F}_n$.

4.4.2 The Euclidean algorithm

The Euclidean algorithm finds the greatest common divisor (GCD) c of two natural numbers a and b, which is the largest natural number that divides both a and b without a remainder.

It proceeds inductively step by step as follows:

$$
\begin{array}{cc}
\text{Step\#} & \text{Operation} \\
0 & r_{-2} = q_0 r_{-1} + r_0, \\
1 & r_{-1} = q_1 r_0 + r_1, \\
2 & r_0 = q_2 r_1 + r_2 \\
\ldots & \ldots \\
N & r_{N-2} = q_N r_{N-1}.
\end{array}
\tag{4.27}
$$

where $r_{-2} = \max(a, b)$, $r_{-1} = \min(a, b)$. Since the remainders $r_n \geq 0$ decrease with every step, the algorithm eventually runs its course and stops. The corresponding GCD $c = r_{N-1}$.

4.4.3 The extended Euclidean algorithm

Theorem 4.1 Theorem: Given nonzero integers A and B, there exist integers α and β such that

$$\alpha A + \beta B = \gcd(A, B). \tag{4.28}$$

This theorem can be proved along the lines outlined above. Detailed analysis is left for the reader as an exercise.

As was mentioned earlier, if $\gcd(b, n) = 1$, then division by b is possible since the extended Euclidean algorithm yields:

$$\alpha b + \beta n = 1. \tag{4.29}$$

Hence

$$\alpha b = 1 \bmod n, \tag{4.30}$$

so that

$$b^{-1} = \alpha. \tag{4.31}$$

We leave it to the reader to show that division by b is not possible when $\gcd(b, n) > 1$.

4.4.4 Fermat's little theorem

Given the above, we can prove the following useful and exciting theorem.

Theorem 4.2 Fermat's little theorem: If p is a prime number and $\gcd(A, p) = 1$ (p does not divide A), then

$$A^{p-1} = 1 \bmod p. \tag{4.32}$$

The proof is straightforward and archetypal for other similar theorems. Modular multiplication by A is one-to-one since A^{-1} exists. Thus, numbers

$$A \times i \bmod p, \quad i = 1, ..., p - 1. \tag{4.33}$$

are all distinct. Accordingly,

$$(A \times 1)(A \times 2)...(A \times (p-2)) \times (A \times (p-1))$$

$$= 1 \times 2... \times (p-2) \times (p-1) \bmod p. \tag{4.34}$$

The product $1 \times 2... \times (p-2) \times (p-1)$ is invertible because every factor is invertible, so that the theorem holds. QED

4.4.5 Euler's theorem

Euler's theorem is an exciting and very useful generalization of Fermat's little theorem.

Theorem 4.3 Euler's theorem: If $\gcd(A, n) = 1$, then

$$A^{\phi(n)} = 1 \bmod n, \tag{4.35}$$

where $\phi(n)$ is the celebrated Euler's totient equal to the number of integers $1 \le m < n$, such that $\gcd(m, n) = 1$.

The proof is straightforward (but an efficient calculation of $\phi(n)$ is anything but simple!). Since $\gcd(A, n) = 1$, A is invertible, A^{-1} exists, and multiplication by A is one-to-one. The set of integers $m, 0 < m < n$, such that $\gcd(m, n) = 1$ is closed with respect to multiplication by A. Denote these integers by $m_1, ..., m_{\phi(n)}$. As before

$$\left(A \times m_1\right)\left(A \times m_2\right)...\left(A \times m_{\phi(n)-1}\right) \times \left(A \times m_{\phi(n)}\right)$$

$$= m_1 \times m_2... \times (\phi(n) - 1) \times \phi(n) \bmod n. \tag{4.36}$$

Since the product $m_1 \times m_2... \times (\phi(n) - 1) \times \phi(n)$ is invertible, we have

$$A^{\phi(n)} = 1 \bmod n, \tag{4.37}$$

as claimed. QED

If n is prime, then $n = p$, $\phi(p) = p - 1$ (why?), hence we recover Fermat's little theorem. Given Euler's theorem, calculations $\bmod n$ in the base are equivalent to calculations $\bmod \phi(n)$ in the exponent. In general, Euler's totient function is tough to compute! However, for cryptographic purposes described below, it is sufficient to consider a semiprime number n that is the product of two prime numbers $p, q, n = pq$. In this case, we have

$$\phi(n) = (p-1)(q-1). \tag{4.38}$$

For a practical application of this theorem, see example in Section 4.5.5.

4.5 Finite ring cryptography

4.5.1 Modular exponentiation

Consider a large natural number n. For a given natural x, $0 < x < n$, we can define its modular exponent as follows

$$y = x^e \bmod n. \tag{4.39}$$

The inverse function can be written in the form:

$$x = y^d \bmod n. \tag{4.40}$$

Of course, for an arbitrary n, finding the corresponding d is extremely difficult. However, it is easy to do in some cases, while in other instances, it is easy to do provided that some additional information about n is known.

4.5.2 A good trapdoor function?

Let's start with $n = p$, where p is a large prime number. Given Fermat's little theorem, we can efficiently perform calculations mod $(p-1)$ in the exponent.

Indeed, if

$$y = x^e \bmod p. \tag{4.41}$$

then

$$x = y^d \bmod p, \tag{4.42}$$

provided that

$$ed = 1 \bmod (p-1). \tag{4.43}$$

It is clear that we have to have $\gcd(e, p-1) = 1$ for Eq. (4.40) to be solvable. In this case, the extended Euclidean algorithm can be used to find $d = e^{-1} \bmod (p-1)$. Too bad! Thus, we cannot use exponentiation mod p as a trapdoor function, since it is too easy to invert! Not a good choice. But the germ of an idea nevertheless. We can try to use a semiprime number $n = pq$.

4.5.3 A good trapdoor function!

Given Euler's theorem, for a given $n = pq$, we can efficiently perform calculations mod $\phi(n)$ in the exponent. In general, $\phi(n)$ is extremely difficult to compute. However, if someone *knows* that $n = pq$, computation of $\phi(n)$ is straightforward, but the problem is that large n cannot be factored efficiently! Thus, solving

$$y = x^e \bmod n, \tag{4.44}$$

is as hard as factoring $n = pq$. Thus we have our trapdoor function, or, at least, we hope so. Of course, if we know the actual factors, we can use Euler's theorem and write x as

$$x = y^d \bmod n, \tag{4.45}$$

where

$$ed = 1 \bmod \phi(n), \quad d = e^{-1} \bmod \phi(n). \tag{4.46}$$

In 1977 three MIT researchers — Rivest, Shamir, Adleman — proposed algorithms using the trapdoor function given by Eq. (4.44); see Rivest et al. (1978). Their algorithm is known as the RSA algorithm.[3]

4.5.4 RSA encryption-decryption algorithm

To start with, we show how the encryption-decryption works in the RSA framework. Denote the message and the cyphertext as m and c, respectively. As was mentioned earlier, we assume that $m, c \in \mathbb{F}_n$. The algorithm proceeds as follows:

(1) Bob chooses secret primes p and q, and sets $n = pq$;
(2) Bob chooses e with $\gcd(e, \phi(n)) = 1$;
(3) Bob computes d so that $de = 1 \mod \phi(n)$;
(4) Bob makes n and e public but keeps p, q, and d secret;
(5) Alice encrypts m as $c = m^e \mod n$;
(6) Alice sends Bob c;
(7) Bob computes $m = c^d$ and recovers Alice's message m.

For RSA to work, secret numbers p and q have to be prime numbers. How to test a given number n for being prime? Use Fermat's little theorem. Take $A \neq 0, \pm 1 \mod n$. If

$$A^{n-1} \neq 1 \mod n, \tag{4.47}$$

then n is composite; otherwise, it is prime with high probability (but not for sure!). Repeat for many As to increase the likelihood of p, q being prime.

Somewhat ironically, RSA is occasionally used to transmit symmetric keys used for AES or DES encryption.

4.5.5 RSA digital signature algorithm

We summarize the digital signature algorithm, along the lines of the original RSA paper:

(1) Alice chooses secret primes p and q, and sets $n = pq$;
(2) Alice chooses e with $\gcd(e, \phi(n)) = 1$;
(3) Alice computes d so that $de = 1 \mod \phi(n)$;
(4) Alice makes n and e public but keeps p, q, and d secret;
(5) Alice signs m as $s = m^d \mod n$;
(6) Alice sends Bob s and $m \mod n$;
(7) Bob computes $m' = s^e$ and compares it with m;

If $m' = m$, then Alice has signed the message properly!

---

[3]In fairness, an identical algorithm was developed earlier by James H. Ellis, Clifford Cocks, and Malcolm Williamson (1970–1974), but kept secret by their employer — the Government Communications Headquarters (GCHQ). One can assume that the NSA and the KGB also developed algorithms along similar lines.

Let us consider a concrete toy example:

$$p = 601, \quad q = 7, \quad n = 4207 = 601 \times 7,$$
$$\phi(n) = 3600 = 600 \times 6, \quad e = 1463, \quad d = 1127, \tag{4.48}$$

where

$$1463 \times 1127 \bmod 3600 = 1. \tag{4.49}$$

To calculate d given e, we use the Euclidean algorithm, and then apply it in reverse. The direct and inverse steps are as follows:

$$
\begin{array}{l|l}
\text{Direct steps} & \text{Inverse steps} \\
3600 = 2 \times 1463 + 674 & 674 = 3600 - 2 \times 1463 \\
1463 = 2 \times 674 + 115 & 115 = -2 \times 3600 + 5 \times 1463 \\
674 = 5 \times 115 + 99 & 99 = 11 \times 3600 - 27 \times 1463 \\
115 = 99 + 16 & 16 = -13 \times 3600 + 32 \times 1463 \\
99 = 6 \times 16 + 3 & 3 = 89 \times 3600 - 219 \times 1463 \\
16 = 5 \times 3 + 1 & 1 = -458 \times 3600 + 1127 \times 1463
\end{array}
\tag{4.50}
$$

Thus, $\gcd(3600, 1463) = 1$, and $1127 \times 1463 = 1 \bmod 3600$.

To start, we use squaring to compute the following binary exponentiation table for $58^l \bmod 4207$, $l = 2^{k-1}$, $k = 1, ..., 12$; see Table 4.1.

This table allows one to compute $58^{1463} \bmod 4207$ by using the Russian-style exponentiation and representing the exponent 1463 in the binary form:

$$1463 = 1024 + 256 + 128 + 32 + 16 + 4 + 2 + 1. \tag{4.51}$$

This method is extremely efficient since it allows one to compute the requisite factors sequentially. It is clear that instead 1462 multiplications, we only need 15. Accordingly, we have

$$58^{1463} \bmod 4207 = 471 \times 268 \times 1054 \times 144$$
$$\times 1815 \times 3873 \times 3364 \times 58 \bmod 4207 = 2937. \tag{4.52}$$

Thus, we can easily compute 58^π for any π. In general, it is hard to calculate $x = y^{1/e} \bmod n$ for a given y; however, in the case in question, we can do it because we know how

Table 4.1 Exponentiation table for $58^l \bmod 4207$.

k	1	2	3	4	5	6
l	1	2	4	8	16	32
58^l	58	3364	3873	2174	1815	144
k	7	8	9	10	11	12
l	64	128	256	512	1024	2048
58^l	3908	1054	268	305	471	3077

Table 4.2 Exponentiation table for $2937^l \bmod 4207$.

k	1	2	3	4	5	6
l	1	2	4	8	16	32
2937^l	2937	1619	200	2137	2174	1815
k	7	8	9	10	11	12
l	64	128	256	512	1024	2048
2937^l	144	3908	1054	268	305	471

to decompose n into a product of two primes. To illustrate this point, let us compute $2937^{1127} \bmod 4207$. Once again, we compute the binary exponentiation Table 4.2.

Since

$$1127 = 1024 + 64 + 32 + 4 + 2 + 1, \tag{4.53}$$

we have

$$2937^{1127} \bmod 4207 = 305 \times 144 \times 1815$$

$$\times 200 \times 1619 \times 2937 \bmod 4207 = 58. \tag{4.54}$$

Thus, assuming that 58 is a plain message, and 2937 is the encoded message, we managed to decipher it! We can confirm that 1127 is the magic number directly. First, we notice that

$$58^{3600} \bmod 4207 = 1. \tag{4.55}$$

Indeed

$$3600 = 2048 + 1024 + 512 + 16, \tag{4.56}$$

$$58^{3600} \bmod 4207 = 305 \times 471 \times 3077 \times 1815 \bmod 4207 = 1, \tag{4.57}$$

and

$$\left(58^{1463}\right)^{1127} \bmod 4207 = 58^1 \bmod 4207, \tag{4.58}$$

since

$$1463 \times 1127 \bmod 3600 = 1. \tag{4.59}$$

4.5.6 Blind digital signature algorithm

The blind RSA signature algorithm is a variation of the standard RSA signature algorithm. At first, the need for a blind signature is hard to appreciate, but several situations can be envisioned when such a construct is useful. Assume that Alice needs to prove that she has created a message m, but wants to keep it secret, and Ben is ready to help him; see the next section for a practical example. This setup usually implies that Ben is in a position of authority — he can be a notary, a bank (see below), or an election official. To start with, Ben creates the usual RSA key pair (e, d) corresponding to a suitable semi-private n, and makes e, n public, while keeping d secret. Alice, who wishes Ben to sign a message m, chooses a random integer k and gives Ben the modified message:

$$\mu = k^e m \bmod n. \tag{4.60}$$

Although the number μ is random to Ben, he signs the message and returns the signed message to Alice:

$$\text{sign}\,(\mu) = \mu^d = k^{ed} m^d = km^d \bmod n. \tag{4.61}$$

Alice now divides by k (which is this possible in probability) and retrieves his message $m^d \bmod n$ signed by Ben!

4.5.7 Chaum's anonymous cash

David Chaum suggested using blind signatures, along the lines just outlined, to build the famous Digicash — anonymous electronic cash — which can be viewed as an early precursor of Bitcoin. To describe Chaum's framework, let Ben be a banker, Alice a depositor in his bank, and Bob a merchant. Alice wishes to send some money to Bob so that her bank cannot detect that it is her who spent the money at Bob's store. To this end, Alice sends Ben L blinded messages:

$$\mu_l = k_l^e m_l \bmod n. \tag{4.62}$$

Every message contains the amount to be paid, say A, and a randomly chosen (very long) number v_l, so that for practical purposes, all v_l are different. Ben asks Alice to unblind $L-1$ randomly chosen messages and agrees to keep one message, say, μ_λ blinded. Provided that $L - 1$ messages are in order, Ben signs the remaining message μ_λ and sends the signed message back to Alice,

$$\text{sign}\,(\mu_\lambda) = \mu_\lambda^d = k_\lambda m_\lambda^d \bmod n. \tag{4.63}$$

Alice extracts the signed message $m_\lambda^d \bmod n$ and sends it to Bob together with the unsigned message m_λ. Bob checks the signature and sends the message and the signature to the bank. Ben checks the signature and the number v_λ. If the number is not in its database of spent coins, Ben credits Bob's account (if it exists) or transfers funds to Bob's account at another bank. If the number is in its database (Alice tries to double spend!), the transaction does not go through. The flowchart is shown in Figure 4.4.

These operations' results are as follows: Alice's balance at Ben's bank is reduced by A; Bob's account at Ben's (or a different) bank is increased by the same amount A (minus some fees); goods are shipped to Alice. The whole point is that Ben does not know who paid Bob. In principle, unless Alice uses his physical address, Bob might not know her identity either. The above algorithm is privacy-preserving. As such, it is naturally open to abuse, similarly to physical cash! If there is only one request to use digital cash, the bank knows a payer's identity. However, if there are many users of this scheme, then a merchant might be paid by someone who requested another payment in the same amount, while Alice's account will be credited to pay another merchant. Such scrambling hides the identity of payers.

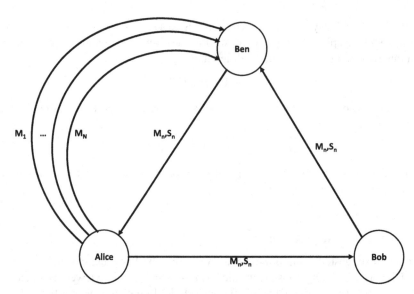

Figure 4.4 Conceptual framework for blind signature algorithm underlying anonymous digital cash. Alice sends N messages for signature to her bank. The bank checks randomly-chosen $N-1$ messages for validity and, if satisfied, signs the remaining message blindly. Alice sends message signed by the bank to Bob, who, in turn, submits it to the bank for payment. Own graphics.

Digicash solves some of the problems Bitcoin wishes to solve, except for one crucial aspect — it is centralized and relies on the trusted third party — Ben the Banker. Alice is the first, and Bob is the second vertex in the corresponding triangle.

4.6 Finite field cryptography

4.6.1 Definitions

So far, we have looked at one possible candidate for a trapdoor function — exponent modulo $n = pq$. Are there other possibilities? Yes, for instance, discrete logarithms mod p are as difficult to compute as factoring of n — same complexity as before! As we shall see shortly, discrete logarithms have some useful features; for example, one can use them for the so-called Diffie-Hellman key exchange. With some additional tricks, one can utilize discrete logarithms to build trapdoor functions.

We start with some useful definitions. Let's define discrete logarithms as solutions of the congruence:

$$\alpha^x = y \bmod p, \quad x = \log_\alpha (y) \equiv L_\alpha (y). \tag{4.64}$$

To ensure that the above equation is solvable, we assume that p is a prime (as the notation suggests), and α is a primitive root $\bmod\, p$. We consider primitive roots in some detail in the following section. Here, briefly mention that α a primitive root $\bmod\, p$, provided that any $q, 0 < q < p$, there is an integer r, such that

$$q = \alpha^r \bmod p. \tag{4.65}$$

Thus, given (p, α, y), we need to find x. To make x unique, we always choose the smallest possible solution. Discrete logarithms are additive (why?):

$$L_\alpha(yz) = \big(L_\alpha(y) + L_\alpha(z)\big) \bmod (p - 1). \tag{4.66}$$

4.6.2 Primitive roots

In general, α is a primitive root modulo n if every number m coprime to n is congruent to a certain power of α, $m = \alpha^{r(m)} \bmod n$. When n is prime, every number m, $0 < m < n$, is a power of α. For example, we show in Table 4.3 that $2, 6, 7, 11$, are primitive roots $\bmod\, 13$.

This table shows that for every $n', 1 \le n' \le 12$ we have a power $r\,(n')$ such that $\alpha^{r(n')} = n'$ for $\alpha = 2, 6, 7, 11$ so that they are primitive roots. It is easy to check that there are no other primitive roots for $p = 13$.

For a given n, finding primitive roots is difficult. One possibility of checking whether α is a primitive root is as follows. First, compute the totient function $\phi(n)$. Second, determine its prime factors $p_1, ..., p_l$. Third, for every p_i, compute

$$z_i = \alpha^{\phi(n)/p_i} \bmod n. \tag{4.67}$$

If $z_i \ne 1$, then α is a primitive root.

For example, $\phi(13) = 12 = 2^2 \times 3$; $6^6 \bmod 13 = 12 \ne 1 \bmod 13$; $6^4 = 9 \ne 1 \bmod 13$; so 6 is a primitive root.

In general, the calculation of discrete logarithms is very involved, although one can quickly find them in some instances. We are not going to discuss details here since we don't need them for what follows. In Figure 4.5, we graphically illustrate why this calculation is hard. In Figures 4.5 (a), (b), we show $5^n \bmod 97$ for $n = 1, ..., 96$, and $3^n \bmod 127$ for $n = 1, ..., 126$, as points on a unit circle according to the mapping

$$n \to \left(\cos\left(2\pi \frac{\alpha_i^n}{p_i} \right), \sin\left(2\pi \frac{\alpha_i^n}{p_i} \right) \right),$$

$$\tag{4.68}$$

$$\big(p_1, \alpha_1\big) = (97, 5), \quad \big(p_2, \alpha_2\big) = (127, 3).$$

Table 4.3 This table demonstrates that 2, 6, 7, 11 are primitive roots $\bmod\, 13$.

r	1	2	3	4	5	6	7	8	9	10	11	12
$2^r \bmod 13$	2	4	8	3	6	12	11	9	5	10	7	1
$6^r \bmod 13$	6	10	8	9	2	12	7	3	5	4	11	1
$7^r \bmod 13$	7	10	5	9	11	12	6	3	8	4	2	1
$11^r \bmod 13$	11	4	5	3	7	12	2	9	8	10	6	1

This figure clearly shows that these points are distributed in a fairly chaotic fashion. For instance, in Figures 4.5 (a), (b), we see the following sequence of points $96, 34, 70, 68, 1, 8, ..., 126, 72, 1, 18, 87, ...$ As a curious aside, in Figures 4.5 (b), (d), we show that by connecting these points consecutively, we get cardioids represented as envelopes of pencils of lines connecting consecutive points.

4.6.3 ElGamal encryption algorithm

The ElGamal encryption, is based on discrete logs rather than exponentiation; see ElGamal (1985). It is more modern and secure than RSA. The algorithm proceeds as follows:

(1) Alice chooses a large prime p and a primitive root $\alpha \bmod p$;
(2) Alice takes a random integer a and calculates $\beta = \alpha^a \bmod p$;

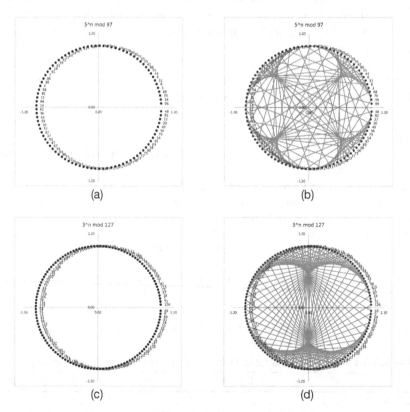

Figure 4.5 (a) Successive powers of α^n, mod p, $p = 97$, $\alpha = 5$. These powers are graphically presented at points on a unit circle. Their distribution is clearly fairly chaotic; (b) Connecting consecutive points, we get a cardioid. Same graphs for $p = 127$, $\alpha = 3$ are shown in Figures (c), (d). Own graphics.

(3) Alice makes the public key that is the triple (p, α, β) public while keeping her secret key a private;

(4) Bob chooses a random integer k and produces the ciphertext as a pair $(\alpha^k, \beta^k m) \bmod p$;

(5) Alice uses the secret power a to decrypt the ciphertext as follows:

$$\left(\alpha^k\right)^{-a}\left(\beta^k m\right) = \left(\alpha^{-ak}\right)\left(\alpha^{ak} m\right) = m \bmod p. \tag{4.69}$$

Thus, ElGamal encryption combines key transmission and OTP encryption/decryption: α^k is used to transmit the "one-time secret" k; β^k is used as a "one-time pad" for m, as discussed above. Since the one-way function is exponentiation $\bmod p$, its security depends on the difficulty of calculating discrete logarithms $L_\alpha(y)$, i.e., finding the solution to $\alpha^x = y \bmod p$. If it was easy to find discrete logarithms, the attacker could extract $L_\alpha(\beta) = a$ and decrypt the message as easily as its intended recipient. ElGamal encryption is generally better off than vanilla RSA because it uses a random k, which helps with short messages.

4.6.4 ElGamal digital signature algorithm

One can complement the ElGamal encryption algorithm with the ElGamal signature algorithm; see ElGamal (1985). The algorithm works as follows:

(1) Alice chooses a large prime p and a primitive root $\alpha \bmod p$;

(2) Alice takes a random integer a and calculates $\beta = \alpha^a \bmod p$;

(3) Alice makes the public key that is the triple (p, α, β) public while keeping her secret key a private;

(4) Alice chooses a random integer $0 < k < p-1$, such that $\gcd(k, p-1) = 1$, and computes $r = \alpha^k \bmod p$;

(5) Alice creates a signature, which is the pair (r, s) where

$$(r, s) = \left(r, (m + ar) k^{-1} \bmod (p - 1)\right). \tag{4.70}$$

The verification algorithm proceeds in two steps:

(1) Bob computes $t = \beta^{-r} r^s \bmod p$ and $u = \alpha^m \bmod p$.

(2) Bob compares t and u.

If the equality holds, the signature is valid. Indeed, for a valid signature, we have

$$\beta^{-r} r^s = (\alpha^a)^{-r}\left(\alpha^k\right)^{(m+ar)k^{-1}} \bmod p$$

$$\tag{4.71}$$

$$= \alpha^{-ar}\alpha^{m+ar} \bmod p = \alpha^m \bmod p.$$

Visualization of the ElGamal DSA is presented in Figure 4.6.

Consider a simple example of the above algorithm. Alice chooses p, α, a, β as follows:

$$p = 997, \quad \alpha = 7, \quad a = 146, \quad \beta = 7^{146} \bmod 997 = 634. \tag{4.72}$$

Here $\alpha = 7$ is the smallest primitive root of the prime $p = 997$. She makes (p, α, β) public and keeps a private. To sign a message $m = 52$, Alice chooses a random $k = 59$ and

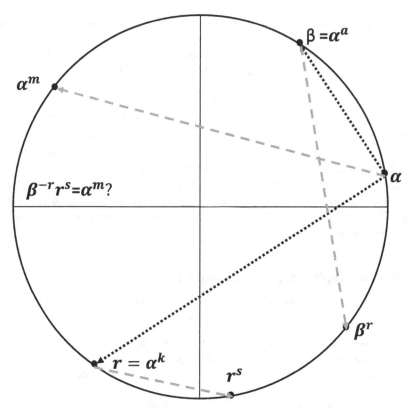

Figure 4.6 Visualization of the ElGamal DSA. Own graphics.

computes

$$r = 7^{59} \bmod 997 = 602,$$

$$s = (52 + 146 \times 602) \times 59^{-1} \bmod 996 = (52 + 146 \times 602) \times 287 \bmod 996 = 292,$$

$$(4.73)$$

which she makes public. The pair $(602, 292)$ is Alice's signature of the message $m = 52$.
To check Alice's signature, Bob computes

$$\beta^{-r} \bmod p = \left(634^{602}\right)^{-1} \bmod 997 = 708^{-1} \bmod 997 = 928,$$

$$r^s \bmod p = 602^{292} \bmod 997 = 356, \qquad (4.74)$$

$$\alpha^m \bmod p = 7^{52} \bmod 997 = 361,$$

and compares $\beta^{-r} r^s \bmod p$ and $\alpha^m \bmod p$ to get

$$928 \times 356 \bmod 997 = 361, \qquad (4.75)$$

so that Alice's signature is valid. Thus, the signing of a message requires one exponentiation, while its verification requires three exponentiations and is expensive.

As always, it is imperative not to reuse the OTP. Accordingly, Alice has to choose the parameter k at random for every new signature. If she fails to do that, the first half of a DSA signature r,

$$r = \alpha^k \bmod p, \tag{4.76}$$

will tell Charlie that k is reused. By knowing this, Charlie can do the following calculation:

$$s_1 = (m_1 + ar)k^{-1} \bmod (p-1),$$

$$s_2 = (m_2 + ar)k^{-1} \bmod (p-1),$$

$$s_2(m_1 + ar) = s_1(m_2 + ar) \bmod (p-1), \tag{4.77}$$

$$a = \left(s_2 m_1 - s_1 m_2\right)\left((s_1 - s_2)\, r\right)^{-1} \bmod (p-1).$$

Thus, Charlie recovers the secret key and decodes the corresponding message as easily as the intended recipient.

4.6.5 Schnorr digital signature algorithm

The Schnorr DSA is an influential and ingenious extension of the ElGamal DSA; see Schnorr (1989). The algorithm chooses suitable primes p, q, and a generator α, such that

$$\alpha^q = 1 \bmod p. \tag{4.78}$$

The triple (p, q, α) is common for all users. The recommendation is to choose 1024-bit prime p, and 160-bit prime q.

The signature algorithm proceeds in four steps:

(1) Alice chooses a random integer a, $0 < a < q$ and calculates $\beta = \alpha^a \bmod p$;
(2) Alice's public key is (p, q, α, β), while her secret key is a;
(3) Alice chooses a random integer $0 < k < q$, and computes $\gamma = \alpha^k \bmod p$;
(4) Alice concatenates the message m and γ, hashes the result $r = h(m||\gamma) \bmod q$, and creates the signature as the pair (r, s):

$$(r, s) = (r, (k + ar) \bmod q). \tag{4.79}$$

The verification algorithm proceeds in two steps:

(1) Bob computes $t = \alpha^s \beta^{-r} \bmod p$ and $u = h(m||t) \bmod q$;
(2) Bob compares r and u.

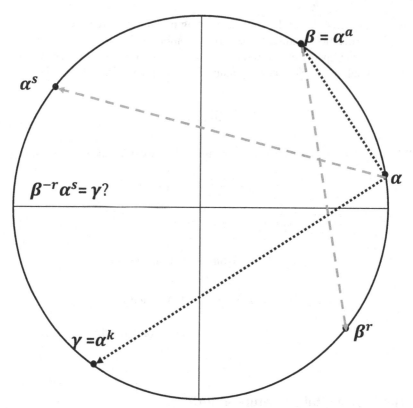

Figure 4.7 Visualization of the Schnorr DSA. Own graphics.

If the signature is valid, r and u should be equal, since

$$\alpha^s \beta^{-r} \bmod p = \alpha^{(k+ar+Z_1 q)} \alpha^{-ar} \bmod p$$

$$= \alpha^{(k+Z_1 q)} \bmod p = \alpha^k \bmod p = \gamma. \tag{4.80}$$

Visualization of the Schnorr DSA is presented in Figure 4.7.

Comparison of Figures 4.6 and 4.7 makes it clear that the Schnorr DSA is more efficient than the ElGamal DSA. Signatures are generated $\bmod q$ instead of $\bmod p$. Thus, the amount of calculations is reduced, and the speed is increased. At the same time, security is based on the difficulty of the discrete log problem $\bmod p$. For verification, we can use two exponentiations rather than three in ElGamal's algorithm, which is naturally faster.

How do we generate all the numbers we need, i.e., p, q, α? To start, we generate q. Then, we create a sequence $p = q\rho + 1$ with random ρ until we hit a prime. To generate α, we first compute a random σ, and then $\alpha = s^\rho \bmod p$. If $\alpha = 1$, which is extremely unlikely, we start again; otherwise, we have our α. In practice, it is faster to use p, q, α, which are pregenerated, since this does not reduce the level of security, because k is random. You can pick the standard values described in the literature.

Consider an example of the above algorithm. Start with $q = 97$, which is a prime number. It is easy to show that $97 \times 40 + 1 = 3881$ is also prime, so that we can choose $p = 3881$. Now choose random $\sigma = 61$. A simple calculation yields

$$61^{40} = 3440 \bmod 3881,$$

$$3440^{97} = 1 \bmod 3881,$$

$$(4.81)$$

so we can use $\alpha = 3440$. Now we randomly choose the secret key $a = 76$ and calculate the corresponding public key β,

$$\beta = 3440^{76} \bmod 3881 = 1568. \tag{4.82}$$

Let us show how Alice, whose secret key is $a = 76$ and public key is $\beta = 1568$, can sign a message $m = 52$. First, she chooses an ephemeral random $k = 22$ and calculates

$$\gamma = 3440^{22} \bmod 3881 = 121. \tag{4.83}$$

Next, Alice needs to compute the hash of the concatenated message $h(m||\gamma)$. We discuss how to calculate suitable hash functions below. Here we make a shortcut. We notice that $\phi = 10$ and $\psi = 21$ are primitive roots of $q = 97$, and define the hash function as follows:

$$h(m||\gamma) = \phi^m \psi^\gamma \bmod q = 10^{52} \times 21^{121} \bmod 97 = 88 \times 74 \bmod 97 = 13. \tag{4.84}$$

Accordingly, the signature has the form:

$$(r, s) = (13, (22 + 76 \times 13) \bmod 97) = (13, 40). \tag{4.85}$$

Bob can verify the validity of Alice's signature as follows. Bob calculates

$$\alpha^s \bmod 3881 = 3440^{40} \bmod 3881 = 554,$$

$$\beta^{-r} \bmod 3881 = 1568^{-13} \bmod 3881 = 1480^{-1} \bmod 3881 = 2305, \tag{4.86}$$

$$554 \times 2305 \bmod 3881 = 121,$$

and after hashing, recovers r, thus confirming Alice's signature.

4.6.6 Diffie-Hellman key exchange

The Diffie-Hellman (DH) key exchange can be based on the exponentiation $\bmod p$, with the following choices for f and α:

$$g(x, y) = x^y \bmod p,$$

$$f(x) = g(\alpha, x) = \alpha^x \bmod p, \tag{4.87}$$

where g is a primitive root $\bmod p$. It is easy to see that the symmetry is satisfied

$$g(f(a), b) = (\alpha^a)^b = (\alpha^b)^a = g(f(b), a) \bmod p. \tag{4.88}$$

The key exchange proceeds as follows:

(1) Alice (or Bob) choose parameters p and α and makes them public;
(2) Alice takes a secret random a and makes $\alpha^a \bmod p$ public;
(3) Bob takes a secret random b and makes $\alpha^b \bmod p$ public;
(4) Both Alice and Bob calculate their shared secret k:

$$k = \left(\alpha^a\right)^b = \left(\alpha^b\right)^a = \alpha^{ab} \bmod p. \tag{4.89}$$

Since the one-way function used in DH exchange is exponentiation $\bmod\, p$, its security depends on the difficulty of calculating discrete logarithms, i.e., finding the solution of the problem (4.64). If discrete logarithms were easy to calculate, an attacker could find both a and b from $\alpha^a \bmod p$ and $\alpha^b \bmod p$ and the whole edifice would collapse. In general, as we discussed, finding discrete logs is a hard problem.

4.7 Elliptic Curve Cryptography

4.7.1 Continuous elliptic curves

An elliptic curve is the set of real points (x, y), which solve the following equation:

$$y^2 = x^3 + a_2 x^2 + a_1 x + a_0. \tag{4.90}$$

In other words,

$$E = \left\{(x, y) : y^2 = x^3 + a_2 x^2 + a_1 x + a_0\right\}. \tag{4.91}$$

It is imperative to understand that the corresponding curves are not ellipses! Below we briefly discuss a fascinating reason why they are called elliptic. Every elliptic curve E is symmetric with respect to reflection $y \to -y$. In other words, if $(x, y) \in E$, then $(x, -y) \in E$, as well. We use this fact below when we define addition on elliptic curves.

A couple of representative examples of elliptic curves

$$y^2 = x^3 - x + 1,$$

$$\tag{4.92}$$

$$y^2 = x^3 - x.$$

Not all elliptic curves are useful for our purposes — we want to avoid points with multiple roots, such as

$$y^2 = x^3 + x^2 - x - 1 = (x - 1)(x + 1)^2. \tag{4.93}$$

Several useful elliptic curves are shown in Figure 4.8 below.

In practice, it is sufficient to consider curves with $a_2 = 0$, since we can always shift x by $-a_2/3$, thus removing the x^2 term. Accordingly, for cryptographic purposes, we can use

$$E = \left\{(x, y) : y^2 = x^3 + ax + b\right\}. \tag{4.94}$$

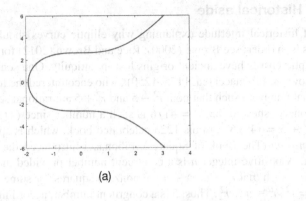

(a)

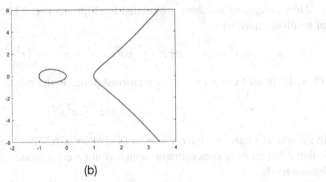

(b)

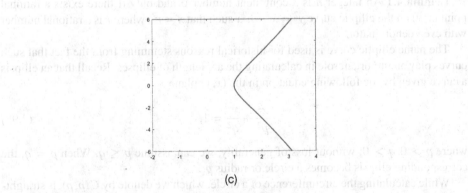

(c)

Figure 4.8 Several representative elliptic curves: (a) $y^2 = x^3 - x + 1$; (b) $y^2 = x^3 - x$; (c) $y^2 = x^3 + x^2 - x - 1$. Own graphics.

It is easy to show that the reduced curve E has multiple roots if and only if its discriminant Δ vanishes:

$$\Delta = -16 \left(4a^3 + 27b^2\right) = -1728 \left(\left(\frac{a}{3}\right)^3 + \left(\frac{b}{2}\right)^2\right) = 0. \qquad (4.95)$$

Before we jump into the pool at a deep end, a brief historical aside is helpful.

4.7.2 Historical aside

A brief historical interlude explaining why elliptic curves should not be confused with ellipses is in order; see Brown (2000); Rice and Brown (2012) for further details.

Elliptic curves have ancient origins lost in antiquity. One source is Leonardo of Pisa, also known as Fibonacci (ca. 1175–1250) , who encountered the following problem: "Find a rational number r such that both $r^2 - 5$ and $r^2 + 5$ are rational squares."

Fibonacci showed that $r = 41/6$ is such a number since $(41/6)^2 - 5 = (31/6)^2$ and $(41/6)^2 + 5 = (49/6)^2$. In his 1225 celebrated book, which is appropriately called *Liber Quadratorum* (The Book of Squares), Fibonacci introduced the concept of a congruent number. A positive integer n is a congruent number provided that there exists a rational number r, such that $r^2 - n$, $r^2 + n$ are nonzero squares for some rational numbers r_-, r_+, $r^2 - n = r_-^2$, $r^2 + n = r_+^2$. Thus, 5 is a congruent number, according to Fibonacci.

How congruent numbers are related to elliptic curves? The product $r_-^2 r^2 r_+^2$ is a square of a rational number s:

$$s^2 = \left(r^2 - n\right) r^2 \left(r^2 + n\right) = r^6 - n^2 r^2 = \left(r^2\right)^3 - n^2 \left(r^2\right). \qquad (4.96)$$

Thus, the point $\left(x = r^2, \ y = s\right)$ is a rational point on the elliptic curve,

$$y^2 = x^3 - n^2 x. \qquad (4.97)$$

In Fibonacci's case, we have $x = r^2 = (41/6)^2$, $y = 62279/216$. An important observation is that r has an even denominator, which is not a coincidence. The following fundamental lemma reads:

Lemma 4.1 An integer n is a congruent number if and only if there exists a rational point (x, y) on the elliptic curve $y^2 = x^3 - n^2 x$ such that $x = r^2$, where r is a rational number with even denominator.

The name elliptic curve is used for historical reasons stemming from the fact that such curves play an important role in calculating the arc length of ellipses. Recall that an ellipsis a curve given by the following equation in the (x, y) plane:

$$\frac{x^2}{p^2} + \frac{y^2}{q^2} = 1, \qquad (4.98)$$

where $p > 0, q > 0$; without loss of generality, we can assume $p \leq q$. When $p = q$, the corresponding ellipsis becomes a circle or radius p.

While calculating the circumference of a circle, which we denote by $C(p, p)$, is straight-forward, $C(p, p) = 2\pi p$, calculating the circumference of an ellipse, which we denote by $C(p, q)$ is very hard, even now. Several great mathematicians, including Isaac Newton (1669), Leonhard Euler (1733), and Colin Maclaurin (1742), expressed the arc length of an ellipse as an infinite series. Subsequently, Adrien-Marie Legendre, Niels Henrik Abel, Carl Gustav Jacobi, and Karl Weierstrass studied the integrals directly. We notice that in contrast with evaluating the circumference $C(p, q)$, finding the area $A(p, q)$ of an ellipse is straightforward:

$$A(p, q) = \pi p q. \qquad (4.99)$$

We can restrict ourselves to a positive quadrant and represent the ellipse parametrically as $x(\theta) = p\cos\theta$, $y(\theta) = q\sin(\theta)$, where θ is the angle. The well-known arc length formula yields

$$C(p,q) = 4\int_0^{\pi/2}\sqrt{p^2\sin^2\theta + q^2\cos^2\theta}\,d\theta$$

$$= 4\int_0^{\pi/2}\sqrt{q^2 - (q^2-p^2)\sin^2\theta}\,d\theta \tag{4.100}$$

$$= 4q\int_0^{\pi/2}\sqrt{1 - k^2\sin^2\theta}\,d\theta,$$

where $k^2 = (1 - p^2/q^2)$, $0 \le k \le 1$. Of course, the nontrivial case is $0 < k < 1$, because an ellipse becomes a circle when $k = 0$, and a straight line when $k = 1$.

To clarify the connection of ellipses with elliptic curves, we introduce $x = \sin^2\theta$ and get

$$d\theta = \frac{dx}{2\sqrt{x(1-x)}}, \tag{4.101}$$

so that

$$C(p,q) = 2q\int_0^1 \frac{\sqrt{1-k^2x}}{\sqrt{x(1-x)}}dx$$

$$= 2q\int_0^1 \frac{(1-k^2x)}{\sqrt{x(1-x)(1-k^2x)}}dx \tag{4.102}$$

$$= 2qk\int_0^1 \frac{\left(\frac{1}{k^2}-x\right)}{\sqrt{x^3 - \left(\frac{1}{k^2}+1\right)x^2 + \frac{1}{k^2}x}}dx.$$

Integrals of the form

$$\int \frac{\text{polynomial}}{\sqrt{\text{cubic with three distinct roots}}}, \tag{4.103}$$

appear in all kinds of applications. Since the calculation of the arc length of an ellipse was the first one, they are called elliptic integrals.

We show a typical ellipse and its associated elliptic curve in Figures 4.9 (a), (b).

Elliptic integrals cannot be evaluated in terms of elementary functions. For completeness, we give the formula for the arc length of an ellipse with semi-axises p, q, see Abramowitz and Stegun (1948), Formula 17.3.10:

$$C(p,q) = 2\pi b F\left(-\frac{1}{2},\frac{1}{2};1;k^2\right). \tag{4.104}$$

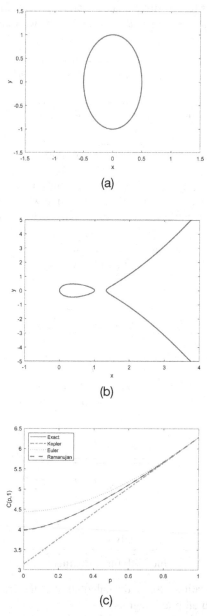

Figure 4.9 (a) The ellipse given by $4x^2 + y^2 = 1$; (b) The corresponding elliptic curve $y^2 = x^3 - 7x^2/3 + 4x/3$; (c) The exact and three approximate expressions for the ellipse circumference. While Kepler's and Euler's approximations fail for elongated ellipses, Ramanujan's approximation holds extremely well even in the limit. Own graphics.

Here $F(a, b; c; z)$ is the celebrated hypergeometric function:

$$F(\alpha, \beta; \gamma; z) = \sum_{n=0}^{\infty} \frac{(\alpha)_n (\beta)_n}{(\gamma)_n} \frac{z^n}{n!}, \tag{4.105}$$

where (α), (β), (γ) are defined as follows:

$$(\alpha)_n = \begin{cases} \alpha(\alpha+1)(\alpha+2)\dots(\alpha+n-1), & n > 0, \\ 1 & n = 0, \end{cases}$$

$$(\beta)_n = \dots, \tag{4.106}$$

$$(\gamma)_n = \dots,$$

are the Pochhammer symbols. It is clear that for the circle, this formula becomes $C(p, p) = 2\pi p$. Alternatively, by using Abramowitz and Stegun (1948), Formula 15.3.17, we can rewrite C in a symmetric form

$$C(p, q) = \pi(p+q) F\left(-\frac{1}{2}, -\frac{1}{2}; 1; \frac{(b-a)^2}{(b+a)^2}\right). \tag{4.107}$$

Several approximate expressions have been developed over centuries by prominent scientists. Johannes Kepler (1609) suggested the following simple formula:

$$C(p, q) \approx \pi(p+q). \tag{4.108}$$

Leonhard Euler (1773) used the approximation of the form:

$$C(p, q) \approx 2\pi \sqrt{\frac{p^2 + q^2}{2}}, \tag{4.109}$$

while Srinivasa Ramanujan (1914) discovered the following remarkable expression:

$$C(p, q) \approx \pi\left(3(p+q) - \sqrt{(p+3q)(3p+q)}\right), \tag{4.110}$$

We show the exact formula and three approximations in Figure 4.9 (c). This simple figure shows the true genius of Ramanujan — his approximation, of the same complexity as the other two, is so accurate that one cannot see the difference with a naked eye.

4.7.3 Addition on elliptic curves

So far, for the RSA purposes, we have considered a simple multiplication mod p, which was our nontrivial group operation. By using elliptic curves, we can introduce a much more exciting group operation. For simplicity, we shall start with the continuous case.

We need to define a group operation on E, which we call addition. Given two points, $P_1 = (x_1, y_1)$ and $P_2 = (x_2, y_2)$, we want to define $P_1 + P_2 = P_3 = (x_3, y_3)$. To do so, we draw a straight line through the two points until it intersects with the curve at a point $Q = (x_3, -y_3)$; see Figure 4.10 below. The reason why such a point exists is due to the celebrated Bézout's Theorem for algebraic curves.

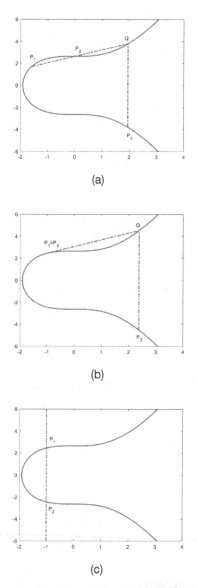

(a)

(b)

(c)

Figure 4.10 A geometric approach to EC addition. We use the curve shown in Figure 4.8 (a): (a) Addition of two distinct points P_1 and P_2 with different abscissas; $P_1 + P_2 + Q = \infty$, $P_1 + P_2 = -Q = P_3$; (b) Doubling of P_1; $P_1 + P_1 + Q = \infty$, $P_1 + P_1 = -Q = P_3$; (c) Addition of two distinct points A and B with identical abscissas; $P_1 + P_2 = \infty$, $P_2 = -P_1$. Own graphics.

Theorem 4.4. Bézout's Theorem: Two algebraic curves of degrees m and n intersect in $m \cdot n$ points, provided that each point is counted with its intersection multiplicity. They cannot meet in more than $m \times n$ points unless they have a common component so that the equations defining them have a common factor.

This theorem was stated by Isaac Newton in his *opus magnum Principia* in 1687, where he demonstrated that for two curves of degrees m and n, the number of intersection points equals the product of their degrees, $m \times n$. Subsequently, this result was published by Étienne Bézout in his book *Théorie générale des équations algébriques* in 1779.

In general, intersections can occur in the complex space, but in our case, they have to be real (why?). Since we are interested in an elliptic curve (of degree 3) and a straight line (of degree 1), the number of intersections should be precisely 3.

We reflect Q in the x-axis and get $P_3 = (x_3, y_3)$, see Figure 4.10. To put it differently, we say that three points P_1, P_2, Q, which belong to the elliptic curve E and the straight line L, add up to the "neutral element" O, which is our zero or neutral element:

$$P_1 + P_2 + Q = O. \tag{4.111}$$

When $P_1 = P_2 = P$, we, naturally, go to the limit, draw a tangent line through P, and get Q and P_3:

$$P + P + Q = O. \tag{4.112}$$

In the particular case: $P_1 = (x_1, y_1)$, $P_2 = (x_1, -y_1)$, the line L through these two points is obviously vertical and will not intersect the curve E again! In this case, we have

$$P_1 + P_2 = O. \tag{4.113}$$

Thus, we can say

$$P_1 + P_2 = -Q = P_3. \tag{4.114}$$

The neutral element has the following properties:

$$O + (x, y) = (x, y), \quad (x, y) + (x, -y) = O. \tag{4.115}$$

Now, we need to find the actual coordinates of (x_3, y_3), where

$$(x_1, y_1) + (x_2, y_2) = (x_3, y_3). \tag{4.116}$$

If $(x_1, y_1) \neq (x_2, y_2)$, then we can use chord and reflection addition

$$(x_3, y_3) = \left(m^2 - x_1 - x_2, m \left(2x_1 + x_2 - m^2 \right) - y_1 \right),$$

$$\tag{4.117}$$

$$m = \frac{(y_2 - y_1)}{(x_2 - x_1)}.$$

If $(x_1, y_1) = (x_2, y_2) = (\bar{x}, \bar{y})$ then we can use tangent and reflection addition

$$(x_3, y_3) = \left(m^2 - 2\bar{x}, m \left(3\bar{x} - m^2 \right) - \bar{y} \right),$$

$$\tag{4.118}$$

$$m = \frac{(3\bar{x}^2 + a)}{2\bar{y}}.$$

Some representative examples of the group operation are shown in Figure 4.10.

We now show how to derive formulas (4.117) and (4.118). Consider the straight line

$$y = m \left(x - x_1 \right) + y_1, \tag{4.119}$$

where m is given by (4.117). Equivalently,

$$L = \{(x, y) : y = m(x - x_1) + y_1\}. \tag{4.120}$$

Thus, (x_1, y_1), (x_2, y_2), and $(x_3, -y_3)$ all belong to both the elliptic curve in question and the straight line. Accordingly, x_1, x_2, x_3 are roots of the following cubic:

$$x^3 + ax + b(m(x - x_1) + y_1)^2 = 0. \tag{4.121}$$

Explicitly

$$x^3 - m^2 x^2 + \left(a + 2m^2 x_1 - 2my_1\right) x + \left(b - (mx_1 - y_1)^2\right) = 0. \tag{4.122}$$

Thus, by Vieta's formula, we get

$$x_1 + x_2 + x_3 = m^2, \tag{4.123}$$

so that

$$x_3 = m^2 - x_1 - x_2, \quad -y_3 = m\left(m^2 - 2x_1 - x_2\right) + y_1. \tag{4.124}$$

After reflection in the x-axis, we get

$$x_3 = m^2 - x_1 - x_2, \quad y_3 = m\left(+2x_1 + x_2 - m^2\right) - y_1, \tag{4.125}$$

as advertised.

When $(x_1, y_1) = (x_2, y_1) = (\bar{x}, \bar{y})$, we calculate the tangent passing through this point via an implicit equation:

$$2\bar{y}dy = (2\bar{x}^2 + a)dx,$$

$$\frac{dy}{dx} = \frac{(2\bar{x}^2 + a)}{2\bar{y}} = m. \tag{4.126}$$

Thus, the point $(x_3, -y_3)$ belongs to the elliptic curve E and the straight line L given by the following equation:

$$y = m(x - \bar{x}) + \bar{y}, \tag{4.127}$$

so that, as before

$$x_3 = m^2 - 2\bar{x}, \quad -y_3 = m\left(m^2 - 3\bar{x}\right) + \bar{y}. \tag{4.128}$$

After reflection, we get

$$x_3^* = m^2 - 2\bar{x}, \quad y_3^* = m\left(3\bar{x} - m^2\right) - \bar{y}, \tag{4.129}$$

as claimed.

In this subsection, we have learned how to add two points on an elliptic curve, including the particular case of adding two identical points.

4.7.4 Projectivization of elliptic curves

In this subsection, we briefly discuss connections between elliptic curves and projective geometry. The fact that the vertical line passing through the points (x, y) and $(x, -y)$ belonging to the elliptic curve intersects with it seemingly at two rather than three points deserves additional discussion.

To elucidate the problem, we need to "lift" elliptic curves from the affine (x, y) plane to the projective $(X : Y : Z)$ plane. This procedure is called the projectivization. We proceed as follows. Consider an affine elliptic curve given by Eq. (4.94). Let $x = X/Z, y = Y/Z$, assumes the form:

$$Y^2Z = X^3 + aXZ^2 + bZ^3, \tag{4.130}$$

in the (X, Y, Z) space. We eliminate the origin $(0, 0, 0)$ to avoid ambiguity since it belongs to every elliptic curve.

Eq. (4.130) is homogeneous of degree 3. Thus, if (X, Y, Z) is a point on the curve, so is $(\lambda X, \lambda Y, \lambda Z)$ for any λ. To avoid having "too many" points, we define two-dimensional projective space as the set of equivalence classes of nonzero points (X, Y, Z), where points (X, Y, Z) and $(\lambda X, \lambda Y, \lambda Z)$ are considered to be equivalent. Intuitively, the three-dimensional space is reduced to two dimensions by treating lines passing through the origin as a single point. We use the notation $(X : Y : Z)$ for a projective point to emphasize this fact. Thus, $(X : Y : Z) = (\lambda X : \lambda Y : \lambda Z), \lambda \neq 0$.

It is clear that there are two types of solutions to the homogeneous equation: if $Z \neq 0$, then we can divide X, Y by Z, consider $x = X/Z, y = Y/Z$ and get a solution of the affine equation.

However, if $Z = 0$, then division by Z is not allowed, so that there are points on the projective curve that do not correspond to points on the affine curve. These points have the form $(0 : y : 0)$, for any $y \neq 0$. Given that points in the projective plane are defined as equivalence classes, we can choose the following point $O = (0 : 1 : 0)$, called the point at infinity or the neutral element. We can intuitively think that O is a point located infinitely high (and low) on the y-axis.

Consider a vertical straight line given by the following equation in the affine plane:

$$x - \bar{x} = 0. \tag{4.131}$$

In the projective plane, it looks as follows:

$$X - \bar{x}Z = 0. \tag{4.132}$$

The point O belongs to this line for any $\bar{x}$. Thus, after projectivization, we find the third point, in addition to $(\bar{x}, \bar{y})$ and $(\bar{x}, -\bar{y})$, belonging to the elliptic curve and the straight line, and this point is the point at infinity, O.

4.7.5 Multiplication on elliptic curves

We are only interested in multiplication by an *integer*, which is defined intuitively as repeated addition. For instance

$$3 \times (x, y) = (((x, y) + (x, y)) + (x, y)). \tag{4.133}$$

Here and below, we use × to denote elliptic curve point multiplication by a scalar. We illustrate this operation in Figure 4.11 below. As usual, the binary decomposition of the multiplier helps to perform multiplication quickly.

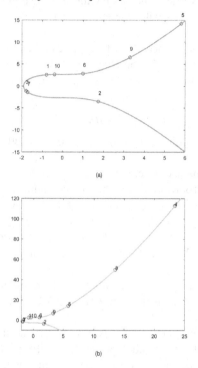

Figure 4.11 Doubling of the point A = (−0.8, 2.55) on the EC $y^2 = x^3 + 7$; (a) The sequence of A, 2A, 4A, 8A,... is shown on a small scale; (b) Same sequence on a large scale. Own graphics.

4.7.6 Discrete elliptic curves

With the above material under our belt, we can now go back to cryptography. Not surprisingly, we have to operate with a finite set of discrete points. Earlier, we considered points $\mod p$ or $\mod n = \mod pq$. Now we wish to use an EC and a large prime p and consider EC $\mod p$. We denote this curve by E_p. Consider coordinates (x, y) as integers $\mod p$, $x = 0, 1, ..., p − 1, y = 0, 1, ..., p − 1$. Thus,

$$E_p = \left\{(x, y) : y^2 = x^3 + ax + b \mod p\right\}. \tag{4.134}$$

Let us assure you that finding points (x, y) on E_p is not that easy!

Assuming that we know the set of points constituting E_p, we can define addition by using formulas (4.117), (4.118), but m, which, of course in an integer, has to be understood as appropriate, namely as a product of the numerator and the inverse of the denominator.

To warm up, let us consider E_p of the form:

$$y^2 = x^3 + 7 \mod p, \tag{4.135}$$

where $p = 2^5 - 1 = 31$, $p = 2^7 - 1 = 127$, $p = 2^{13} - 1 = 8191$, which are the so-called Mersenne primes, i.e., prime numbers that are one less than a power of two.[4] We shall see below that using quasi-Mersenne primes for constructing ECs has considerable advantages. These curves are shown in Figure 4.12. They contain 21, 127, and 8011 points, respectively. Since 21 is not a prime, E_{31} is not a viable candidate for a cryptographically useful EC, while E_{127} and E_{8191} are (127 and 8011 are both primes), if not for the fact that they are way too small.

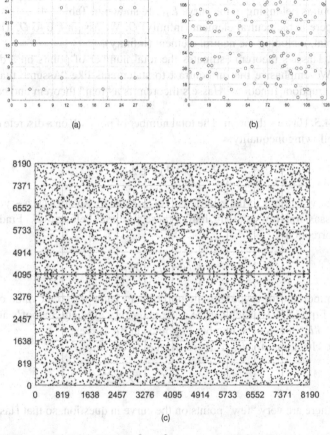

(a)

(b)

(c)

Figure 4.12 Three discrete ECs $\{E_p : y^2 = x^3 + 7 \bmod p\}$: (a) $p = 31$, $m = 21$; (b) $p = 127$, $m = 127$; (c) $p = 8191$, $m = 8011$. Here $m = \#E_p$ is the total number of points on E_p. The "neutral element" O is not shown. Own graphics.

---

[4]These primes are named after French monk Marin Mersenne (1588–1648), who stated in his *Cogitata Physica-Mathematica* (1644) that the numbers $2^n - 1$ with $n \leq 257$ were prime for $n = 2, 3, 5, 7, 13, 17, 19, 31, 67, 127$, and 257. While $n = 2, ..., 19$ were known, $n = 67, 257$ are wrong, and $n = 61, 89, 107$ are omitted, the name Mersenne primes became standard.

Table 4.4 Finite points belonging to the curve E_{31}.

x	0	0	1	1	4	4	5	5	7	7
y	10	21	15	16	3	28	15	16	3	28
#	1	2	3	4	5	6	7	8	9	10
x	11	11	20	20	24	24	25	25	27	27
y	6	25	3	28	6	25	15	16	6	25
#	11	12	13	14	15	16	17	18	19	20

All finite points belonging to the curve E_{31} are shown in Table 4.4.

Besides, there is the neutral element (infinity) O. We assign # 0 to O. Thus, there are 21 points in total, including the neutral element (infinity).

In general, Hasse's theorem estimates the total number of points on a discrete elliptic point E_p. We emphasize that in contrast to statements like "Assume that f is a good cryptographic trapdoor function," Hasse's theorem is a "real" theorem and can be proven rigorously.[5]

Theorem 4.5. Hasse's theorem: The total number of points N on a discrete elliptic curve satisfies the following inequality:

$$|N - (p + 1)| < 2\sqrt{p}. \tag{4.136}$$

If a point (x, y) belongs to the elliptic curve, $(x, y) \in E_p$, so does $(x, p - y)$. There are p possible abscissas, roughly speaking, half of these abscissas belong to E_p. Finding the exact value of the correction

$$t = N - (p + 1), \tag{4.137}$$

is extremely important for cryptographic purposes to know the exact value of the correction, given by Eq. (4.137), which can be done via a very efficient Schoof's algorithm; see Schoof (1985). $\#E_p$ denotes the number of points on the curve.

It is easy to check that for E_{31} Hasse's inequality is valid:

$$|21 - 32| = 11 < 2 \times 5.57. \tag{4.138}$$

Interestingly, there are very "few" points on the curve in question, so that Hasse's theorem barely holds.

Table 4.5 shows how to add points on E_{31}.

---

[5]However, it is a conditional statement of the form "If p is prime, then ... ."

Table 4.5 Addition table for E_{31}. Points on the curve are numbered as in Table 4.4.

+	0	1	2	3	4	5	6	7	8	9	10	11	12	13	14	15	16	17	18	19	20
0	0	1	2	3	4	5	6	7	8	9	10	11	12	13	14	15	16	17	18	19	20
1	1	2	0	16	6	3	15	20	14	17	11	18	9	7	19	4	5	12	10	8	13
2	2	0	1	5	15	16	4	13	19	12	18	10	17	20	8	6	3	9	11	14	7
3	3	16	5	9	0	12	11	18	13	2	4	20	17	7	5	8	19	15	1	6	10
4	4	6	15	0	10	2	12	1	7	6	15	19	14	16	11	18	13	8	3	17	8
5	5	3	16	12	2	18	0	15	11	5	7	17	5	14	2	10	6	20	19	9	18
6	6	15	4	11	12	0	18	9	5	13	16	3	9	3	15	5	19	1	8	7	14
7	7	20	13	18	1	15	9	16	8	6	10	6	11	17	14	8	2	2	6	9	3
8	8	14	19	13	7	11	5	8	12	0	5	15	7	18	11	2	15	7	12	13	18
9	9	17	12	2	6	5	13	6	0	15	7	20	14	16	5	9	10	11	16	13	14
10	10	11	18	4	15	7	16	10	5	7	0	14	2	9	10	8	6	1	19	3	17
11	11	18	10	20	19	17	3	6	15	20	14	19	0	9	4	3	15	2	8	16	12
12	12	9	17	14	5	8	3	11	6	18	2	0	20	4	3	10	16	7	1	11	15
13	13	7	20	11	14	10	19	16	1	17	9	4	3	18	7	15	18	1	1	5	5
14	14	19	8	15	12	20	9	15	11	5	11	5	10	7	18	16	6	2	4	20	10
15	15	4	6	2	18	1	10	14	9	5	7	20	16	8	17	11	0	3	13	12	19
16	16	5	3	17	1	9	2	10	13	8	6	15	19	18	7	0	12	14	4	20	11
17	17	12	9	8	16	19	5	4	11	20	1	2	7	15	18	3	14	13	0	10	6
18	18	10	11	15	7	6	20	12	3	2	19	8	1	17	16	13	4	0	14	5	9
19	19	8	14	7	9	13	17	2	4	18	3	16	11	1	6	12	20	10	5	15	0
20	20	13	7	10	8	18	14	3	1	4	17	12	15	5	2	19	11	6	9	0	16

We illustrate results shown in the addition table graphically in Figure 4.13.

To this end, consider two randomly chosen points on E_{31}: $A = (4, 28)$, $B = (24, 6)$. Denote by L the straight line passing through these two points, which is given by the following equation

$$L = \{(x, y) : y = \alpha\,(x - 4) + 28 \bmod 31\}, \tag{4.139}$$

where the coefficient α is an integer between 0 and 30, such that

$$20\alpha + 28 = 6 \bmod 31. \tag{4.140}$$

The Euclidean algorithm yields $\alpha = 2$ so that the equation for the straight line has the form:

$$L = \{(x, y) : y = 2\,(x - 4) + 28 \bmod 31\}. \tag{4.141}$$

As expected, the graph shows that there is one more point $C = (7, 3)$, which simultaneously belongs to E_{31} and L. Accordingly, $A + B = D$, where $D = (7, 28)$. The process is shown in Figure 4.13 (a).

Similarly, for $A = (5, 15)$, $B = (20, 3)$ we have

$$L = \{(x, y) : y = \alpha\,(x - 5) + 15 \bmod 31\}, \tag{4.142}$$

where

$$15\alpha + 15 = 3 \bmod 31. \tag{4.143}$$

Accordingly, $\alpha = 24$ and L is given by

$$L = \{(x, y) : y = 24(x - 5) + 15 \bmod 31\}, \tag{4.144}$$

so that $C = (24, 6)$ and $A + B = D = (24, 26)$. The process is shown in Figure 4.13 (b).

Let us choose a point of E_{31}, say $G = (7, 3)$ and compute $G, 2 \times G, 3 \times G, \ldots$. This sequence is shown in Figure 4.14. This figure clearly shows that this sequence is chaotic so that knowing $A = kG$, it is very hard to find the actual k. We define G's integer order as a number n such that $n \times G = O$, where O is the identity element. The cofactor h is the ratio N/n, where N is the total number of points on E_{31}. We have $n = 21$, $h = 1$.

4.7.7 The discrete log problem on elliptic curves

We are ready to formulate the elliptic curve version of the discrete log problem (4.64). To this end, we need to be a bit more specific and add a few details to the previous construct. Consider an EC, which is characterized by a sextuple $T = (p, a, b, G, n, h)$, and denoted by E_T. Here p, a, b are as before, while G is the so-called base point, which is used to generate the set of points of the form $\mathfrak{S}_G = \{k \times G\}$, n is the order of G, i.e., the cardinality of the set $\mathfrak{S}_G$, $n = \#\mathfrak{S}_G$, and h is the cofactor of G, $h = \#E_T/\#\mathfrak{S}_G$.[6] For cryptographic purposes, n has to be prime. A good base point should have $h = 1$, which means that the set $\mathfrak{S}_G$ of all the points covers the entire curve. Thus, G is similar to a primitive root considered above.

---

[6]Lagrange's lemma, which states that for any finite group G and every subgroup H of G, the order (number of elements) of H divides the order of G, guarantees that h is an integer.

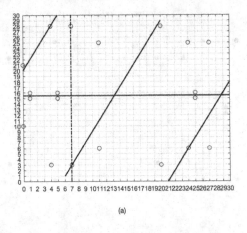

(a)

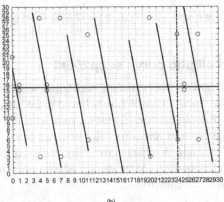

(b)

Figure 4.13 A geometric approach to producing the addition table: (a) (4, 28) + (24, 6) = (7, 28); (b) (5, 15) + (20, 3) = (24, 25). Own graphics.

The discrete log problem requires finding x for a given triple (p, α, y). We can formulate its elliptic curve version as follows. Given EC_T, and a point $A \in \mathfrak{S}_G$, find k, such that

$$A = k \times G = (((G + G) + G) + \dots + G). \tag{4.145}$$

When $\#E_T = \#\mathfrak{S}_G$, the problem becomes particularly to formulate (but, in general, very difficult to solve!). Given a point $A \in EC_T$, find k, such that $A = k \times G$.

In many respects problems (4.64) and (4.145) are similar! Given this fact, we can develop the elliptic curve version of the ElGamal signature and the Diffie-Hellman key exchange.

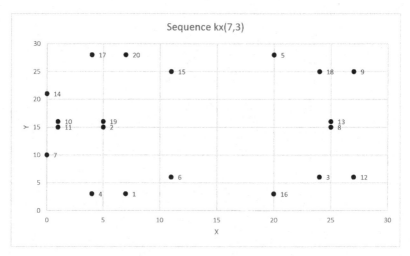

Figure 4.14 The sequence $k \times G$, $G = (7, 3)$. $21 \times G = O$. Own graphics.

4.7.8 The standard elliptic curve secp256r1

First, we describe the U.S. government standard, which uses secp256r1, the "r" in the penultimate position stands for "random," which is believed to be secure. In the spirit of mistrusting the government, Bitcoin deliberately uses a somewhat different curve, secp256k1, the "k" in the penultimate position stands for "Koblitz," see below.

The curve secp256r1 itself is characterized by the sextuple $T = (p, a, b, G = (G_x, G_y), n, h)$. The corresponding parameters, which are described in Certicom Research (2009), *inter alia*, have the form:

$$
\begin{aligned}
p &= \begin{array}{l} \texttt{ffffffff00000001000000000000000000} \\ \texttt{00000000ffffffffffffffffffffffff} \end{array} \\
a &= \begin{array}{l} \texttt{ffffffff00000001000000000000000000} \\ \texttt{00000000fffffffffffffffffffffffc} \end{array} \\
b &= \begin{array}{l} \texttt{5ac635d8aa3a93e7b3ebbd55769886bc} \\ \texttt{651d06b0cc53b0f63bce3c3e27d2604b} \end{array} \\
G_x &= \begin{array}{l} \texttt{6b17d1f2e12c4247f8bce6e563a440f2} \\ \texttt{77037d812deb33a0f4a13945d898c296} \end{array} \\
G_y &= \begin{array}{l} \texttt{4fe342e2fe1a7f9b8ee7eb4a7c0f9e16} \\ \texttt{2bce33576b315ececbb6406837bf51f5} \end{array} \\
n &= \begin{array}{l} \texttt{ffffffff00000000ffffffffffffffff} \\ \texttt{bce6faada7179e84f3b9cac2fc632551} \end{array} \\
h &= \qquad\qquad 1
\end{aligned}
\tag{4.146}
$$

It is clear that the corresponding p can be written as follows:

$$ p = 2^{256} - 2^{224} + 2^{192} + 2^{96} - 1, \tag{4.147} $$

it is a generalized Mersenne prime.

Parameters a, b were chosen verifiably at random from the seed S, chosen randomly and fixed afterward. The idea of selecting parameters from seed is simple, although its implementation is complex; we do not cover it in this book, referring the reader to Certicom Research (2009). In a nutshell, the algorithm proceeds as follows:

$$S = \text{c49d360886e704936a6678e1139d26b7819f7e90},$$

$$(4.148)$$

$$H = h(S), \quad a = \alpha(H), \quad b = \beta(H).$$

where $h(.)$ is a cryptographic hash function described in the next section, which is easy to compute but impossible to invert. The choice of G is less important. For the moment, it is sufficient to know that such a function is straightforward to compute but tough to invert. The functions $\alpha(.)$, $\beta(.)$ are known and can be inverted.

Thus, if one were to find a particular H, which would make the corresponding EC vulnerable, it can be written as:

$$H = \alpha^{-1}(a) = \beta^{-1}(b).$$

$$(4.149)$$

However, finding a seed S, such that

$$S = h^{-1}(H),$$

$$(4.150)$$

is practically impossible.

4.7.9 Elliptic curve digital signature algorithm

We can now collect all pieces and formally describe the all-important elliptic curve digital signature algorithm (ECDSA), being the Bitcoin protocol's heart.

The ECDSA is similar to the ElGamal DSA described earlier. The main difference is that the former employs elliptic curves instead of discrete logarithms, as does the latter. Choose the EC E_T with parameters $T = (p, a, b, G, n, h)$. Pick up a random integer k and compute

$$A = k \times G.$$

$$(4.151)$$

The public key (or address) is the point A, and the secret key is a single integer k. Note that Eq. (4.151) is practically impossible to invert by the existing methods because of the enormous number of points on secp256k1, or any other EC suitable for cryptographic purposes.

The ECDSA proceeds as follows. To sign a message m, which is an integer number less than p, we choose a random integer l, which is coprime with n, $\gcd(l, n) = 1$, so that $l^{-1} \bmod n$ exists, calculate

$$B = l \times G,$$

$$(4.152)$$

and represent the signature as a pair:

$$(r, s) = \left(r, (m + kr) \, l^{-1}\right),$$

$$(4.153)$$

where $r = B_x$. It is worth mentioning that B_y is not used by the algorithm and is not revealed, although it is trivial to calculate it modulo parity. Two points are in order: (a) one should keep l secret (same as the secret key k); (b) one should never reuse l. Violation of one of these conditions makes the secret key k recoverable.

To verify the signature, we need to compare $P_3 = P_1 + P_2$ and B, where

$$P_1 = ms^{-1} \times G, \quad P_2 = rs^{-1} \times A. \tag{4.154}$$

If the signature is correct, we get

$$ms^{-1} \times G + rs^{-1} \times A = (m + rk)\, s^{-1} \times G$$

$$\tag{4.155}$$

$$= (m + rk)(m + kr)^{-1}\left(l^{-1}\right)^{-1} \times G = l \times G = B.$$

Visualization of the ECDSA is presented in Figure 4.15.

Some sizes used in ECDSA: secret key — 256 bits; public key — 512 bits; message — 256 bits; signature — 512 bits.

A simple example is in order. Consider E_{127} with $G = (45, 33)$, as above. Choose a secret key $k = 48$, then the corresponding public key (or address) is

$$A = k \times G = (96, 51). \tag{4.156}$$

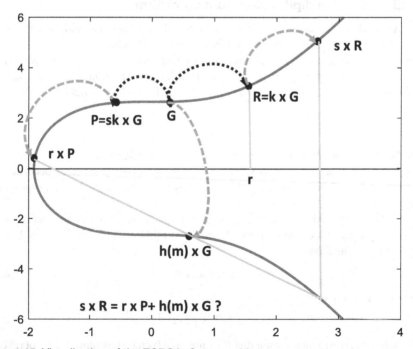

Figure 4.15 Visualization of the ECDSA. Own graphics.

Let a message be $m = 66$. Choose random $l = 83$. It is easy to check that:

$$l^{-1} = 101,$$
$$B = l \times G = (122, 3),$$
$$s = (m + kr)\, l^{-1} = 79 \bmod 127,$$
$$s^{-1} = 82,$$
$$ms^{-1} = 78 \bmod 127, \qquad (4.157)$$
$$P_1 = ms^{-1} \times G = (85, 77),$$
$$rs^{-1} = 98 \bmod 127,$$
$$P_2 = rs^{-1} \times A = (18, 88),$$
$$P_3 = P_1 + P_2 = (122, 3) = B.$$

Since $P_{3,x} = B_x$, the signature is valid.

4.7.10 Elliptic curve Schnorr digital signature algorithm

EC Schnorr DSA has some very convenient properties and can potentially replace the ECDSA for the Bitcoin blockchain protocol.

Visualization of the EC Schnorr DSA is presented in Figure 4.16.

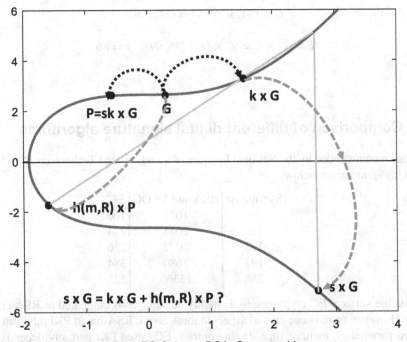

Figure 4.16 Visualization of the EC-Schnorr DSA. Own graphics.

4.7.11 Diffie-Hellman key exchange

The elliptic curve based Diffie-Hellman key exchange (ECDH) follows the same reasoning as above; see Eqs (4.22) and (4.23). As before, the key exchange uses two one-way functions, f, g. Not surprisingly, elliptic curve multiplication of the base point G by secret keys can do the trick because

$$d \times (c \times G) = c \times (d \times G).$$ (4.158)

As with the discrete log version, this cannot be used for encryption or signing but can be used for the key exchange.

Given public parameters $T = (p, a, b, G, n, h)$, the ECDH proceeds as follows:

(1) Alice chooses a secret random c and makes $C = c \times G$ public;
(2) Bob chooses a secret random d and makes $D = d \times G$ public;
(3) Both create the *same* key, which is the x coordinate of $E = d \times C = c \times D$:

$$k = E_x = (d \times (c \times G))_x = (c \times (d \times G))_x.$$ (4.159)

To make the above construct more concrete, we consider a simple example. For obvious reasons, using cryptographically suitable numbers is out of the question, so instead, we consider the curve E_{127} shown in Figure 4.12 (b). As was mentioned earlier, the number of points on $\#E_{127} = 127$, which is prime, so we can choose any point on the curve as the base point G, say $G = (45, 33)$. The order of this point is 127. Let Alice choose a random secret $c = 97$, and Bob choose $d = 34$. Then

$$C = (94, 30), \quad D = (51, 109),$$

$$E = 34 \times C = 97 \times D = (86, 89), \quad k = 86.$$ (4.160)

4.8 Comparison of different digital signature algorithms

Key sizes recommended by the National Institute of Standards and Technology (NIST), see NIST (1999), are given below:

Symmetric	RSA and DLOG	EC
80	1024	160
112	2048	224
128	3072	256
192	7680	384
256	15360	521

Thus, on the surface, EC cryptography is highly advantageous compared to RSA cryptography. However, there is one crucial aspect to think about. RSA-based PKI relies on user-supplied parameters, namely, (n, e, d). In contrast, EC-based PKI partially depends on a specific EC, *say* secp256r1, with parameters given by Eq. (4.146) that are provided by a third party, and partially on a random secret key k, chosen by the user. Although highly likely, there is no guarantee that the third party holds nothing-up-my-sleeve and has not inserted backdoors, allowing to find secret keys associated with public addresses faster than via brute-force attack. Provided that there are enough parameters to be selected to build an EC, theoretically, it might be possible to scan the set of parameters and find a curve that allows the insertion of a backdoor; see Bernstein et al. (2015). The Bitcoin protocol addresses this issue with brutal efficiency by choosing the secp256k1 instead of secp256r1. The former curve is so simple, $a = 0$, $b = 7$, that the existence of a backdoor is highly unlikely.

4.9 Cryptographic hash functions

4.9.1 Hash functions

Hash functions are somewhat different from one-way functions, but not completely. Recall that a one-way function is easy to compute but hard to inverse. A hash function is easy to compute but difficult to find a preimage for.[7] Thus, it is easy to calculate $y = h(x)$ from x. However, it is hard to find x' such that $h(x') = y$, even though there are plenty of possible candidates.

By its very nature, every hash function with longer inputs than outputs necessarily has collisions. For example, a hash function such as SHA-256 produces 256 bits of output from any input (which can be up to $2^{64} - 1$ bits long). Thus, SHA-256 generates 2^{256} outputs, for the *much larger* set of $2^{2^{64}-1}$ inputs, so that some inputs will hash to the same output (the so-called pigeonhole principle). Collision resistance means that collisions are hard to find; see, e.g., Goldwasser and Bellare (2008). We show how to build collision-resistant functions, which are very important for cryptography, next.

4.9.2 Collision-resistant hash functions

A strongly collision-resistant hash function is a function, which is easy to compute but hard to find any collision. In other words, it is tough to find x and x', $x \neq x'$, such that

$$h(x) = h(x'). \tag{4.161}$$

A weakly collision-free hash function is a one-way function easy to compute but hard to find a second preimage for a given x. Thus, given x, it is hard to find x', $x \neq x'$, such that

$$h(x) = h(x'). \tag{4.162}$$

We call such $h(.)$ second preimage resistant functions.

The main difference between strongly-resistant and weakly-resistant hashes is that we are free to choose x and x', $x \neq x'$ arbitrarily for the former. At the same time, for the latter,

---

[7]In general, the inverse for a hash function does not exist because there are many preimages for a given image by construction.

we restrict ourselves to a particular choice of x. As their names suggest, strong collision resistance is more challenging to achieve than weak collision resistance.

Strongly-resistant hash functions come handy if one wants to create hash handles to the entries in an extensive database and be sure that no two entries have the same handle. Weakly-resistant hash functions can be used to ensure that a particular password, stored in a database of hashed passwords, cannot be easily broken.

Somewhat disappointingly, we do not know whether either strongly or weakly collision-resistant hash functions exist at all! Moreover, there is no real evidence that one can construct such hash functions. Thus, although hash functions are widely used, their collision-resistance is more a matter of faith than rigorous mathematical proof.

4.9.3 Signatures and hash functions

As mentioned earlier, it is impractical to sign the whole message. Fortunately, it is sufficient to sign its hash. However, signing a hash adds another possible vector of attack and poses the following problem. Suppose that Charlie has seen the pair $(m, \text{sign}(h(m)))$ signed by Alice, and wants to pretend that she signed his message m' instead, which would be easy to do if $h(m') = h(m)$, because, obviously,

$$\text{sign}\,(h\,(m)) = \text{sign}(h(m')). \qquad (4.163)$$

Thus, a good hash function $h\,(m)$ should make it difficult to find messages m' such that $h(m') = h(m)$, i.e., it has to be a weakly collision-resistant hash function.

4.9.4 Signatures and birthday attacks

The birthday problem arises in probability theory. The problem is to find the probability that, in a set of n randomly chosen people, there is a pair of them with the same birthday. In its present form, the problem was formulated by Richard von Mises; see Frank et al. (1964). Occasionally, this problem is also called the birthday paradox, not because it is a paradox *per se*, but rather because the result seems to be very counterintuitive at first. This probability is clearly 100% for $n = 367$ since only 366 birthdays are possible (we include February 29 in the set). Below, with all due respect to leaplings, we disregard leap years and assume that all days of the year are equally probable. It turns out that 50% probability is reached with just 23 people, $n = 23$, and 99.9% probability is reached with 70 people, $n = 70$.

The solution to the birthday problem is simple. Consider the set of N people, $N \leq 365$. We need to calculate the probability that two (or more) of them have the same birthday, which we denote by P_N. As is often the case, it is easier to calculate the complementary probability $P'_N = 1 - P_N$ than no two people have the same birthday. We arbitrarily number the people from 1 to N. We say that "Outcome 2" takes place when person 2 does not have the same birthday as person 1, the "Outcome 3" takes place when person 3 does not have the same birthday as either person 1 or person 2, and so on. It is clear that P'_N is the same as the probability of "Outcome N," which can be calculated by using conditional probability as follows:

$$P'_2 = 1\frac{364}{365} = 1\left(1 - \frac{1}{365}\right), \quad P'_3 = P'_2\frac{363}{365} = P'_2\left(1 - \frac{2}{365}\right),$$

(4.164)

$$P'_N = P'_{N-1}\frac{365 - (N-1)}{365} = P'_{N-1}\left(1 - \frac{N-1}{365}\right),$$

so that

$$P'_N = \frac{365!}{365^N(365-N)!},$$

(4.165)

where ! is the factorial operator. Accordingly,

$$P_N = 1 - P'_N.$$

(4.166)

In particular, $P_{23} = 50.7\%$, $P_{70} = 99.9\%$. By using the fact that $1 - x \approx e^{-x}$, when $x \ll 1$, it is not hard to see that P_N can be approximated as follows:

$$P_N \approx 1 - e^{N(N-1)/730}.$$

(4.167)

Cryptographic consequences are significant. The birthday attack algorithm allows us to find a collision of a hash function in $2^{L/2}$ attempts, where L is the number of bits defining the classical preimage resistance security. Here is a simple example.

Assume that Charlie knows that Alice is prepared to sign a document d. He wants to design a different document d' and claim that Alice signed it. For instance, instead of "Pay to the order of Bob," he wants the signed document to read "Pay to the order of Charlie." Of course, merely changing Bob's name to Charlie's would not do the trick because the hashes of the corresponding documents will be completely different due to the hash function's collision resistance. However, Charlie can proceed as follows.

Charlie takes the genuine contract d and produces small variations by adding spaces at the line ends, changing the wording slightly, or making other innocuous changes. Charlie takes the fraudulent contract d' and produces similar small changes in it. K changes will give 2^K different documents, $d_1, ..., d_{2^K}$, and similarly, 2^K different documents $d'_1, ..., d'_{2^K}$. He now attempts to find a match for the hash value of $h(d_k)$ and $h(d'_l)$, for an arbitrary pair (k, l), $1 \leq k, l \leq 2^K$. If there is a match $h(d_k) = h(d'_l)$, the same signature will be valid for those two contracts. Charlie presents the acceptable version, d_k, to Alice for signing. After Alice has signed, Charlie takes the signature, attaches it to the altered contract d'_l, and claims that Alice signed it. If the hash values are shorter than 2K bits, then a match's probability is very high due to the birthday paradox.

4.9.5 The random oracle hash

A random oracle hash function (ROHF) is a function that produces a fixed-length random output for any input. When it receives a particular input for the first time, it gives a random output. If it gets the same input again, it will reproduce the same output. A ROHF is an ideal strongly collision-resistant hash function. It can be used to prove the security of encryption/signing provided that the assumption that the functions used behave like a random oracle is valid. However, to show that a given hash function is ROHF usually provides strong evidence rather than real proof. Still, if a hash function fails a ROHF test, we should look for new candidate hash functions.

4.9.6 A practical hash function

While ROHF is more of a theoretical construct, it is crucial to design hash functions that one can successfully use in practice. A practical hash function transforms messages of any length to a fixed-length digest. It has to depend on every bit of the message so that changes of one bit will result in changes in the hash. In the simplest case, the algorithm has to be completely transparent, so that anyone who wants should be able to check the validity of the hash. In general, a practical hash function could depend on a secret parameter, which restricts possible verification to the group of participants who know the secret.

A general hash function has to have the following properties:

(1) Input is a string of *any* size, say n-bit, where n is a large number;
(2) Output is a string of *fixed* size, say 256-bit;
(3) Computation is efficient and requires $O(n)$ operations.

A *cryptographic* hash function has to have some additional properties:

(1) Strong collision-resistance — it is infeasible to find two values x and z, such that $x \neq z$ and $H(x) = H(z)$;
(2) Obfuscation — given y, it is infeasible to find x, such that $y = H(x)$;
(3) Binding — it is infeasible to find x, x', such that

$$H(r||x) = H\left(r'||x'\right), \tag{4.168}$$

where $||$ stands for concatenation of inputs.

4.9.7 Search puzzles and hash functions

A search puzzle can be formulated as follows: given a hash function, H, a value x from a high min-entropy distribution, i.e., a sufficiently random distribution, and a target set Z, find a solution y, such that

$$H(x||y) \in Z, \tag{4.169}$$

where $||$ denotes concatenation. A hash function $H(.)$ is puzzle-friendly, if, for an x, chosen from a high min-entropy distribution, it is infeasible to find y, such that

$$H(x||y) \in Z, \tag{4.170}$$

faster than in $O(2^n)$ time. For a puzzle-friendly hash function, there is no better strategy than a random try. As we shall see in the next chapter, such functions are ideal instruments for the so-called proof-of-work consensus algorithms.

Here is a simple example. Let H be SHA-256 considered in Section 4.10.3 and $x = \text{aaa}$ be a hexadecimal number. We wish to find a nonce y such that $H(\text{aaa}||y) \in Z$, where Z consists of hexadecimal numbers starting with two zeros. The only way of finding a y is to check possible solutions one-by-one, starting with $y = 1$. The corresponding calculations are as follows:

$$
\begin{aligned}
\text{aaa} &\Rightarrow \text{d7bbbd87ccad846b9851316fedf49753...} \\
\text{aaa1} &\Rightarrow \text{f582ba7e93825419ade3d56273f91ef9...} \\
\text{aaa2} &\Rightarrow \text{5c5ac118de3c6c054bcec887953a2cbf...} \\
\text{aaa3} &\Rightarrow \text{cf4bcbea91fcb437dcff9d8697c98303...} \\
\text{aaa4} &\Rightarrow \text{94a7969d0d3f3d83b01ba6a8b02f75be....} \qquad (4.171) \\
\text{aaa5} &\Rightarrow \text{1192bd655bf008aa64954575246a6927...} \\
\text{aaa6} &\Rightarrow \text{d3bbd8b64ce60159c171186a228ee9d2...} \\
\text{aaa7} &\Rightarrow \text{86036a264ec88ee6fa0c3c8ca337d335...} \\
\text{aaa8} &\Rightarrow \text{002c5879349404d5b88419400043e4c3...}
\end{aligned}
$$

Thus, $y = 8$ is the nonce we need. There are infinitely many others, of course.

4.9.8 Other uses of cryptographic hash functions

Since cryptographic hash functions hide the input needed to produce a given output, they are useful for password storage and other similar purposes. In a sense, they provide the called zero-knowledge proof (ZKP) identification. Indeed, if a password pwd is hashed, and its image is $hpwd$, you can prove that you know the password, without revealing the password itself, by calculating $H(pwd)$ and showing that:

$$H(pwd) = hpwd. \qquad (4.172)$$

That is why, if this setup is implemented in earnest, your bank or another security provider can truthfully say that they do not have your password.

Other applications of hash functions in computer science include hash tables and verification of file integrity. Requirements for such functions are often weaker than requirements for functions used for cryptographic purposes. In many respects, modern techniques used to check file integrity can be traced to the old ones, used 2000 years ago when scribes verified the validity of the newly copied Old Testament scrolls; see, e.g., Tov (2012).

4.9.9 Iterative hash function with padding

Initially, constructing a mapping of an arbitrary-length input into a fixed-length output is complicated, almost magical, and, in a sense, it is. However, upon reflection, we can convince ourselves that a simple way of doing so is to use an iterative hash function, which divides the message into blocks of fixed length n, and hashes them iteratively. If the length of the message, N, is not a multiple of the block lengths, $N \neq tn$, where t is an integer, the message is padded by adding zeros and ones according to a predefined rule. This construct, known as the Merkle-Damgård transform, is illustrated graphically in Figure 4.17 .

In each iteration, since we deal with fixed-length blocks of length n, there is a good chance of finding a collision once we have searched approximately $2^{n/2}$ messages. Thus, for t blocks (corresponding to a padded message of length tn), we need to search approximately $t2^{n/2} = 2^{\log t + n/2}$ messages to generate 2^t collisions. For a useful hash function, to find a

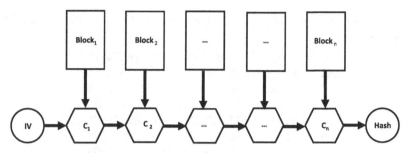

Figure 4.17 Sequentional construction of a hash function known as the Merkle–Damgård transform. Own graphics.

multi-birthday event, corresponding to k collisions, we should go through approximately $2^{n(k-1)/k}$ messages.

4.10 Secure hash algorithms

4.10.1 Background

Although designing a reliable, secure hash algorithm (SHA) is difficult, there are numerous candidates. Bitcoin and many other systems use SHA-256 and RIPEMD-160 (RIPE Message Digest). As its name suggests, SHA-256 maps an arbitrary string into a 256 long digest.

Below we briefly look as SHA-256; however, for practical purposes, it is sufficient to note that SHA-256 uses the Merkle-Damgård transform and turns a fixed-length collision-resistant compression function into a function with arbitrary-length inputs by applying it sequentially. The input is padded as necessary to make its length a multiple of 512 bits.

Several hashing algorithms are used in practice, such as MD5, SHA1, SHA2. However, older algorithms, such as MD5, are broken and can no longer be used for cryptographic purposes (even though occasionally they are still used for less sensitive tasks). It is a cautionary tale — a combination of human ingenuity and computer power can break seemingly unbreakable cryptographic primitives.

In 2007 NIST started a contest for future replacements for SHA1 and SHA2. The proposed algorithms' quality was judged based on their speed, security, depth of analysis, and other considerations. Five SHA3 candidates were chosen for the final evaluation: BLAKE, Grøstl, JH, Keccak, and Skein. In 2012 NIST announced Keccak as the winner.

4.10.2 MD5

MD5, which was invented by Ronald Rivest in 1992, converts a variable-length message into a fixed-length output of 128 bits (=16 bytes); see Rivest and Dusse (1992). The input message is divided into chunks of 512-bit blocks, viewed as 16 32-bit words. If the need occurs, the message is padded so that its length is divisible by 512. When processing the first 512-bit input, the output is initialized to a given set of eight 32-bit initial values. When processing subsequent 512-bit inputs, the output is initialized to the set of four 32-bit values obtained from the previous calculations.

Thus, it is enough to understand how one 512-bit input is transferred to a 128-bit hash. The input is divided into 16 32-bit words, denoted by $W_0, ..., W_{15}$. The set of 16 words in expanded to the set of 64 words $M_{i,j}$, $1 \leq i \leq 4, 0 \leq j \leq 15$, according to the following rule:

$$M_{1,j} = W_j, \ M_{2,j} = W_k, \ k = (5j+1) \bmod 16,$$

(4.173)

$$W_{3,j} = W_k, \ k = (3j+5) \bmod 16, \ W_{4,j} = W_k, \ k = 7j \bmod 16.$$

Since there are 64 words, there are 64 rounds. The flowchart for MD5 is shown in Figure 4.18.

The transformation of a message block consists of four stages or rounds. Each round is composed of 16 similar sub-rounds using every one of 16 sub-messages; they rely on non-linear functions $F_1, ..., F_4$ operating on bits as follows:

$$F_1(B,C,D) = (B \wedge C) \vee (\neg B \wedge D), \ F_2(B,C,D) = (B \wedge D) \vee (C \wedge \neg D),$$

(4.174)

$$F_3(B,C,D) = B \oplus C \oplus D, \ F_4(B,C,D) = C \oplus (B \vee \neg D).$$

Here $\oplus, \wedge, \vee, \neg$ are XOR, AND, OR, NOT bitwise operations.

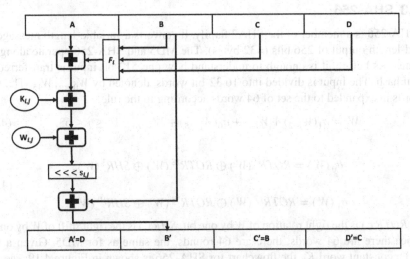

Figure 4.18 Flow chart for MD5. Own graphics. Source: Wikipedia.

The $i.j$-th round, $i = 1, ..., 4, j = 0, ..., 15$, proceeds as follows:

$$A_{i,j} = D_{i,j-1}, \quad C_{i,j} = B_{i,j-1}, \quad D_{i,j} = C_{i,j-1},$$

$$B_{i,j} = ROTL^{S_{i,j}} \left(A_{i,j-1} + F_i \left(B_{i,j-1}, C_{i,j-1}, D_{i,j-1} \right) \right) \tag{4.175}$$

$$+ K_j + W_j) + B_{i,j-1}.$$

Here $ROTL(.)$ is left rotation of W by one bit.

Some examples of MD5 hashes are shown in Eqs (4.176)–(4.178).

$$
\begin{gathered}
\text{Mary had a little lamb.} \\
\Downarrow \\
\texttt{ca964b1677d5476ea11eed1e1837c342}
\end{gathered}
\quad , \tag{4.176}
$$

$$
\begin{gathered}
\text{Mary had a little lamb} \\
\Downarrow \\
\texttt{e946adb45d4299def2071880d30136d4}
\end{gathered}
\quad , \tag{4.177}
$$

$$
\begin{gathered}
\text{Mary had a little lamb!} \\
\Downarrow \\
\texttt{ecb0d9a4d79da7379d6a070fb7216ff4}
\end{gathered}
\quad . \tag{4.178}
$$

These examples clearly show that "small" changes in the input make the output unrecognizable, as intended.

At present, MD5 is broken and should not be used in practice; see Wang and Yu (2005). Yet, it serves as a prototype for other, more complicated, and secure algorithms.

4.10.3 SHA-256

The SHA-256 is a member of the SHA2 family. It converts a variable-length message into a fixed-length output of 256 bits (=32 bytes). The MD5 and SHA-256 operate along similar lines. As before, it is enough to understand how one 512-bit input is transferred to a 256-bit hash. The input is divided into 16 32-bit words, denoted by $W_0, ..., W_{15}$. The set of 16 words in expanded to the set of 64 words according to the rule

$$W_j = \sigma_1(W_{j-2}) + W_{j-7} + \sigma_0(W_{j-15}) + W_{j-16}, \quad 16 \le j \le 63. \tag{4.179}$$

Here

$$\sigma_0(W) = ROTR^7(W) \oplus ROTR^{18}(W) \oplus SHR^3(W),$$

$$\sigma_1(W) = ROTR^{17}(W) \oplus ROTR^{19}(W) \oplus SHR^{10}(W), \tag{4.180}$$

where $ROTR(.)$ is the right rotation of W by one bit, $SHR(.)$ is the right shift of W by one bit.

Since there are 64 words, there are 64 rounds, the same as for MD5. Given a word W_i and a constant word K_i, the flowchart for SHA-256 is shown in Figure 4.19; see, e.g., Barker (2002); Sanadhya and Sarkar (2008).

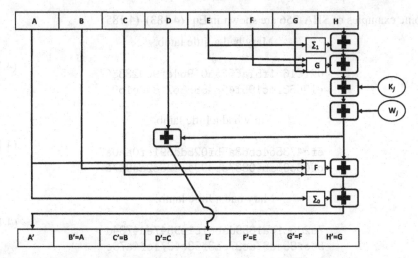

Figure 4.19 Flow chart for SHA2. Own graphics. Source: Wikipedia.

The j-th round proceeds as follows:

$$A_j = \Sigma_0(A_{j-1}) + F(A_{j-1}, B_{j-1}, C_{j-1}) + H_{j-1}$$

$$+ \Sigma_1(E_{j-1}) + G(E_{j-1}, F_{j-1}, G_{j-1}) + K_j + W_j,$$

$$B_j = A_{j-1}, \quad C_j = B_{j-1}, \quad D_j = C_{j-1},$$

$$E_j = D_{j-1} + H_{j-1} + \Sigma_1(E_{j-1})$$

(4.181)

$$+ G(E_{j-1}, F_{j-1}, G_{j-1}) + K_j + W_j,$$

$$F_j = E_{j-1}, \quad G_j = F_{j-1}, \quad H_j = G_{j-1}.$$

Here

$$F(A, B, C) = (A \wedge B) \oplus (\neg A \wedge C),$$

$$G(A, B, C) = (A \wedge B) \oplus (B \wedge C) \oplus (C \wedge A),$$

(4.182)

$$\Sigma_0(W) = ROTR^2(W) \oplus ROTR^{13}(W) \oplus ROTR^{22}(W),$$

$$\Sigma_1(W) = ROTR^6(W) \oplus ROTR^{11}(W) \oplus ROTR^{25}(W).$$

Some examples of SHA-256 are shown in Eqs (4.183)–(4.185).

Mary had a little lamb.

$$\Downarrow$$

d2fc16a1f51a653aa01964ef9c923336
e10653fec195f493458b3b21890e1b97 (4.183)

Mary had a little lamb

$$\Downarrow$$

efe473564cb63a7bf025dd691ef0ae0a
c906c03ab408375b9094e326c2ad9a76 (4.184)

Mary had a little lamb!

$$\Downarrow$$

7e2dbc1ca1859dabe1c1e9547ed4734d
56ef85ec87ae87ea3f63c4371cf4a79e (4.185)

Once again, we see that "small" changes in the input make the output unrecognizable, as intended.

It is essential to understand that, in some cases, the SHA-256 algorithm can interpret the input in two different ways: (a) as a text; (b) as a hexadecimal number. The outputs are entirely different, of course. In Eqs (4.186) and (4.187), we show some examples of SHA-256 applied to text strings vs. hexadecimal numbers.

efe473564cb63a7bf025dd691ef0ae0a
c906c03ab408375b9094e326c2ad9a76

$$\Downarrow$$

f8c72ab0790e80e9191af68a37659e80
33c3de935f68be0df8ed04bfe35ac3c5 (4.186)

efe473564cb63a7bf025dd691ef0ae0a
c906c03ab408375b9094e326c2ad9a76

$$\Downarrow$$

a34e5a180726e23e5ea2aea4b16ce101
99331eb897003e419c947619426e4da0 (4.187)

We note that in the Bitcoin protocol, and, by extension, most other crypto protocols, which predominantly deal with hexadecimal numbers rather than strings, inputs are always hashed twice:

$$x \rightarrow y = SHA256(x) \rightarrow z = SHA256(y). \qquad (4.188)$$

The second hashing is used as a safety mechanism against the so-called "length-extension" attack; see Ferguson et al. (2010). A length extension attack allows an attacker who knows the length of a message $m1$ and its Hash($m1$) to calculate the hash Hash($m1\|$ $m2$) of the message $m1$ concatenated with an attacker-chosen message $m2$, without knowing $m1$. Algorithms, relying on the Merkle–Damgård construction, such as MD5, SHA-1, and SHA-2, can be vulnerable against such an attack. The newer SHA-3 algorithms are secure in this respect.

4.10.4 RIPEMD-160

RIPEMD-160 was developed in 1992 and improved in 1996 by Hans Dobbertin, Antoon Bosselaers and Bart Preneel; see Dobbertin et al. (1996).[8] It converts a variable-length message into a fixed-length output of 160 bits (= 20 bytes). This algorithm operates on five 32-bits words. The flowchart for RIPEMD-160 is shown in Figure 4.20. For details, we refer the reader to the original paper.

Once again, in the Bitcoin protocol, and by extension, many other crypto protocols, inputs are hashed twice so that:

$$x \to y = SHA256(x) \to z = RIPEMD160(y). \tag{4.189}$$

Some examples of RIPEMD-160 in action applied to a SHA256 of a hexadecimal number are shown in Eqs (4.190) and (4.191). We see that a small local change of the input — the first digit changed from a to b — results in a global change in the output.

$$
\begin{array}{c}
\texttt{a34e5a180726e23e5ea2aea4b16ce101} \\
\texttt{99331eb897003e419c947619426e4da0} \\
\Downarrow \\
\texttt{c73e70858495648065bf86a58d7cd30783b3f86d}
\end{array} , \tag{4.190}
$$

$$
\begin{array}{c}
\texttt{b34e5a180726e23e5ea2aea4b16ce101} \\
\texttt{99331eb897003e419c947619426e4da0} \\
\Downarrow \\
\texttt{289f9b3a5479cfeea34e9c14d7ea6d4291c89561}
\end{array} . \tag{4.191}
$$

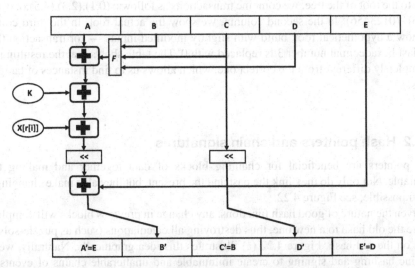

Figure 4.20 Flow chart for RIPEMD160. Own graphics. Source: Wikipedia.

---

[8]RIPEMD stands for RACE (Research in Advanced Communications Europe) Integrity Primitives Evaluation Message Digest.

These examples clearly show that "small" changes in the input make the output unrecognizable, as intended.

4.11 Merkle trees and hash pointers

4.11.1 Merkle trees

The idea of a hash, combined with padding for efficiency, is beneficial for making data simultaneously immutable and compressed. Suppose we have a set of N data blocks. We can pad them to make the total number of blocks a multiple of 2, say 2^n. Then we can hash them sequentially until we reach the root node; see Merkle (1987). These constructs are called Merkle trees and are very useful in cryptography in general and blockchains in particular. In the next chapter, we show how Bitcoin and Etherium protocols (as examples) use Merkle trees.

A typical Merkle tree is shown in Figure 4.21 (a). By construction, Merkle trees are tamper-proof, in a sense that changes in any leaf along the path to the root will change the root itself beyond recognition. This property is helpful if, for example, we wish to verify that a particular data point, say a transaction, belongs to a particular Merkle tree. Using Merkle trees, we can verify membership of a transaction in a set in $O(\ln(N))$ steps. This idea is illustrated in Figure 4.21 (b).

In Table 4.6, we use hashes of the first seven transactions from the Bitcoin Block 600,000 and build a balanced tree, instead of a padded tree to speed up the process. In other words, to get to the root of the tree, we combine transactions as follows: (0,1),(2,3),(4,5),6; (0123), (4,5,6); (0123456). In the second column, we show the actual root. In the third column, we show a hypothetical root, build with slightly modified inputs — for transaction 0, the very last hexadecimal number 3 is replaced with 0. The table shows that the resulting root is completely different from the correct one, which allows us to find instances of tampered inputs.

4.11.2 Hash pointers and chain signatures

Hash pointers are beneficial for chaining blocks of data together and making them immutable. Not only do they link the past and the present, but they also make changing the past impossible, see Figure 4.22.

Given the nature of good hash functions, any change in previous blocks will completely change the old hash to a new one, thus destroying all calculations (such as puzzle-solving) based on the old hash. Figure 4.22 (a) illustrates this idea graphically. Naturally, we can combine hashing and signing to create immutable and unalterable chains of events. By signing a block of data and hashing the pair $(m, \text{sign}(h(m)))$ we create a hash pointer. In turn, this pointer becomes an integral part of the next block. Thus, we create a blockchain shown in Figure 4.22 (b).

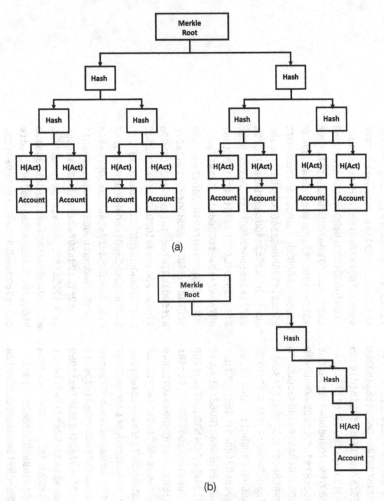

(a)

(b)

Figure 4.21 (a) Merkle tree is used to compress the information about multiple transaction into a single Merkle root; (b) Merkle tree can be used to obtain proof of inclusion allowing the user to verify that their account is included and their deposit is correct. For privacy, account information itself is hashed. The leaf node represents an account. If need occurs, it is revealed and all the nodes on the path from the leaf to the Merkle root are traced. Own graphics.

4.11.3 Chain signatures combined with Merkle trees

Above, we have shown how to create blockchains by combining hash pointers and signatures. When blocks are non-uniform, we can add Merkle trees into the framework and consider the root of a large block to be signed and chained. This method helps create non-alterable records containing numerous transactions; see Figure 4.23.

Table 4.6 Calculation of the Merkle tree root.

Round	Actual hash	Modified hash
0	93955d40d918d014903843d258aada5c 720a5d37afac7889268f459a97b148a3	93955d40d918d014903843d258aada5c 720a5d37afac7889268f459a97b148a0
1	a8178a722337241ac060b4bba4b33b8 b4847a756fa715af7fd11bfd143d5	a8178a722337241ac060b4bba4b33b8 b4847a756fa715af7fd11bfd143d5
2	efb3f60304532ebc80163b5f375fa8a9 4a39a8b0807b99703b6b646c1f7af5bf	efb3f60304532ebc80163b5f375fa8a9 4a39a8b0807b99703b6b646c1f7af5bf
3	a070eda356c87a7af9bff22eab3b3c38 460605eb00938c84a86a1d6d3c608078	a070eda356c87a7af9bff22eab3b3c38 460605eb00938c84a86a1d6d3c608078
4	b96b516295b8e4f5452405db8213ca56 cde630b7a30c2a400c4991b9a17f072a	b96b516295b8e4f5452405db8213ca56 cde630b7a30c2a400c4991b9a17f072a
5	d8974a3a6596fbd86bd1f794a221639c 37d7633394bf03fddd61b48a1505f8b1	d8974a3a6596fbd86bd1f794a221639c 37d7633394bf03fddd61b48a1505f8b1
6	d87a5a7ea8a8fb566d81605d1ef9ab11 32462f7812dab97e24b32d667aa1f959	d87a5a7ea8a8fb566d81605d1ef9ab11 32462f7812dab97e24b32d667aa1f959
01	8b038b740bb16a4e54593367f0bc48f bde0c17518821f2efd8489daa0720849	0f87ad6a8497309300d00fc0c12d2af74 a2ff901b27a7cd61efba01db3817b09d
23	17efe128dcaede1600101af5ef2223d8 51578f06ab89725733fa2aac31e972d0	17efe128dcaede1600101af5ef2223d8 51578f06ab89725733fa2aac31e972d0
45	9e27b512cb2ec01cc0eb68df6a683f67 4c6c2b6c50bdcf9a2ee3eaeb9575ea4d	9e27b512cb2ec01cc0eb68df6a683f67 4c6c2b6c50bdcf9a2ee3eaeb9575ea4d
6	d87a5a7ea8a8fb566d81605d1ef9ab11 32462f7812dab97e24b32d667aa1f959	d87a5a7ea8a8fb566d81605d1ef9ab11 32462f7812dab97e24b32d667aa1f959
0123	b11872bc8ca21a33c52efd1fe1fbd161 8203cb181c52db3f22e0b70bb2c165ba	3e40422b5f6080810a187564b6e67da2 a2e612e36ccf6b9ee410d56dbc356eb3
456	b50bffbc033369a859af8a757cf9adfe 6334e2c28aedb912eadab99b8986f12e	b50bffbc033369a859af8a757cf9adfe 6334e2c28aedb912eadab99b8986f12e
01234567	1bfedd44486da71aa3d2da322108bbb2 31d3f59077af88175ae065a0e5e64518	06659e1edeb39df2e1867f0d1f8d0e2 c91e18d66c5dceecbede07d9f67ca1f2

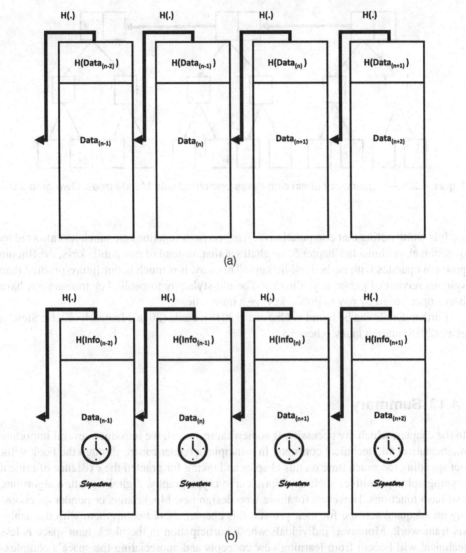

Figure 4.22 (a) A sequence of hash pointers. (b) A sequence of chain signatures. Own graphics.

4.12 Quantum-resistant cryptography

The public key infrastructure framework considered in this chapter relies on assuming that specific calculations, such as factoring a large number, solving the discrete log problem, or its EC analog, are hard. These assumptions are valid, given the state of the current hardware. However, the rapid, if uneven progress achieved in quantum computing, could dramatically change the situation.

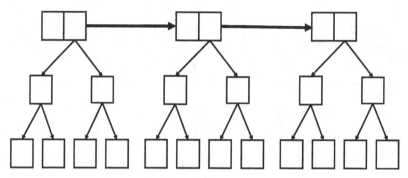

Figure 4.23 A sequence of chain signatures combined with Merkle trees. Own graphics.

It is worth noting that computations relying on hash functions are much less affected by quantum algorithms. In Chapter 5, we shall see that, instead of raw public keys, the Bitcoin protocol operates with hashed public keys. This way, it is much better future-proofed than systems relying of public keys directly. The old-style pay to public key transactions have been superseeded by pay to public key hash transactions.

Further details can be found in Gao et al. (2018); Ikeda (2018); Jentzsch (2016); Stewart et al. (2018) among many others.

4.13 Summary

In the chapter, which, by necessity, is somewhat technical, we introduce several important mathematical and technical concepts. In principle, a reader can go through the book without spending too much time on this chapter and taking for granted the existence of critical cryptographic primitives such as elliptic curve cryptography, digital signature algorithm, and hash functions. However, for those who design new blockchains or ponder on choosing an adequate scheme for their project, this chapter should equip them with the analysis framework. Moreover, individuals whose participation in the blockchain space is less technical will benefit from learning new concepts and appreciating the space's complexity. By its very nature, we show that cryptography is a natural mechanism for maintaining the integrity of computerized record keeping. This fact naturally makes it the foundational tool for building financial and non-financial applications. While symmetric cryptography is essential (for securely storing data in databases and other similar applications), it is secondary to transact over the Internet. At the same time, asymmetric cryptography is the primary instrument in this respect.

PKI, which includes public/secret key pair creation, distribution, and usage, provides a robust framework for building distributed applications, such as cryptocurrencies and blockchains. For transactional purposes, a digital signature is the main instrument. Hash functions are particularly useful because we can compress messages efficiently by using a

Merkle tree of hashes. With these instruments at our fingertips, we are now ready to understand how to build peer-to-peer digital currencies and distributed ledgers.

Of the multitude of books and lecture notes on cryptography, we found the following particularly useful: Goldwasser and Bellare (2008); Katz and Lindell (2014); Larsson (2018); Trappe and Washington (2006).

4.14 Exercises

1. Decrypt the message m_1, encrypted with an affine cipher,

$$m_1 = \begin{array}{l} \text{ZLDDRVL W FD LEBYHOILGXFIA} \\ \text{SFVLELYL PLH JQ ULEVIA QJNY} \end{array}.$$

2. Decrypt the message m_2, encrypted with a Vigenère cipher by using a hint from m_1,

$$m_2 = \begin{array}{l} \text{ORMFGCBWQ VVOQI URFZFQF WVQOKJRZ} \\ \text{WPU Z SFWIVFQ AEF DZU UKALTDZU} \end{array}.$$

3. Solve the problem described in the encrypted message m_2.
4. Design a physical system along the lines of Chaum's DigiCash for an anonymous absentee voting system.
5. Describe the birthday paradox. Calculate how many people need to be in the room so that the probability of two of them have the same birthday (day and month) be 99%. What about 99.9%? Build the graph of this probability as a function of the number of people. What does it mean for the security of a 512-bit hash function $y = f(x)$? How many people need to be in the room to guarantee that the probability of three of them having the same birthday be 95%?
6. Consider an elliptic curve

$$E_p = \left\{(x, y) : y^2 = x^3 + 7\right\} \bmod p,$$

where $p = 127 = 2^7 - 1$ is the fourth Mersenne prime. How many integer points are on this elliptic curve? Compare your answer with Hasse's estimate. Prove that the point $\alpha = (19, 32)$ belongs to E_p and construct a sequence of $\beta_n = n\alpha, n = 1, ..., 100$. Explain in your own words whether this sequence is suitable for encryption purposes, and if so, how.
7. Use the MD5 algorithm as an inspiration and construct your own hash function $y = h(x)$, which maps an input x of any length into an output y with the length of 32 bits. Is it a good hash function? If not, can you produce a collision, i.e., find two distinct messages x, x' such that $h(x) = h(x')$?
8. Design an efficient implementation of the Random Oracle hash function of your own with 32-bit long output. Can you produce a collision, i.e., find two distinct messages x, x' such that $h(x) = h(x')$?
9. Explain how one can use a good hash function to protect users' passwords. How would you try to break it? What is a dictionary attack?
10. Use the standard SHA-256 hash function and find a nonce that produces a three 000 starting with a string LiptonTreccani.

5 | Bitcoin — A Deep Dive

"I've been working on a new electronic cash system that's fully peer-to-peer, with no trusted third party — The main properties: Double-spending is prevented with a peer-to-peer network. No mint or other trusted parties. Participants can be anonymous. New coins are made from Hashcash style proof-of-work. The proof-of-work for new coin generation also powers the network to prevent double-spending."

Satoshi Nakamoto, 2008, Cryptography Mailing List

5.1 Introduction

By now, there are thousands of different blockchain protocols and associated coins and tokens. However, the so-called Big Three — Bitcoin, Ethereum, Ripple — retain their lead.

This chapter discusses the pros and cons of Bitcoin — the undisputed champion of the crypto world, which inspired hundreds of competing alternative coins, also known as altcoins. Bitcoin itself experienced multiple hard forks, including Bitcoin Cash, Bitcoin SV, etc. Therefore, an often repeated statement that only a limited number of Bitcoins will be mined is not strictly speaking true.

The Bitcoin protocol rests on the giants' shoulders. All the necessary cryptographic ideas were developed in the 20th century; see, e.g., Narayanan and Clark (2017). Much like the goddess Aphrodite of Greek mythology, the Bitcoin protocol was born fully formed and ready to go. There is a striking contrast between Bitcoin and numerous other technologies such as telephone, television, computer, etc., which underwent a very long period of gradual improvements before becoming as impressive as they are. Still, on several levels, the Bitcoin protocol failed to achieve its objectives. It is not as decentralized as claimed or believed by some. In particular, mining is highly centralized; see discussion below. In addition, the Bitcoin protocol governance is centralized as well; see Tariq and Jamison (2019).

Below we differentiate between Bitcoin as a protocol and bitcoin, or BTC, as the corresponding native token. BTC price dynamics and Google trends are presented in Figure 5.1. This figure clearly shows that public interest in Bitcoin is correlated with the BTC price.

The chapter, central to the whole book, is organized as follows. In Section 5.2 we present a detailed description of Bitcoin addresses — their meaning, how to construct them, and how to use them. In Section 5.3, we study Bitcoin transactions, which boil down to disposing of all the bitcoins held at a particular address, with a desired amount of BTCs sent from

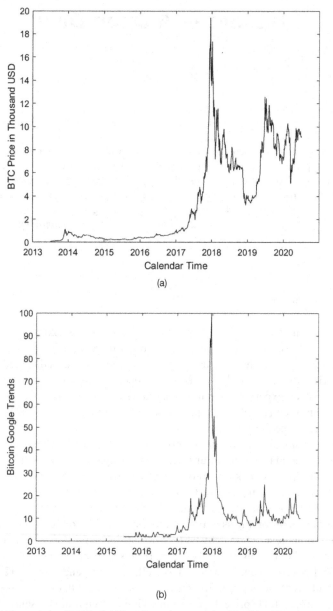

Figure 5.1 (a) The price of BTC; (b) The Google trend for Bitcoin. The corresponding time series are strongly correlated. Own graphics. Sources: (a) coinmarketcap.com, (b) Google.

the payer to the payee and the rest returned to the payer as change. We dedicate Section 5.4 to the analysis of the Bitcoin ecosystem and describe the relevant ingredients, including nodes, wallets, miners, exchanges, and other details. Section 5.5 deals with consensus and mining. It explains how miners reach the proof-of-work consensus, what they do in practice, and what are their incentives. In Section 5.6, we study a particular block, specifically Block 600,000, to illustrate the ideas presented in the earlier sections. We summarize Bitcoin pros and cons in Section 5.7 and draw our conclusions in Section 5.8.

5.2 Bitcoin addresses

As was mentioned in Subsection 4.7.9, the elliptic curve digital signature algorithm (ECDSA) is usually based on a U.S. government standard. This standard uses secp256r1, which is believed to be secure. However, for obvious reasons, Bitcoin does not trust government recommendations and uses secp256k1 instead. To put it differently, Bitcoin rejects secp256r1 in favor of secp256k1 precisely because secp256r1 is NIST-backed and fails to demonstrably satisfy the nothing-up-my-sleeve principle. Lack of proof may or may not suggest that the sextuple characterizing the 256r1 EC has been selected to favor some form of backdoor vulnerability; see Section 4.8.

Given that both secret and public keys are, basically, very large integers, we need to find a way to represent them in a relatively compact form. We have several options: (a) binary — numerals 0 to 1 (this is how numbers are internally represented in computers); (b) decimal — numerals 0 to 9 (this is what we use in our daily lives); (c) hexadecimal — numerals and letters a to f[1]; (c) Base64 — 26 lowercase letters, 26 capital letters, numerals 0 to 9, and two more characters such as ""/; (d) Base58 — a subset of Base64 — lowercase and capital letters and numbers without the four easy-to-confuse symbols (0, O, l, I). We emphasize that hexadecimal numbers are case-insensitive, while both Base58 and Base64 numbers are case sensitive.

The elliptic curve domain parameters over $\mathbb{F}_p$ for the curve secp256k1 are specified by the sextuple $T = (p, a, b, G = (G_x, G_y), n, h)$, of the form:

$$
\begin{aligned}
p = &\ \texttt{ffffffffffffffffffffffffffffffff} \\
&\ \texttt{fffffffffffffffffffffffffeffff\,c2f} \\
a = &\ \texttt{00000000000000000000000000000000} \\
&\ \texttt{00000000000000000000000000000000} \\
b = &\ \texttt{00000000000000000000000000000000} \\
&\ \texttt{00000000000000000000000000000007} \\
G_x = &\ \texttt{79be667ef9dcbbac55a06295ce870b07} \\
&\ \texttt{029bfcdb2dce28d959f2815b16f81798} \\
G_y = &\ \texttt{483ada7726a3c4655da4fbfc0e1108a8} \\
&\ \texttt{fd17b448a68554199c47d08ffb10d4b8} \\
n = &\ \texttt{fffffffffffffffffffffffffffffffe} \\
&\ \texttt{baaedce6af48a03bbfd25e8cd0364141} \\
h = &\ 1
\end{aligned}
\tag{5.1}
$$

---

[1]Such numbers are frequently written with a prefix 0x to indicate that the usage of the hexadecimal base.

The corresponding p can be rewritten as follows:

$$p = 2^{256} - 2^{32} - 2^{9} - 2^{8} - 2^{7} - 2^{6} - 2^{4} - 1, \tag{5.2}$$

and represents a generalized Mersenne number, see Subsections 4.7.6 and 4.7.9. This sextuple was discovered by Niel Koblitz, hence the "k" in the penultimate position stands for "Koblitz."

Object sizes used in the Bitcoin ECDSA framework are as follows: Secret key — 256 bits; Public key — 512 bits; Message to sign — 256 bits; Signature — 512 bits. It is worth noting that these are the sizes in the so-called raw form. Later, we shall see that typically signatures are around 568–576 bits in length due to padding required to satisfy various representation standards.

As explained in Section 4.7, a secret key, k, can be converted into a public key, K, but a public key, K, cannot be converted back into a secret key, k because the transformation $k \rightarrow K$ is a one-way function. We have to choose secret key k, which is just a very long number (256 bits in binary code), at random.

Our randomly chosen raw hexadecimal key has the form:

$$k_{HEX} = \frac{\text{d6ba4be53fb53e6dffd5aac6ed7f7023}}{\text{f2eca5fa9ddc050ed40c7e1463d0adb7}}. \tag{5.3}$$

Of course, we can write k in different formats. Various representations of the same secret key k, which must be hidden at all times, are shown in Eqs (5.4)–(5.6):

$$k_{BIN} = \begin{matrix} \text{1101011010101110100100101111100101} \\ \text{0011111110110101001111100110110101101} \\ \text{1111111110101011010101011000110} \\ \text{1110110101111111011100000100011} \\ \text{1111001011101100101001011111111010} \\ \text{1001110111011100000000101000001110} \\ \text{1101010000001100011111110000010100} \\ \text{0110001111010000101011011011011011111} \end{matrix}, \tag{5.4}$$

$$k_{DEC} = \frac{\text{9712410696238081644394820180010104877}48}{\text{50879409010076863372728208030912523703}}, \tag{5.5}$$

$$k_{WIF} = \frac{\text{5KSrWhmfdd9uBLe5QfypzEY1jN}}{\text{HdkAtJPtCg5T8uxGDzASpEBJL}}. \tag{5.6}$$

Given that the secret key's knowledge is tantamount to the ownership of bitcoins associated with the corresponding public key, it is imperative to devise formats that would handle the key as safe as possible. Bitcoin uses the so-called Wallet Import Format (WIF), also known as Wallet Export Format (WEF), to encode a secret ECDSA key and make it easier to copy. We illustrate the corresponding steps graphically in Figure 5.2.

Once k given by Eq. (5.3) is chosen, we find the associated public key (address) by calculating the product $k \times G$:

$$K = k \times G = (x, y). \tag{5.7}$$

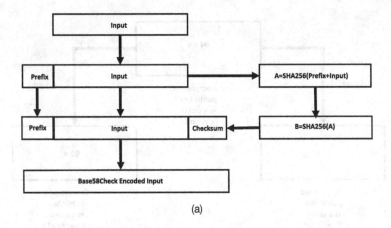

(a)

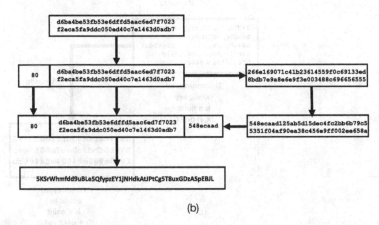

(b)

Figure 5.2 (a) Flow chart of the Base58Check Encode algorithm; (b) Algorithm applied to a randomly chosen secret key, which is error-checked and written in WIF form. Own graphics.

Here G is given by Eq. (5.1) and

$$x = \frac{579afad03cf11bedd612f7f156bd8c8e}{e7ab063eecf855a989e84604efa488cb},$$
$$y = \frac{1c6909c43ed944d7fb55d4ea5763f482}{d96a8705f31668a30664861fdae2855d}. \tag{5.8}$$

As before, this key can be written in different formats, determined by the corresponding prefixes, as shown in Figure 5.3.

In the associated Figure 5.3, we present K in two equivalent forms: (a) in the uncompressed form, with the prefix 04; (b) in the compressed form, here with the prefix 03, because y is odd. For even y, the corresponding prefix is 02.

The idea is that we can represent any point $(x, y) \in E_T$ either in the full form, as a pair (x, y), or in the abridged form, as its x coordinate, and the parity of the y coordinate. It is

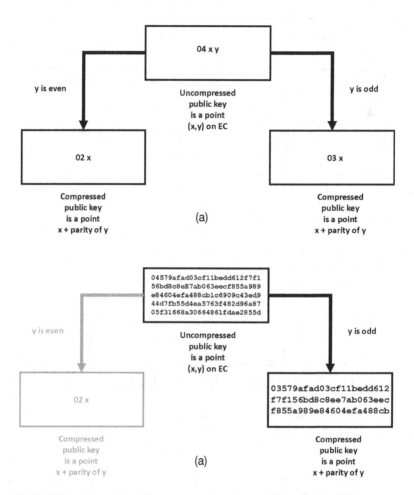

Figure 5.3 Public keys can be left uncompressed (as points (X, Y) on EC) or compressed (as abscissas X on EC plus an indication of the parity of the ordinate Y), which comes in handy in many situations: (a) Generic flowchart; (b) Specific example of compression. Own graphics.

clear that both forms describe the same point on the EC secp256k1, but, as expected, the compressed form is half the length of the uncompressed one. However, as is shown below, this fact does not affect their Bitcoin address representations, which are of similar length.

Both uncompressed and compressed keys can be conveniently hashed (twice) and error-checked; see Table 5.1.

We illustrate the above operations graphically in Figures 5.4–5.6. In Figure 5.4, we show the flow chart of the general Base58Check Encoding procedure, while in Figures 5.5

Table 5.1 Different representations of the same public key K.

Uncompressed Public Key

04579afad03cf11bedd612f7f156bd8c8e
e7ab063eecf855a989e84604efa488cb
1c6909c43ed944d7fb55d4ea5763f482
d96a8705f31668a30664861fdae2855d

SHA256

28bcaff01234071645bbb9af5da1e698
1b9a4bfe317488b33e111845991cbe36

RIPEMD160

4bcc96fe592c2e4d59c6b9facbc5d57c6150b312

Base58 Encoded Bitcoin Address

17unmY9kEdgSuYB5xd4paUBP1Ubzo3EgtL

Compressed Public Key

03579afad03cf11bedd612f7f156bd8c8e
e7ab063eecf855a989e84604efa488cb

SHA256

0c73c5e00bcbdbdbb7092b29fdcf8671
f53fb2fffc1a823cac4ba990d108a9be

RIPEMD160

581617b45032f35772705c9c0688efaf2ea11a30

Base58 Encoded Bitcoin Address

192kvuHdhNgJ9h1YG3E8AKxWSyRHdH9Tqt

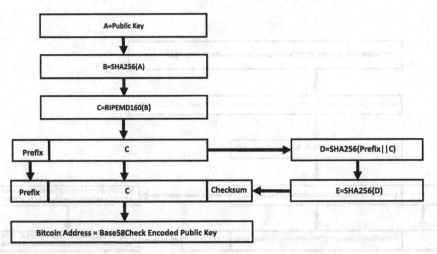

Figure 5.4 Both uncompressed and compressed keys are hashed twice for convenience and Base58Check Encoded. Two different representations of the same address are produced as a result. Own graphics.

and 5.6, we illustrate how this procedure is applied to the specific public key we are interested in.

Because both uncompressed and compressed keys undergo several hashing procedures, the resulting keys are approximately the same length when written in the Base58 format.

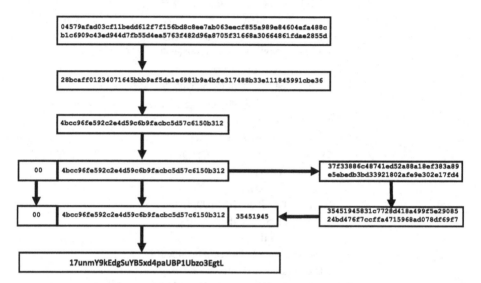

Figure 5.5 An uncompressed public key is hashed twiced and Base58Check Encoded by using the algorithm illustrated in Figure 5.4. Own graphics.

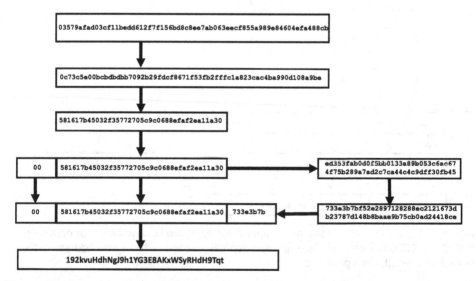

Figure 5.6 A compressed public key is hashed twiced and Base58Check Encoded by using the algorithm illustrated in Figure 5.4. Own graphics.

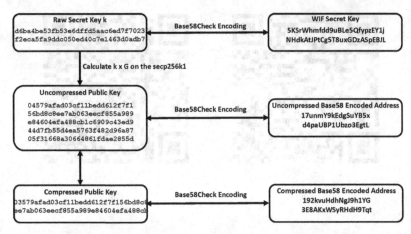

Figure 5.7 Mappings of a raw secret key into the WIF secret key and Base58 Encoded Bitcoin addresses. Own graphics.

We summarize all the steps starting with the choice of a random secret key in Figure 5.7. Figure 5.8 shows the secret key, compressed public key, and uncompressed public key in the convenient computer-readable QR format.

5.3 Transactions

5.3.1 Distributed ledger accounting

The distributed ledger allows us to have a consensus-based view of the amount of bitcoins associated with a particular address. It is important to emphasize that consensus does not help in any way to calculate the dollar value of the corresponding bitcoins. This task is strictly outside of the relatively limited confines of the Bitcoin protocol itself. This value naturally changes every time a new transaction associated with the address in question is accepted as valid by the community.

To be consistent with the general rules of accounting, we need a proper ledger. Essentially, we have two options. We can construct a rudimentary account-based ledger shown in Table 5.2.

In this table, TF stands for cumulative transaction fees paid by all the parties whose transactions are included in the block. This account-based system requires a fairly complicated mechanism for tracing balances at all times. Building such a mechanism is not impossible. As described in the next chapter, Ethereum does exactly that. Still, Bitcoin designers, being trailblazers, have chosen a simple approach. Specifically, they have built a transaction-based ledger. The way it operates is shown in Table 5.3.

In this table, we distinguish between the cumulative transaction fee, TF, paid to Alice as a miner of a block, and an individual transaction fee, tf, paid by the relevant parties per transaction. At present, TF is roughly 1,000 times larger than tf.

We can think about the inputs of a transaction as coins destroyed and outputs as coins created. It is imperative to understand that in a transaction-based system, every transaction has to spend the entire amount associated with a particular address, even if only a fraction of

(a)   (b)

(c)

Figure 5.8 QR codes: (a) raw secret key; (b) compressed public key; (c) uncompressed public key. Source: https://www.qr-code-generator.com/.

Table 5.2 A rudimentary account-based ledger.

Transactions	Signed by
Create 12.5 bitcoins + TF and credit to Alice	Miners
Send 8 bitcoins from Alice to Bob, tf to miners	Alice
Send 4 bitcoins from Bob to Charlie, tf to miners	Bob
Send 3 bitcoins from Charlie to Alice, tf to miners	Charlie
Send 7 bitcoins from Alice to David, tf to miners	Alice

Table 5.3 A rudimentary transaction-based ledger.

#	Inputs	Outputs	Signed by
1	⊘	12.5+TF->Alice	Miners
2	1[0]	8->Bob,(4.5+TF-tf$_2$)->Alice,tf$_2$->Miners	Alice
3	2[0]	4->Charlie,(4-tf$_3$)>Bob,tf$_3$->Miners	Bob
4	3[0]	3->Alice,(1-tf$_4$)->Charlie,tf$_4$->Miners	Charlie
5	1[0],3[0]	7->David,(0.5+TF-tf$_2$- tf$_5$)->Alice,tf$_5$->Miners	Alice

Table 5.4 Moving bitcoins between Alice and Bob.

	Scenario$_1$	Scenario$_2$
State$_n$	{Alice: 50, Bob: 20}	{Alice: 50, Bob: 20}
TX	{send 10 BTC from Alice to Bob, send 40 BTC from Alice to Alice}	{send 60 BTC from Alice to Bob, send -10 BTC from Alice to Alice}
Result	Success	Failure
State$_{n+1}$	{Alice: 40, Bob: 30}	{Alice: 50, Bob:20}

the amount is sent to the payee's address, while the remainder goes to an address controlled by the payer. Many authors suggest that, for security purposes, one should not be reuse addresses. Thus, the output amount is split into the payment itself, change, and transaction fees. For instance, Alice has initially received $12.5 + TF$ bitcoins. To pay 8 bitcoins to Bob, she completely wipes out her account by sending the required 8 bitcoins to Bob, and the balance of $4.5 + TF - tf$ bitcoins to herself.

A typical transition, shown in Table 5.4, is conceptually similar to the case of moving money between two deposits in the same bank in Table 2.6.

As we shall see below, the transaction-based ledger artificially inflates the number of bitcoins transacted per block. However, given that most users do not value their privacy that much, it is relatively easy to split most of the transactions into the actual payment and change.

5.3.2 Bitcoin transactions and mathematical induction

The concept behind Bitcoin protocol is pure mathematical. In particular, transactions are organized in chains, traced back to the so-called Coinbase transactions, or even further to the genesis block.

Recall that, to prove a statement by induction, we need to perform two steps. First, we need to verify the statement for $n = 1$. This verification is the genesis step. Second, assuming that the statement holds for any given n, we must show that it also holds for $n+1$. This demonstration is the induction step. Together, these two steps prove the statement for every natural number n.

Here is a simple example. Let us prove the following identity:

$$1^3 + 2^3 + \ldots + n^3 = \left(\frac{n(n+1)}{2} \right)^2. \tag{5.9}$$

The genesis step yields

$$1^3 = \left(\frac{1 \times 2}{2} \right)^2. \tag{5.10}$$

The induction step yields

$$1^3 + 2^3 + \ldots + n^3 + (n+1)^3 = \left(\frac{n(n+1)}{2} \right)^2 + (n+1)^3$$

$$= \left(\frac{(n+1)}{2} \right)^2 (n^2 + 4n + 4) = \left(\frac{(n+1)(n+2)}{2} \right)^2, \tag{5.11}$$

as claimed. Hence, Eq. (5.9) holds.

Bitcoin transactions are organized along very similar lines. Every transaction moves its unspent transaction outputs (UTXO) to the next one. Given the nature of the protocol, these transitions are cryptographically secured, as shown in Figure 5.9.

In Figure 5.9, we give three complementary views of a transaction chain. Figure 5.9 (a) shows a sequence of three transactions in detail; Figure 5.9 (b) illustrates the continuity between transactions by emphasizing the fact that the output of a transaction becomes the input of the next one; finally, Figure 5.9 (c) considers a particular sequence of actual transactions.

Because UTXOs move from one Bitcoin address to the next, it is not surprising that the stated goal of achieving anonymity of transactions is challenging to achieve. Analytical tools for blockchain analysis help tracing transactions, and, under certain conditions, deanonymize the owners of public keys; see, e.g., Möser and Böhme (2017). Moreover, Bitcoin transactions can be visualized by using powerful graphics tools, which helps to uncover owners of the addresses of interest; see, e.g., Baumann et al. (2014); McGinn et al. (2016).

5.3.3 Transaction structure

By necessity, the Bitcoin transaction's structure has to be highly standardized, with its general format shown in Table 5.5.

A properly formed transaction is hashed (twice, as usual) and bit-reversed to get the hash called transaction ID (TXID). These IDs are always 32-bytes long.

Transactions can have many inputs and outputs, as presented in Figure 5.10.

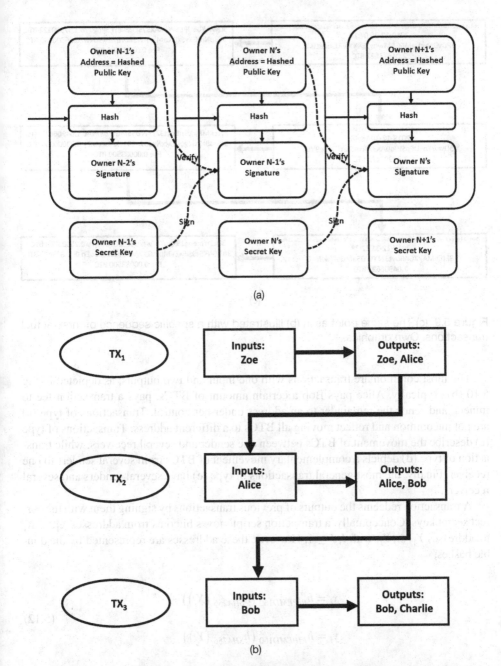

(a)

(b)

Figure 5.9 Bitcoin transactions are chained. They represent movements of value from one public key to the next. The current owner uses her secret key to sign a change of ownership message, while miners use her public key to verify that the signature is legitimate: (a) Transfer of value in a sequence of three transactions; (b) A row of three generic transactions with emphasis on how the previous transaction outputs become the next one's inputs;

(*Continued*)

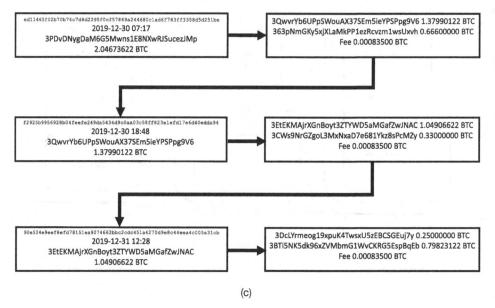

(c)

Figure 5.9 (c) The same point as in (b) illustrated with a specific sequence of three actual transactions. Own graphics.

The most common are transactions with one input and two outputs, as depicted in Fig 5.10 (b). Typically, Alice pays Bob a certain amount of BTCs, pays a transaction fee to miners, and sends the remainder to an address under her control. Transactions of type (a) are not uncommon and reflect moving all BTCs to a different address. Transactions of type (c) describe the movement of BTCs between one sender and several receivers, while transaction of type (d) depicts a complementary movement of BTCs from several senders to one receiver. Finally, the most general transactions of type (e) have several senders and several receivers.

A transaction redeems the outputs of previous transactions by signing them with the correct secret keys. Conceptually, a transaction script moves bitcoins from addresses $X_1, ..., X_I$ to addresses $Y_1, ..., Y_J$, with $I, J \geq 1$. However, these addresses are represented by the double hashes,

$$x_i = h_{RIPEMD160}\left(h_{SHA256}\left(X_i\right)\right),$$

$$y_j = h_{RIPEMD160}\left(h_{SHA256}\left(Y_j\right)\right).$$

(5.12)

Thus, the script moves bitcoins from addresses with hashes x_i to addresses with hashes y_j.

Table 5.5 The general format of a Bitcoin transaction.

Field	Size	Description
Version	4 bytes	currently 1
Flag	2 bytes or none	either 0001 or none
In-counter	1 - 9 bytes	# inputs
List of inputs	as needed	inputs
Out-counter	1 - 9 bytes	# outputs
List of outputs	as needed	outputs
List of witnesses	as needed or none	witnesses
Lock_time	4 bytes	shows when to consummate TX

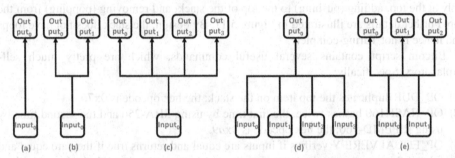

Figure 5.10 Bitcoin transactions can have a different number of inputs and outputs: (a) one input, one output — value moving from one address to the next; (b) one input, two outputs — a portion of the value transferred to a new owner, the rest returned to the original owner as change; (c) one input, several outputs — value splitting; (d) multiple inputs, one output — value combination; (e) multiple inputs, multiple outputs. The most common one is a transaction with one input and two outputs. Own graphics.

5.3.4 Transaction scripts

Bitcoin operates on a pushdown stack, processed from left to right. A pushdown stack is a container of objects inserted and removed according to the last-in-first-out (LIFO) principle. Such a stack allows only two operations: pushing an item into the stack and popping an item out of the stack. Thus, a stack is a limited access data structure because one can add or remove elements only at the top.

A physical analogy is a stack of books. There are two options to handle such a stack: either add a new book to the top or remove the top book. Push and pop do precisely what their names suggest. They push items, such as public keys or signatures, into the stack and

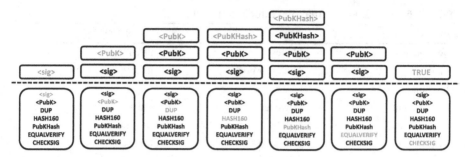

Figure 5.11 Bitcoin transactions being verified in a series of steps on a stack. The stack allows only two operations — push and pop. Own graphics. Source: Narayanan et al. (2016).

pop them out of the stack. Thus, in a limited access data stack, items are added and removed only at the top, adding (pushing) to the top of the stack, and removing (popping) from the top. Further details are illustrated in Figure 5.11. By design, the stack does not have loops and hence is not Turing-complete.

Bitcoin script contains several useful commands, which are pretty much self-explanatory. Specifically:

(1) OP_DUP duplicates the top item on the stack; the hex opcode is 0x76.
(2) OP_HASH160 hashes twice, the first time by using SHA-256 and the second time by using RIPEMD-160; the hex opcode is 0xa9.
(3) OP_EQUALVERIFY verifies if inputs are equal and returns true if they are equal and false if they are not; in the latter case, it marks transaction invalid; the hex opcode is 0x88.
(4) OP_CHECKSIG checks that the input signature is a valid signature corresponding to the input public key for the hash of the current transaction; the hex opcode is 0xac.[2]

The basic Bitcoins script are Pay to PubKey (P2PK) (obsolete),[3] Pay to PubKey Hash (P2PKH), Pay to Script Hash (P2SH), and Pay to MultiSig. Replacing P2PK with P2PKH makes the protocol future-proof since hashing protects public keys from being broken via quantum computers.

Bitcoin allows very complicated scripts in the form of P2SH. Thus, although seldomly emphasized, Bitcoin can be viewed as the first operational example of programmable money, provided that one is prepared to view Bitcoin as money.

---

[2]OP_CHECKMULTISIG checks that m signatures out of possible M signatures are valid; the hex opcode is 0xae.

[3]This is the earliest transaction script, used in the early days of Bitcoin before hashing of public keys became standard.

5.3.5 Transaction verification

In this book, we consider a relatively straightforward script — P2PKH. For a detailed analysis of more complicated scrips, the reader can consult several useful sources; see, e.g., Antonopoulos (2017). The P2PKH script has the following form

<sig>

<pubKey>

OP_DUP

OP_HASH160

<pubKeyHash'>

OP_EQUALVERIFY

OP_CHECKSIG

Here, quantities in brackets denote the inputs. Recall that in the modern version of the Bitcoin protocol, an address is a hash (<pubKeyHash>) of the actual public address (<pubKey>), so the sender has to send UTXO to a hashed address. For obvious reasons, the recipient, who wishes to redeem these outputs and send them to someone else, has to provide the signature and the public key, which produces the correct hashed address. Assuming that the right signature and address are provided, the stack$\mathbb{S}$ undergoes the following push and pop transformations:

$$\mathbb{S}_1 = \text{<sig>},$$

$$\mathbb{S}_2 = \text{<sig> <pubKey>},$$

$$\mathbb{S}_3 = \text{<sig> <pubKey> <pubKey>},$$

$$\mathbb{S}_4 = \text{<sig> <pubKey> <pubKeyHash>}, \quad\quad (5.13)$$

$$\mathbb{S}_5 = \text{<sig> <pubKey> <pubKeyHash> <pubKeyHash>},$$

$$\mathbb{S}_6 = \text{<sig> <pubKey>},$$

$$\mathbb{S}_7 = \text{TRUE}.$$

Here, <pubKeyHash'> is the hash of the payee's address. Otherwise, the output is FALSE, and the transaction is invalid. The push and pop steps, executed from left to right, are presented graphically in Figure 5.11.

5.3.6 Representative transactions

One cannot provide a comprehensive survey of Bitcoin transactions without mentioning the very first transaction,

$$\text{TXID} = \frac{\texttt{f4184fc596403b9d638783cf57adfe4c}}{\texttt{75c605f6356fbc91338530e9831e9e16}}. \quad\quad (5.14)$$

This transaction was completed on 2009-01-11 at 21:30 and resulted in 50 BTC being sent from Satoshi Nakamoto to Hal Finney in two tranches — 10 and 40 BTC. It is the mother of all subsequent transactions. Another famous transaction is the so-called Pizza transaction, which paid 10,000 BTC for a pizza,

$$\text{TXID} = \frac{\text{a1075db55d416d3ca199f55b6084e211}}{\text{5b9345e16c5cf302fc80e9d5fbf5d48d}} . \tag{5.15}$$

To be concrete, consider a simple TX from Block 400,000,

$$\text{TXID} = \frac{\text{928c4275dfd6270349e76aa5a49b355e}}{\text{efeb9e31ffbe95dd75fed81d219a23f8}} . \tag{5.16}$$

This TX is as simple and as fundamental as it gets — P2PKH transaction with one input and one output. It is a straightforward transfer of 3.20900000 BTC from the following input address

$$\text{IA}_0 = \text{1Jyt2ydxc9uTzHuj8DR8egqbLEwLa6kQPL}, \tag{5.17}$$

which contains 3.21000000 BTC, to the output address

$$\text{OA}_0 = \text{15KMFpvaHJLf26y7DkTFKjxijfYvAAdU2s}, \tag{5.18}$$

with 0.01000000 BTC paid to miners as transaction fee. 3.21000000 BTC is the fifth UTXO of the previous TX,

$$\text{TXID} = \frac{\text{1d321f93349cdc71e56477a665ac1645}}{\text{796615ca3cf0103d70875715f4ff8051}} . \tag{5.19}$$

To prove that the input address's owner is legitimate, she provides the actual Bitcoin address, which has the right hash and digitally signs this transaction.

The TX is detailed in Table 5.6 in human-readable form.

Table 5.6 Details of a typical transaction in Block 400,000. Source: https://www.blockchain.com/btc/tx/.

1	TXID	928c4275dfd6270349e76aa5a49b355e efeb9e31ffbe95dd75fed81d219a23f8
2	Time	2016-02-25 10:24
3	Size	191 bytes
4	Weight	764
5	Block	400,000
6	Total input	3.21000000 BTC
7	Total output	3.20900000 BTC
8	Fees	0.00100000 BTC
9	Fee per byte	523.560 satB
10	Fee per weight	130.890 satWU
11	Input index	0
12	Prev. TXID	5180fff4155787703d10f03cca156679 4516ac65a67764e571dc9c34931f321d
13	Prev. output index	5

(Continued)

Table 5.6 Details of a typical transaction in Block 400,000. Source: https://www.blockchain.com/btc/tx/. (*Continued*)

14	Input address	1Jyt2ydxc9uTzHuj8DR8egqbLEwLa6kQPL
15	Value	3.21000000 BTC
		OP_DUP
		OP_HASH160
16	PubKeyScript	c53c08858049dccc686358e05a0371211fc77c5b
		OP_EQUALVERIFY
		OP_CHECKSIG
17	SigScript	3044022100a98648381f405a6882989faa500147c7cb9f4c
		e03e912d18529fb3609e243a47021f798214efe634e8c47e
		158edae534f5652b98ce1bd3693fa95dcdd2c699d9870102
		bd63ab2a6215bdd16d554ea3fd5d83843bff9c76e0b8c6c1
		50d58ee6ca7ea525
18	Output index	0
19	Output address	15KMFpvaHJLf26y7DkTFKjxijfYvAAdU2s
20	Value	3.20900000 BTC
21	PubKeyScript	OP_DUP
		OP_HASH160
		2f58e6245481be77894d5f0f0e2641decdafc447
		OP_EQUALVERIFY
		OP_CHECKSIG

Let's go through Table 5.6 line by line:

(1) TXID — the unique transaction identifier, it is 32 bytes long and is represented by 64 hex symbols (2 hex symbols per byte). TXID is computed only after the transaction is fully formed, see discussion related to Tables 5.7 and 5.8.

(2) Time — the time when the block containing this TX was mined. In this particular case, it is the time when Block 400,000 was mined. It is important to emphasize that the time stamp does not reflect the time when TX was broadcasted to the network.

(3) Size — the size of transaction in the serialized form, see Table 5.8.[4]

(4) Weight — a quantity related to size and pertinent to modern transactions that use the Segregated Witness (SegWit) method, such as the one described in Table 5.10. For older transactions, such as the one in question, weight is four times size, $764 = 4 \times 191$.

(5) Block — the block number into which the transaction is included.

(6) Total input — the sum of all the UTXO s that are the inputs for the transaction.

(7) Total output — the sum of all outputs for the transaction.

(8) Fees — the difference between the total input and output, it is always positive.

---

[4] Keep in mind that 1 byte = 8 bits = 2 hex numbers.

(9) Fee per byte — the meaning is self-explanatory.

(10) Fee per weight — the meaning is self-explanatory.

(11) Input index — the sequential number of the input. Since our TX has one input, somewhat confusingly, the index is 0.

(12) Previous TXID — the ID of the transaction whose outputs are used as inputs of our TX. The mathematical induction analogy is clear.

(13) Previous output index — shows which output of the previous transaction is used as input for ours.

(14) Input address — shows the address from which the input is coming from.

(15) Value — the meaning is self-explanatory.

(16) PubKeyScript — shows how to unlock the input value.

(17) SigScript — the signature used to unlock the input.

(18) Output index — the sequential number of the output. Since our TX has one output, similarly to Input index (11), it is zero.

(19) Output address — shows where the bitcoins are sent.

(20) Value — the meaning is self-explanatory.

(21) PubKeyScript — shows how to unlock the output value in the future.

Table 5.7 Bitcoin transaction in the parsed form.

1	Version	01000000
2	Input count	01
3	Prev. TXID	5180fff4155787703d10f03cca156679 4516ac65a67764e571dc9c34931f321d
4	Prev. output number	05000000
5	SigScript size	6a
6	Data size	47
7	DER sequence identifier	30
8	Data size	44
9	Int identifier	02
10	Data size	21
11	r value	00a98648381f405a6882989faa500147c 7cb9f4ce03e912d18529fb3609e243a47
12	Int identifier	02
13	Data size	1f
14	s value	798214efe634e8c47e158edae534f56 52b98ce1bd3693fa95dcdd2c699d987
15	sighash	01
16	Data size	21
17	PubKey	02bd63ab2a6215bdd16d554ea3fd5d838 43bff9c76e0b8c6c150d58ee6ca7ea525
18	Sequence number	ffffffff

(Continued)

Table 5.7 Bitcoin transaction in the parsed form. (*Continued*)

19	Output count	01
20	Value	a08b201300000000
21	SciptPubKey size	19
22	OP_DUP	76
23	OP_HASH160	a9
24	Data size	14
25	PubKey hash	2f58e6245481be77894d5f0f0e2641decdafc447
26	OP_EQUALVERIFY	88
27	OP_CHECKSIG	ac
28	Locktime	00000000

Let's study Table 5.7 in detail[5]:

(1) Version — shows which version of the Bitcoin protocol is used.

(2) Input count — indicates the number of inputs.

(3) Previous TX ID — the meaning is self-explanatory.

(4) Previous output number — shows the sequential number of the output of the previous transaction used as an input for the current one. Written as a hexadecimal number in the little-endian form.

(5) ScriptSig size — shows the size in bytes of the SigScript size. It is $0x6a$ in hex or, equivalently, 106 in decimals.

(6) Data size — the total size of all the relevant data.

(7) Distinguished Encoding Rules (DER) sequence identifier — $0x30$.

(8) A one-byte length of what follows — 44 (hex) = 68 (dec).

(9) A header byte indicating an integer — $0x02$.

(10) A one-byte length of r —21 (hex) = 33 (dec).

(11) r value — this is the r part of the pair (r, s) DSA introduced in Section 4.7.9.

(12) A header byte indicating an integer — $0x02$.

(13) A one-byte length of s — 1f (hex) = 31 (dec).

(14) s value — this is the s part of the pair (r, s) DSA introduced in Section 4.7.9.

(15) Sighash — indicates which part of the transaction is signed by the ECDSA signature.

(16) A one-byte length of PubKey — 21 (hex) = 33 (dec).

(17) PubKey — the public key associated with the fifth output address of the previous TX.

(18) Sequence number — the sequence number $0xffffffff$ has no effect on the transaction.

(19) Output count — shows the number of outputs.

(20) Value — the number of bitcoins sent to the output (as shown in Table 5.6, item #20) expressed in hex, in the little-endian format.

(21) A one-byte length of PubKeyScript — 19 (hex) = 25 (dec).

---

[5]Recall that 0x indicates that the number is in the hexadecimal base.

(22) OP_DUP — hex opcode.

(23) OP_HASH160 — hex opcode.

(24) A one-byte length of PubKey hash — 14 (hex) = 20 (dec).

(25) PubKey hash — the HASH160 hash of the public key to which bitcoins are sent.

(26) OP_EQUALVERIFY — hex opcode.

(27) OP_CHECKSIG — hex opcode.

(28) Locktime — indicates the time of TX consummation. The lock time of 00000000 means that TX can be included in the block immediately without any delay.

Of course, under the hood, this transaction looks differently. In the so-called scripted, serialized, or raw form, it is a set of 382 hexadecimal symbols (191 bytes), which compactly and decisively encapsulates the transaction for all time; see Table 5.8.

Table 5.8 Bitcoin transaction in the serialized form.

Raw TX data
01000000015180fff4155787703d10f03cca1566794516ac65a67764e571dc9c
34931f321d050000006a473044022100a98648381f405a6882989faa500147c7
cb9f4ce03e912d18529fb3609e243a47021f798214efe634e8c47e158edae534
f5652b98ce1bd3693fa95dcdd2c699d987012102bd63ab2a6215bdd16d554ea3
fd5d83843bff9c76e0b8c6c150d58ee6ca7ea525ffffffff01a08b2013000000
001976a9142f58e6245481be77894d5f0f0e2641decdafc44788ac00000000
First hash
73c22f3837d62228d41158b35d24694f5b37212f7efd1734360842f958d5114f
Second hash
f8239a211dd8fe75dd95beff319eebef5e359ba4a56ae7490327d6df75428c92
TXID= byte-reveresed second hash
928c4275dfd6270349e76aa5a49b355eefeb9e31ffbe95dd75fed81d219a23f8

This table also shows that after hashing the serialized content twice and byte-reverting the second hash, we recover the correct TXID given by Eq. (5.16). As we know, hashes change beyond recognition even if one character is wrong, so the fact that we get the right TXID is a conclusive proof that our serialized form is correct. Given that Block 400,000 is buried underneath more than 250,000 blocks, this TX is irreversible under any circumstances.

It is imperative to emphasize that the serialized Bitcoin transaction contains only the information necessary and sufficient to verify it properly and nothing else. For instance, it includes the number of output bitcoins but not the amount of input bitcoins since the latter is recoverable from the TXID of the previous transaction it is referring to.

It is easy to check that Public Key, shown as part of SigScript in line 17 of Table 5.6 and explicitly in line 17 of Table 5.7, is the right public key for the hashed Public Key shown as part of PubKeyScript in line 16 of Table 5.6. The corresponding steps are shown in Table 5.9.

Thus, we recover the correct PubKeyHash and the right Bitcoin input address.

Table 5.9 PubKey verification.

Compressed
02bd63ab2a6215bdd16d554ea3fd5d838
43bff9c76e0b8c6c150d58ee6ca7ea525
SHA256
65664f366f7c3195de961e6730d6acc2
27cfc7e530fd98569f294ce7a5ae69d6
RIPEMD160
c53c08858049dccc686358e05a0371211fc77c5b
Base58 Encoded
1Jyt2ydxc9uTzHuj8DR8egqbLEwLa6kQPL

Now we consider a more modern transaction from Block 600,000,

$$\text{TXID} = \frac{\text{d8974a3a6596fbd86bd1f794a221639c}}{\text{37d7633394bf03fddd61b48a1505f8b1}}. \tag{5.20}$$

Details of this transaction are given in Table 5.10.

Table 5.10 Details of a typical transaction in Block 600,000. Source: https://www.blockchain.com/btc/tx/.

1	Hash	d8974a3a6596fbd86bd1f794a221639c
		37d7633394bf03fddd61b48a1505f8b1
2	Time	2019-10-19 00:04
3	Size	249 bytes
4	Weight	669
5	Block	600,000
6	Total Input	4.27887989 BTC
7	Total Output	4.27871189 BTC
8	Fees	0.00016800 BTC
9	Fee/byte	67.470 sat/B
10	Fee/WU	25.112 sat/WU
11	Index	0
12	Input Address	3M92sq9ssFaNbEwF47uteVKJsbw125juS7
13	Value	4.27887989 BTC
		OP_HASH160
14	Pkscript	d55600283b297e12a0a8e1a92da7c03c0bcb6c52
		OP_EQUAL

(Continued)

Table 5.10 Details of a typical transaction in Block 600,000. Source: https://www.blockchain.com/btc/tx/. (*Continued*)

15	Sigscript	0014f8513401ea5e9dcd57597e8a736f162572d19079
		304402207851731c11bea40b3d5171c63520a3a5f96f5d79
		b72900926d56b0803b6f895c02202f16dc19ed6866d226c0
16	Witness	0300115b6ebc6e9a578655638362de4c94e8ae0865660103
		7ef7c159605d43e78d4c1ad53b53e60e46bcc504ad9bca6c
		33fd889fcc324eea
17	Index	0
18	Output Address	1KwCFuU3Y9T4qF3Xt55SbDzWbkdQhMkW9i
19	Value	1.00000000 BTC
		OP_DUP
		OP_HASH160
20	Pkscript	cfb21f760729f5b213201c384da7319db3c47a04
		OP_EQUALVERIFY
		OP_CHECKSIG
21	Index	1
22	Output Address	3M92sq9ssFaNbEwF47uteVKJsbw125juS7
23	Value	3.27871189 BTC
		OP_HASH160
24	Pkscript	d55600283b297e12a0a8e1a92da7c03c0bcb6c52
		OP_EQUAL

The transaction in question is more meaningful since it (presumably) sends one bitcoin to a new owner and returns the rest as change. It has one input address

$$IA_0 = 3M92sq9ssFaNbEwF47uteVKJsbw125juS7,$$

and two output addresses

$$OA_0 = 1KwCFuU3Y9T4qF3Xt55SbDzWbkdQhMkW9i,$$

$$OA_1 = 3M92sq9ssFaNbEwF47uteVKJsbw125juS7.$$

Initially, IA_0 address contains UTXO=4.27887989 BTC. TX moves 1.00000000 BTC to OA_0, returns 3.27871189 BTC to $OA_1 = IA_0$, and pays 0.00016800 BTC to miners as transaction fees. It is interesting to note that this transaction fee is a hundred times less than in the previous one. It is digitally signed by the secret key associated with IA_0. Notice that this transaction uses SegWit, which somewhat reduces the digital print of the corresponding transaction. SegWit is a relatively new modification of the original Bitcoin protocol proposed in Bitcoin Improvement Proposal BIP141. SegWit upgrade was activated at block 477,120. Hence, it is not present in TX with TXID given by Eq. (5.16) from Block 400,000.

5.3.7 Transaction broadcast

Once the originator of a transaction is ready to execute it, she broadcasts it to the entire network, see Figure 5.12.

The transaction stays in the memory pool until it is included in blocks the miners are working on; see Figure 5.13.

In this figure, open circles represent nodes, and filled circles represent miners; see the next section for further discussion. The transaction stays in the so-called memory pool until it is selected by one (or several) of the miners. If the payer does not offer an attractive enough

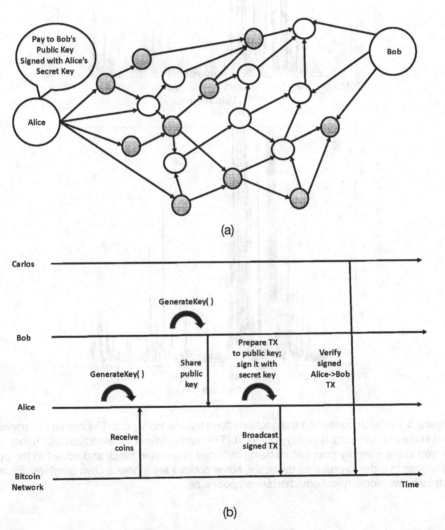

Figure 5.12 (a) Alice signs a transaction and broadcasts it to the Bitcoin network; (b) Alice's actions in more detail. Carlos is a representative miner. Own graphics.

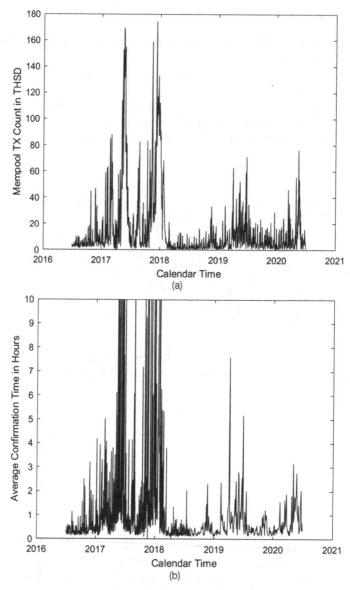

Figure 5.13 (a) Unconfirmed transactions form the memory pool. The number of transaction in the memory pool is highly volatile; (b) The average time a transaction with miner fees stayed in the memory pool before being included in a mined block and added to the public ledger. In order to preserve the scale, some outliers are ignored. Own graphics. Source: https://www.blockchain.com/charts/mempool-size.

transaction fee, her transaction can stay in the pool for a very long time, and potentially might not be processed at all. This fact is the strengths and weaknesses of the Bitcoin protocol at the same time. On the one hand, it prevents tiny transactions, which can overwhelm the system, thus being an anti-spam measure. On the other hand, when block rewards are sufficiently diminished, it will significantly burden participants, who will be forced to pay substantial fees for their transactions to be processed.

It is imperative to understand that all transactions are cryptographically secured by digital signatures and do not require further efforts to ensure that they are valid. However, nothing prevents the owner of a particular address from trying and spending his or her money twice — the celebrated double spend problem. Hence, to ensure that the Bitcoin protocol is self-consistent and double spending is not allowed, it is necessary to have a particular class of participants in the Bitcoin protocol. These participants are called miners — they listen to the network for upcoming transactions, assemble these transactions into blocks, and participate in a competition to have their block added to the system, as explained below.

5.3.8 The Coinbase transaction

Assuming for a moment that mining does what it is supposed to do, the Bitcoin blockchain grows one block at a time. We illustrate a Bitcoin block header's general format in Table 5.11.

The corresponding blocks form a hash chain, with some hashes linking blocks, and other hashes compressing the details of all transactions within a block by calculating the corresponding Merkle root; see Figure 5.14.

Every block has a special transaction that rewards successful miners for their efforts. Like any other, block 600,000 contains the so-called Coinbase transaction, the first one in the list. Details of the Coinbase transaction are shown in Table 5.12.

Table 5.11 The general format of a Bitcoin block header. Notice conversion from the big-endian to the little-endian format. BE = Big-endian, LE = Little-endian.

Field	Size	Format
Version	4 bytes	LE
Previous Block Hash	32 bytes	BE
Merkle Root	32 bytes	BE
Time	4 bytes	LE
Bits Target	4 bytes	LE
Nonce	4 bytes	LE

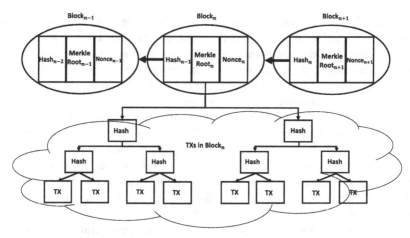

Figure 5.14: The Bitcoin blockchain contains both internal (within a block), and external (between blocks) hash chains. Own graphics.

Table 5.12 Details of the Coinbase transaction for Block 600,000. Source: https://www.blockchain.com/btc/tx/.

1	Hash	93955d40d918d014903843d258aada5c
		720a5d37afac7889268f459a97b148a3
2	Time	2019-10-19 00:04
3	Size	200 bytes
4	Weight	692
5	Block	600,000
6	Total Input	0.00000000 BTC
7	Total Output	12.53764047 BTC
8	Fees	0.00000000 BTC
9	Fee/byte	0.000 sat/B
10	Fee/WU	0.000 sat/WU
11	Value	99,726.27 USD
12	Index	0
13	Input Address	N/A
14	Value	N/A
15	Pkscript	N/A
16	Sigscript	c02709
		OP_0
		ff52aa5d
		2b050830

(Continued)

Table 5.12 Details of the Coinbase transaction for Block 600,000. Source: https://www.blockchain.com/btc/tx/. (*Continued*)

		a6e7b35c6e202a00
		2f426974667572792f
17	Witness	0000000000000000000000000000000000
		0000000000000000000000000000000000
18	Index	0
19	Output Address	3KF9nXowQ4asSGxRRzeiTpDjMuwM2nypAN
20	Value	12.53764047 BTC
		OP_HASH160
21	Pkscript	c08e030911ba85f4a3c324ec6aa6d6722250be74
		OP_EQUAL

The Coinbase transaction is responsible for the process of generating new coins. Details of this TX are as follows:

$$\text{TXID} = \frac{93955d40d918d014903843d258aada5c}{720a5d37afac7889268f459a97b148a3}. \tag{5.21}$$

This transaction sends 12.53764047 BTC to the successful miner's address,

$$OA_0 = \text{3KF9nXowQ4asSGxRRzeiTpDjMuwM2nypAN}. \tag{5.22}$$

The total fee of 12.53764047 BTC is split into mining and transaction fees:

$$12.53764047 \text{ BTC} = 12.5000000 \text{ BTC} + 0.03764047 \text{ BTC}, \tag{5.23}$$

where 12.5 BTC is the mining fee, and 0.038 BTC is the transaction fee. The Coinbase transaction does not need to be signed by the sender (since there is none); instead, it is collectively signed by all miners by accepting Block 600,000 (as assembled and approved by the original miner) as valid.

We will return to the analysis of Block 600,000 in Section 5.6. Here, we mention that given that the total processed volume in Block 600,000 is about 3,398 BTC, see Figure 5.34 in Section 5.6, the overall transaction cost is 0.37%, which is superficially attractive. However, if we consider that the actual number of bitcoins changing hands is estimated to be 450 BTC, the actual transaction costs become as high as 2.80%. The mining fee dwarfed the transaction fees at the time the block was minted. It is unknown what would happen when the mining fee, which, as explained below, halves at periodic intervals, will diminish in size to become irrelevant. One can guess that the transaction fees will have to increase dramatically, at least by a factor of hundred, which will become a severe drag on the protocol in the future.

5.4 The Bitcoin ecosystem

5.4.1 Full node

Full nodes, which we introduced in Chapter 3, perform a vital function in maintaining the network's integrity. They accept transactions and blocks from other full nodes, validating their UTXO, signatures, and hashes and verifying that none of the inputs have been previously spent, and relay validated transactions and blocks further to other full nodes. Full nodes also assist lightweight clients, described in the next subsection, by transmitting their transactions to the network and notifying them about relevant transactions affecting their wallet. The verification function requires full nodes to maintain a complete local copy of the blockchain. Figure 5.15 illustrates changes in the total number of full nodes in the Bitcoin ecosystem over time.

Currently, the number of nodes hovers around 10,000. Table 3.1 shows their geographical distribution. Chapter 9 provides further essential details.

5.4.2 Simplified payment verification node

In contrast with full nodes, Simplified Payment Verification (SPV) nodes use just a small part of the complete Bitcoin transaction database, which is sufficient to verify transactions pertinent to them. As noted in the original white paper, a node can verify bitcoin payments without keeping a full copy of the database; see Nakamoto (2008). The corresponding

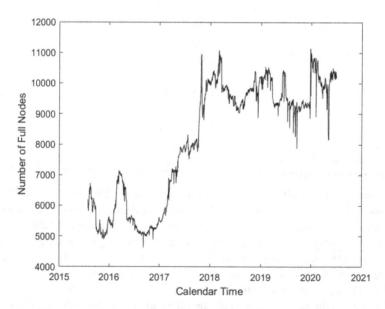

Figure 5.15 Number of full nodes in the Bitcoin ecosystem. Own graphics. Source: coin.dance.

method is called the SPV method, and wallets using this method are called SPV wallets. Instead of the full database, an SPV wallet only maintains a copy of the longest chain block headers, updated by continuously querying full network nodes. The wallet finds the Merkle branch linking the transaction it wishes to verify to its block header. By finding the transaction's place in the longest chain, the wallet confirms that it is valid. In short, in contrast to full nodes maintaining the entire database of transactions, SPV nodes only carry block headers, thus following the original approach for time stamping of digital documents proposed by Stuart Haber and W. Scott Stornetta; see Haber and Stornetta (1990, 1997). Of course, by their very nature, SPVs have to rely on third parties in the form of full nodes, rather than themselves, to keep track of transactions they need.

5.4.3 Wallet

A software program storing a secret key corresponding to a particular Bitcoin address is called a Bitcoin wallet. Such a wallet, used for sending and receiving BTCs, establishes ownership of the BTC balance, associated with a given Bitcoin address, to the wallet's owner. The reader should appreciate that the wallet does not contain any BTCs, just the secret keys. There are several types of Bitcoin wallets, namely, desktop, mobile, web, and hardware. We present a detailed description of Bitcoin wallets in Chapter 9.

5.4.4 Solo miners

Individual miners are critical members of the Bitcoin community at large. Miners assemble transactions into blocks, confirm that every transaction in the block is legitimate, and find a suitable nonce for the block, which allows the block to be included in the blockchain. Miners use application-specific integrated circuit (ASIC) hardware designed for the sole purpose of mining. Successful miners receive block rewards via Coinbase transactions. They are responsible for all the costs associated with mining, such as electricity and hardware. Since miners compete among themselves, getting the reward is infrequent and random; hence, a sole miner's life is hard unless she assembles a large mining rig. Further details are given in Section 5.5.

5.4.5 Mining pools

Since solo mining is financially hazardous, miners are forced to pool resources and organize mining pools. As members of a pool, miners share their processing power over the network and receive rewards proportionally to the amount of work they contributed to the collective efforts. Given the probabilistic nature of the mining game and the fact that pool members hide behind their public addresses, their contributions to the common good are measured by the number of partial proof-of-work blocks they produce over a given period. Mining pools became common and, eventually, dominant when the mining difficultly increased so much that individual miners' rewards became very infrequent. Thus, in exchange for receiving a portion of the block reward consistently, miners agreed to sacrifice some upside.

In the beginning, when mining difficulty was low and there was no need for pools, Satoshi Nakamoto and a few colleagues did all the mining. Currently, a few large pools dominate the industry, much like Oil Majors in the 20th century. Pools are run by managers, who receive a portion of mining rewards for their troubles. All in all, practical considerations initiated an inevitable march toward centralization and away from the Bitcoin protocol founding principles. Further details are presented in Section 5.5.

5.4.6 Exchanges

In contrast to the original theoretical claim of being a fully decentralized peer-to-peer protocol, in practice, Bitcoin operates in a highly centralized fashion. Specifically, the lion share of all the transactions takes place on centralized exchanges. For the vast majority of economic agents, the only way to acquire bitcoins is by becoming an exchange's customer and buying and selling bitcoins as if it were a stock or other financial instrument. All the cryptographic aspects of cryptocurrencies are abstracted away, and all the issues related to centralized finance come to the fore. There are two alternatives, which are much less prevalent than buying BTC on exchanges — (a) an agent can accept BTC for their goods and services; (b) an agent can become a miner and earn BTCs in the form of block rewards.

5.4.7 Bitcoin tumblers or mixers

Given the inherent aspects of the Bitcoin protocol, it does not provide real anonymity of transactions. While at first glance, it would seem this is an issue only for those engaging in some nefarious activity, a little reflection indicates that honest participants might require enhanced anonymity too. For instance, if a payee has a known public address, all its competitors can get a valuable business insight about the payee's financial well-being by monitoring its transactions on a public blockchain. Thus, we need to ensure a reasonable level of anonymity for legitimate reasons.

While some approaches involve building distributed ledger designed around enhanced anonymity such as Z-cash, which we describe in Chapter 12, we find them too radical to use in a regulatory-compliant framework. A suitable mixing service is more appropriate, see, e.g., Meiklejohn and Mercer (2018). Even though peer-to-peer tumblers are not unheard of, third parties do most of the bitcoin mixing — yet another reminder that the Bitcoin ecosystem is not genuinely decentralized.

Although details vary, the principle behind Bitcoin tumblers is straightforward. A bitcoin owner, interested in mixing her bitcoins, sends them to the mixer's address and provides new addresses where the mixer should send mixed bitcoins. The mixer assembles several customers in a pool, charges her fees (of the order of 1%), and distributes bitcoins as requested. The danger is obvious — the mixer can renege on her promises and keep the bitcoins she is supposed to mix. Usually, it does not happen because mixers have to protect their business reputation, but the danger is real.

5.5 Mining

5.5.1 Consensus

As was mentioned in Chapter 3, a distributed ledger, such as Bitcoin, is worth its salt only if it can maintain its integrity via a proper consensus algorithm. In other words, all participants should be able to agree on the validity of all the transactions except for the most recent ones. Of course, it is a difficult task in a distributed system, which was viewed as practically impossible for many years. However, by its very existence, the Bitcoin protocol has shown that such a consensus is indeed possible, albeit challenging and expensive to achieve; see Pass et al. (2017).

One can build a simplified consensus algorithm based on the assumption of a random, but fair, node selection. Such a selection cannot give every node an equal chance because it will make the whole system vulnerable to the so-called Sybil attacks, since acquiring a node, represented by an address, is practically costless, see Subsection 3.6.1. Thus, the consensus by the majority of nodes does not guarantee the system's integrity. Accordingly, an opportunity to be selected as a validator should be proportional to the node's resources spent participating in the game.[6] Of course, miners are not selfless. Their objective is to collect rewards paid by the system for validating transactions and all the transaction fees they can get.

As is shown in Figure 5.12, participants broadcast new transactions to the entire Bitcoin ecosystem. Every mining node collects some of the recent transactions into a block. Due to the natural delays in receiving new transactions, and for other reasons, these blocks are not identical. It is possible to have a situation when a miner decides to mine an empty block to get just the block-mining reward. More importantly, if a particular transaction does not pay an attractive transaction fee, it can be left dangling in a memory pool for a very long time, or be dropped altogether, which is a feeble aspect of the protocol.

Achieving consensus by mining is done along the following lines, see also Subsection 3.6.2. Once the previous block is included in the blockchain, and new transactions announced, miners assemble them into blocks (roughly 2,000 transaction per block), hash them, and form a new block header. They need to solve a conceptually simple but computationally involved puzzle, similar to the one described in Eq. (4.169). The competition winners (there might be several) broadcast their blocks, including the nonce, and the header, to the rest of the community. Other nodes (both miners and tracking nodes) quickly check all the attributes of the block's transactions (i.e., inputs exceed outputs, signatures are valid, etc.), and the block's header (a proper nonce). Although it requires approximately 10 minutes and lots of energy to solve a puzzle, checking its correctness takes only milliseconds. If the block ticks all boxes, it might, but does not necessarily have to, be accepted by miners.

Miners express acceptance by including the hash of the accepted block as part of the blocks they are working on. It is essential to recognize that for the first 60 minutes after a block is mined, there are several branches, all of which could be eligible to become the

---

[6]This is not dissimilar to the shareholder democracy in public companies. While, in theory, every shareholder can speak at the general shareholders meeting, their votes are counted proportionally to the number of shares they possess, which, somewhat loosely correlates with the amount of money they have spent on buying their shares.

main chain. As time elapses, one of the chains grows faster, and the longest-chain-wins rule, mentioned in Subsection 5.4.2, preserves miners' collective self-interest. Scenarios, when this rule doesn't hold and leads to forks, are described in Subsection 5.5.7 below.

Miners provide proof-of-work (PoW) by finding nonces, which solve the requisite hash puzzles. They need to find two nonces since a single one is too short for satisfying the difficulty requirements; as described in Subsection 5.5.3 below. Usually, in distributed systems and computer security, the assumption is that a certain percentage of nodes are honest (robust), and the rest are malicious (faulty). The objective of the system's designer is to show that it will operate as intended provided that this assumption holds. For the Bitcoin protocol, we define the percentage of robust and faulty nodes by the amount of hash power they control. As was mentioned earlier, this is not dissimilar to how shareholders of public companies make their decisions.

However, an alternative way of thinking about this problem is to assume that nodes are neither honest nor malicious, and design a system that promotes honest behavior. We don't know if the honest conduct of nodes is a Nash equilibrium for Bitcoin. In general, PoW-based consensus is robust, provided that more than 50% of the hashing power is controlled by "honest" miners, or, to be more precise, mining pools. If this condition is violated, malicious miners can launch the so-called 51% attack and rewrite the history of blockchain transactions, see also Subsection 3.6.8.[7] In the context of the Bitcoin protocol, an attack is highly unlikely since it will result in a precipitous drop in the BTC dollar price. This drop would go against miners' self-interest.

An interesting historical example of PoW is the celebrated Sultan Suleiman the Magnificent Tughra (or seal) shown in Figure 5.16. So much effort went into the production of a Tughra that its forgery was practically impossible. The fact that a forger, if uncovered, would immediately meet a horrible death, helps to preserve the integrity of communications as well.

Bitcoin security is sequential. Once a block containing a particular transaction is buried deep in the blockchain, i.e., many (no less than six, but, preferably a few more) are built on top of it, the transaction becomes immutable. The idea is illustrated in Figure 5.17.

Bitcoin transactions leave no physical records, which are hard to duplicate, while electronic records' duplication is trivial. Therefore, Alice can try to double spend bitcoins she controls. One can envisage the following scenario. Alice creates two transactions: transaction (B), which sends bitcoins from Alice to Bob plus some change back to Alice, and transaction (C), which sends the same bitcoins from Alice to Charlie plus the remainder to Alice. Only one transaction can be valid since they spend the same bitcoins. To protect himself, Bob and Charlie, or, for that matter, any other provider of goods and services, have to wait and see which transaction is included in a blockchain with several confirms (no less than six). Only after that, one of them will send his merchandise to Alice. A double spend attack and a double spend defense are shown in Figure 5.18, where the arrows are pointers from a block to its predecessor.

---

[7]Some authors argue that a successful attack can be launched by a group of miners who control about 33% of the hashing power; see Eyal and Sirer (2014). We feel that their claims are greatly exaggerated — the very fact that Bitcoin protocol has not collapsed in more than 10 years is the counterexample to their arguments.

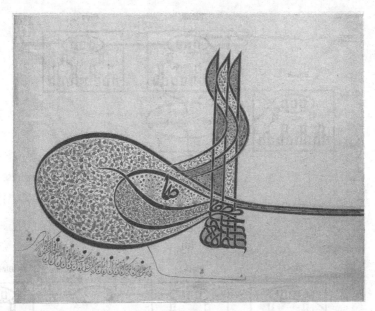

Figure 5.16 Tughra (Insignia) of Sultan Suleiman the Magnificent (r. 1520–66) ca. 1555–60. So much effort goes into producing it, that falsification of the official documents becomes impossible. Source: The Met. Public Domain.

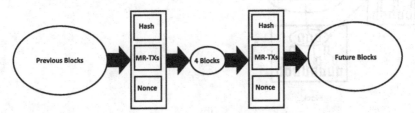

Figure 5.17 Safety in numbers: blockchain security is sequential and is built over time. Blocks buried deep into the chain are much more secure than the recent ones. Even occasional empty blocks play a role in enhancing security of the blockchain. Own graphics.

5.5.2 How Bitcoin rewards miners?

Bitcoin's monetary policy of fixing the total number of BTCs in circulation at 21,000,000 is simple — some view it as a great advantage, while others think of it as superficial. Every time a miner finds a successful block, she receives a reward via the Coinbase transaction. This reward plus all the block's transaction fees are the ultimate reason why miners mine. The size of block rewards started with 50 BTC per block, is halved after 210,000 new blocks are mined (it happens about every four years). The genesis block was mined on January 3, 2009. The first halving occurred on November 28, 2012; the second occurred on July 9, 2016, the most recent halving occurred on May 11, 2020; see Figure 5.19. The current reward is 6.25 BTC. The total number of BTCs in circulation is limited to 21,000,000.

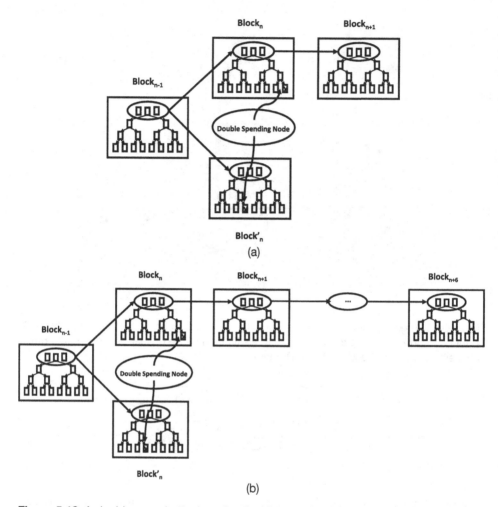

Figure 5.18 A double spend attack and a double spend defense. (a) Alice tries to spend her BTC twice. (b) Eventually, the longest chain wins, so, after a while, only one transaction succeeds and stays in the blockchain. Own graphics.

Indeed, disregarding the fact that BTC is not infinitely divisible, we have

$$Total = \sum_{n=0}^{\infty} \frac{210,000 \times 50}{2^n} = 210,000 \times 50 \times \frac{1}{1 - \frac{1}{2}} = 2,100,000. \tag{5.24}$$

Figure 5.19 (a) shows the growth of the total number of BTCs over time, which looks regular on average. Figure 5.19 (b) illustrates the fact that on small scales, it is somewhat irregular, although the downward jumps, occurring after the rewards' halving, are visible.

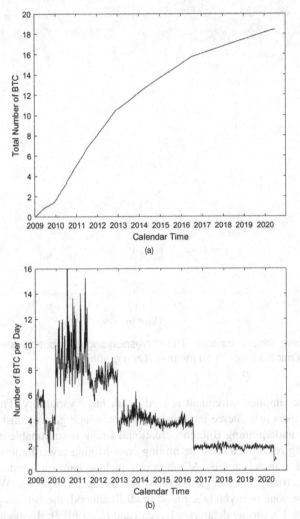

(a)

(b)

Figure 5.19 Mining rewards are halved every four years. In the limit, 21 million BTC will be mined. The most recent halving occurred on May 11, 2020. In the early years, mining rewards were rather haphazard. Eventually, their volatility has diminished; (a) Total number of BTC in existence, in millions; (b) The number of BTC mined per day, in thousands. Own graphics. Source: https://www.blockchain.com/charts/total-bitcoins.

The mining time is random and has the exponential distribution, with the expectation of 10 minutes, so that it takes on average 10 minutes for someone to mine a block; see Figure 5.20.

For an individual miner, who controls a given fraction of the total hash power, the average time she needs to wait before receiving her reward has the form:

$$\text{average time} = \frac{10 \text{ min}}{\text{fraction of hash power}}. \tag{5.25}$$

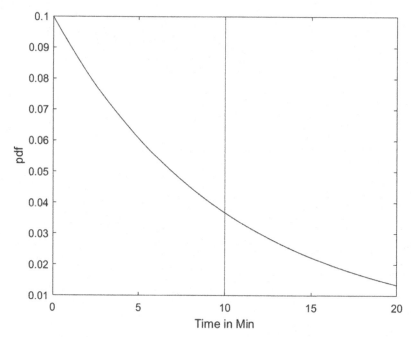

Figure 5.20 Mining time is random. The corresponding distribution is exponential. The expected time to mine a block is 10 minutes. Own graphics.

Small hash power implies infrequent rewards with huge variability. Thus, simple economics forces miners to coalesce into gigantic pools, which goes against the grain of the distributed ethos underpinning Bitcoin's creation. Mining is sustainable in the long term, only if the mining reward exceeds the mining cost. Mining reward consists of the block reward and block transaction fees. Mining costs include hardware costs and operational costs, such as electricity, labor, legal, security, and incidental expenses. What will happen when rewards are gone is anybody's guess. In all likelihood, the Bitcoin protocol will be severely impacted. For further details see Beccuti and Jaag (2017); Bissias et al. (2018); Prat and Walter (2018); Qin et al. (2018); Romiti et al. (2019).

Various aspects of Bitcoin blockchain growth are exhibited in Figures 5.21–5.24.

Figure 5.21 (a) illustrates how the Bitcoin blockchain's size grows over time, while Figure 5.21 (b) shows the cumulative number of confirmed transactions, which at the moment is above the half-billion mark. Regardless of any other consideration, the fact that the Bitcoin protocol managed to process so many transactions in a decentralized fashion without any issues is a magnificent technological achievement in itself.

The Bitcoin blockchain grows block by block. Figure 5.22 (a) demonstrates how the block's size oscillates in time, while Figure 5.22 (b) shows how many transactions constitute a block. Both quantities are highly volatile, but, on average, a block is about 1 MB and contains around 2,000 TXs.

Figures 5.23 (a) and 5.23 (b) complement each other. The first figure illustrates the number of TXs per day, TpD, while the second one shows the number of TXs per second,

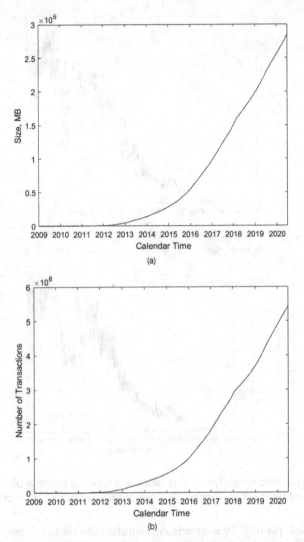

Figure 5.21 Bitcoin blockchain gradually grows with time: (a) the size of the blockchain; (b) the total number of processed transactions. Own graphics. Source: blockchain.com.

TpS. On average, the protocol processes about 3 TpS, and consequently, about $3 \times 60 \times 60 \times 24 \approx 260000$ TpD.

Figures 5.24 (a) and 5.24 (b) show two related numbers, which are of great interest to anyone who wishes to understand the efficiency of Bitcoin and, in particular, estimate BTC daily volume. Figure 5.24 (a) illustrates that the total output of all transactions is 1–2 million, but this volume includes BTCs returned to the payers as change. Figure 5.24 (b) demonstrates the actual transaction volume, which is about 0.2 million BTC. Given that the total number of mined bitcoin in circulation is approximately 18.5 million, about 1% of the

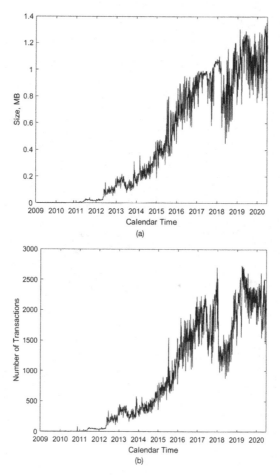

Figure 5.22 Bitcoin blockchain grows one block at a time: (a) average block size in MB; (b) average number of transactions per block. Own graphics. Source: blockchain.com.

total changes hands per day. By comparison, a similar ratio for the Apple stock (AAPL) is about 0.7%.

5.5.3 How do miners mine?

At the heart of mining is the search for a valid nonce that is a 32-bit (4-byte) field whose value is adjusted by miners until the hash of the block is less than or equal to the current target difficulty of the block, see Subsections 3.6.2 and 5.5.4. Since it is unlikely that there is a 32-bit nonce resulting in a hash below the target, the miners have to operate with two nonces self-consistently: (a) nonce in the block header; (b) nonce in the Sigscrip of the Coinbase transaction described earlier in Subsection 5.3.8.[8] Search within the self-consistent

---

[8] Students of quantum chemistry can recognize a similar construct of the foundational Hartree–Fock self-consistent field.

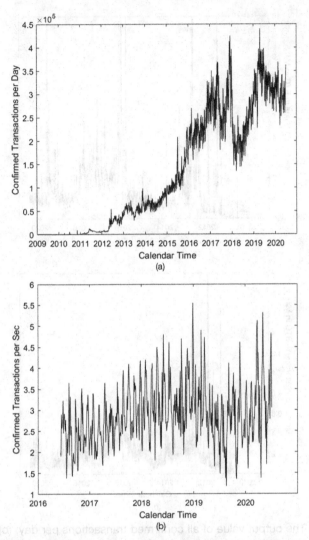

Figure 5.23 Bitcoin blockchain growth is slow compared to many other systems: (a) the number of confirmed transactions per day; (b) the number of confirmed transactions per second. For comparison sake, Visa processes around 2,000 transactions per second. Own graphics. Source: blockchain.com.

framework starts with choosing zero-valued nonce for the Coinbase transaction. The corresponding block header, which includes the hash of the previous block, Merkel tree with hashes from all transactions in the current block, and the block header nonce's first choice, is consequently hashed. If its value satisfies the current difficulty level, the problem is solved, and the miner broadcasts the block to the network. However, if the target hash is not low enough, the miner continues to iterate over all available block header nonces. Assuming

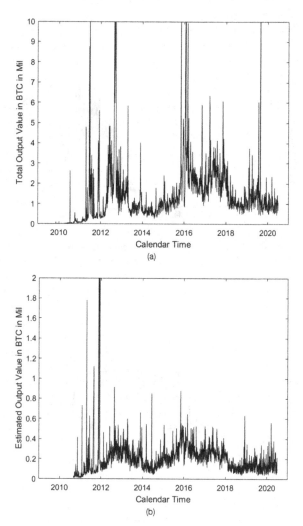

Figure 5.24 (a) The output value of all confirmed transactions per day; (b) the estimated value of all non-change transactions per day. Own graphics. Source: blockchain.com.

that all permutations with the initial value of the Coinbase nonce do not produce a satisfactory result, the miner changes the Coinbase nonce, which propagates to the root, and repeats the process, as demonstrated in Figure 5.25.

Symbolically, the task of a miner is to find two nonces, y_1, y_2, such that

$$H\left(x||y_1, y_2\right) \in Z, \tag{5.26}$$

where Z is a suitable target set. Assuming that finding a combination of nonces producing a proper hash can only be done by brute force, miners must perform a costly calculation

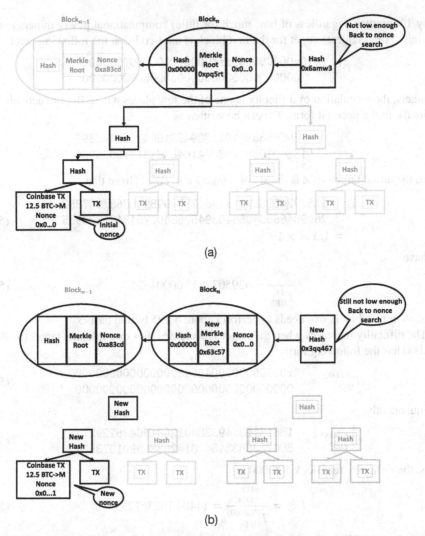

Figure 5.25 (a) A miner starts with nonces containing all zeros and adjusts the block nonce sequentially; (b) A miner changes Coinbase nonce with changes propagating to the root and starts adjusting the block nonce again. Own graphics.

requiring enormous computational resources. Thus, by presenting a self-consistent block that hashes properly, the winning miner proves that she has done the necessary amount of work to earn the block reward.

5.5.4 Mining difficulty

One of the feats of engineering making Bitcoin such an impressive achievement is a mechanism that changes mining difficulty automatically so that, in expectation, blocks are mined every 10 minutes, regardless of how much (or little) computational power miners deploy. The initial (and largest) target for the first block is hard-coded at the following level

$$T_{HEX}^{(1)} = \frac{\text{0x00000000ffff0000000000000000000000}}{\text{000000000000000000000000000000000000}}.$$ (5.27)

Strangely, the calculation of difficulty is one of the few places where Bitcoin actively uses decimals. In the decimal form, $T^{(1)}$ can be written as

$$T_{DEC}^{(1)} = \frac{26959535291011309493156476344723991}{336010898738574164086137773096960}.$$ (5.28)

Thus, the initial target set Z is the set of integers $z \le T_{HEX}^{(1)}$. Given that

$$2^{256} = \frac{115792089237316195423570985008687907853}{269984665640564039457584007913129639935}$$
$$\approx 1,157 \times 10^{77},$$ (5.29)

we have

$$\frac{2^{256}}{T_{HEX}^{(1)}} = 4295032833.00001525.$$ (5.30)

Thus, to mine a block, one needs to perform about 4.295 billion hashes.

The difficulty increases when the target T decreases. For example, the target for Block 100,000 has the following form

$$T_{HEX}^{(100000)} = \frac{\text{0x000000000004864c0000000000000000}}{\text{000000000000000000000000000000000000}},$$ (5.31)

or, equivalently

$$T_{DEC}^{(100000)} = \frac{186131131498380012681564362292723}{0076368334845814253369901973504}.$$ (5.32)

Thus, the difficulty for Block 100,000 is

$$Diff = \frac{T_{DEC}^{(1)}}{T_{DEC}^{(100000)}} = 14484.16236123,$$ (5.33)

so that about 62,210 billion hashes are required to find a winning block.

A block's target is baked-in its header in the form of bits target — an extraordinarily clever bit (punt intended) of engineering. For instance, for Block 600,000, the corresponding bits target in hexadecimal notation is equal to $1715a35c$; see Table 5.13 in Section 5.6. Its interpretation is as follows. The first byte, called the exponent, allows one to calculate how many zeros the target should start with, given the overall 40(hex) or 64(dec) positions. In our case, the exponent equals to 17(hex), or 23(dec); the number of zeros is equal to $40 - 2e = 12$(hex), or, equivalently, $64 - 46 = 18$(dec). The next three bytes are called the coefficient. The coefficient shows the first three nontrivial bytes of the target. Thus, the target for Block 600,000 looks like that:

$$T_{HEX}^{(600,000)} = \frac{\text{0x00000000000000000015a35c00000000}}{\text{000000000000000000000000000000000000}}.$$ (5.34)

It determines the difficulty of finding a valid hash. It moves from left to right, reducing T and increasing mining difficulty.[9] The difficulty adjusts every 2,016 blocks (or approximately two weeks) according to the formula

$$\text{next\_diff} = \frac{\text{prev\_diff} \times 2016 \times 10\text{min}}{\text{time to mine last 2016 blocks}}. \tag{5.35}$$

An astute reader should notice that this idea is similar to the one behind a centrifugal governor.[10]

Since the difficulty goes up and down depending on the miners' (or, more precisely, mining pools') combined hash power, bitcoin should not be viewed as digital gold, as is often claimed. The reason is apparent. On the one hand, if some miners decide to stop working, new bitcoins' production costs will go down. On the other hand, if some gold mine operators decide to shut their mines, other operators' production costs will not go down. Thus, no mechanism limits the bitcoin price in dollars from below, while the gold price in dollar terms is limited below by the production costs.

Given Bitcoin mining allure, the miners' collective hash rate grows exponentially. However, a 10 minute-per-block rule is preserved by automatically increasing the difficulty to offset any gains in hash rate. The process is illustrated in Figure 5.26.

Figure 5.26 (a) shows difficulty and hash rate as functions of time, while Figure 5.26 (b) demonstrates hash rate as a function of difficulty. A simple linear regression produces the following results

$$\text{hash rate} = p_1 \times \text{difficulty} + p_2,$$

$$p_1 = 7.257 \times 10^{-6}(7.222 \times 10^{-6}, 7.291 \times 10^{-6}), \tag{5.36}$$

$$p_2 = 1.859 \times 10^5(3.022 \times 10^4, 3.415 \times 10^5),$$

with $R^2 = 0.9917$, which captures the situation succinctly. As usual, the 95% confidence bounds are shown in parenthesis. The difficulty is a non-dimensional scaling parameter; the hash rate is measured in Terahash per Sec (trillion hashes per second). Let's have a look at another measure, time-to-find-a-block. It fluctuates with difficulty, but stays between eight and 12 minutes; see Figure 5.27.

Once a successful miner finds a winning nonce, she broadcasts it to the network at large, much like Alice broadcasts her desire to send some bitcoins to Bob. For the same reasons as before, it takes time until the broadcast reaches its intended recipients. Several groups monitor the delay time distribution for the broadcast; see, e.g., Decker and Wattenhofer (2013).

---

[9]The nonzero part of T is somewhat similar to the sliding weights for a good old weigh beam eye-level physician scale.

[10]Based on the principle of proportional control, a centrifugal governor is a feedback system that regulates the flow of steam to an engine to maintain a nearly constant speed. Christian Huygens invented centrifugal governors in the 17th century. In the 18th century, James Watt adapted one to regulate steam admission to his steam engine, thus controlling its rotational speed. A simple governor cannot maintain an exact speed; it keeps a speed range instead. Mining difficulty adjustment, accommodating changes in the amount of hashing power used by miners, plays the same role as a governor regulating a steam engine's operation. The principle is simple: more hashing power — greater difficulty, less hashing power — lower difficulty.

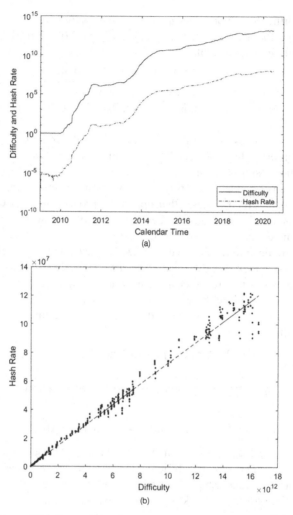

Figure 5.26 Mining difficulty and hash rate over time — an arms race among the miners. (a) Hash rate and difficulty as functions of time; (b) Hash rate as a function of difficulty. Own graphics. Sources: https://www.blockchain.com/charts/hash-rate, https://www.blockchain.com/charts/difficulty.

In Figure 5.28 we show the probability density function (PDF) for the corresponding distribution. The majority of nodes is reached within 600 msec.

5.5.5 Pooling of resources

As described above, the simple economics of Bitcoin mining forces individual miners to coalesce into gigantic pools directed by pool managers. The reason is not hard to

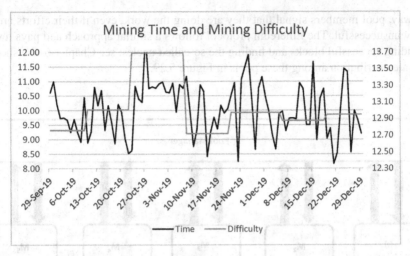

Figure 5.27 Time to find a block and difficulty of doing so shadow each other. Own graphics. Source: blockchain.com

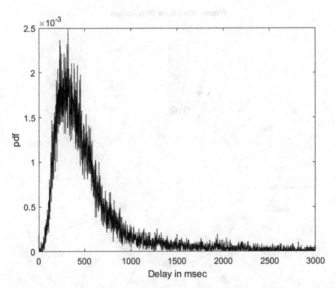

Figure 5.28 Block propagation delay distribution. Own graphics. Source: https://dsn.tm.kit.edu/bitcoin/.

fathom — an individual miner faces a somewhat uncertain stream of income since wins in the mining lottery are proportional to the hashing power she possesses. Hence, it is much safer to pool resources with others. A pool's operator receives successful blocks, publishes them, and distributes rewards among the pool members. Participants communicate successful blocks and blocks that are almost successful, i.e., have hashes starting with several zeros.

This way, pool members signal that they are doing the work, even if their efforts are ultimately unsuccessful. The Ethereum protocol follows a similar approach and pays rewards for finding successful blocks and finding the so-called uncles; see Chapter 6. We show the hash rate distribution among these pools in Figure 5.29.

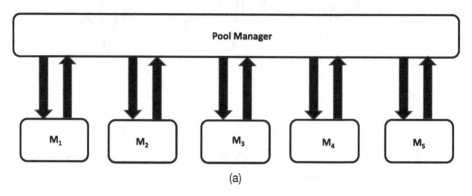

(a)

Bitcoin Hash Rate Distribution

(b)

Figure 5.29 (a) Miners organize themselves in gigantic pools governed by managers; (b) The number of pools responsible for the vast majority of hash rate is embarassingly small. To put it differently — Bitcoin is not sufficiently decentralized. Own graphics. Source: (b) coin.dance.

The emergence of the pools inevitably results in the centralization of Bitcoin protocol and defies its very purpose of being truly decentralized "people's" currency. Several researchers have studied various aspects of this process; see, e.g., Gencer et al. (2018).

Profits earned by miners are presented in Figure 5.30. They demonstrate strong correlation with transaction fees expressed in dollar terms.

5.5.6 Malicious attacks

Malicious actors might perpetrate numerous attacks on the system, including the 51% and 33% attacks; see Eyal and Sirer (2014); Gervais et al. (2016); Shalini and Santhi (2019).

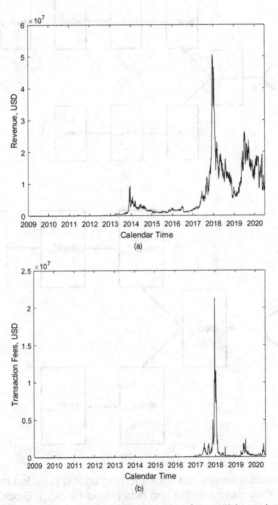

Figure 5.30 (a) Total miners' revenue; (b) Transaction fees paid to miners. Own graphics. Source: blockchain.com.

Two of the most common ones are malicious forking and selfish mining attacks, illustrated in Figure 5.31.

In Figure 5.31 (a), we show a 51% attack by malicious miners. They manage to overwhelm honest miners, create an alternative transaction history, depicted as primed blocks, and double spend their bitcoins despite both transactions buried under six or more blocks. Of course, in the case of Bitcoin protocol, this is too costly an attack to entertain; however, such attacks are not unheard of for smaller blockchains. In Figure 5.31 (b), we show a miner who finds a successful block B'_{n+1} but does not broadcast it right away. Instead, she toils in

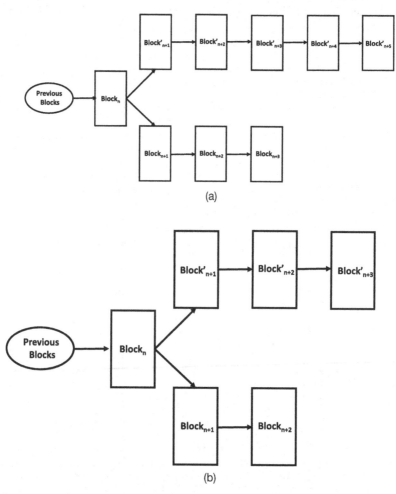

Figure 5.31 (a) Malicious miners can start a forking attack, provided they control enough hashing power; (b) They can be selfish and hide blocks Block$'_{n+1}$, Block$'_{n+2}$, Block$'_{n+3}$, they had found. They publish it when ready both reaping rewards and ruining work of others. Own graphics.

secret and finds several more blocks, B'_{n+2}, B'_{n+3}, and then broadcasts the entire chain. As a result, the work of honest miners is wasted. These attacks, while not particularly frequent, do occur from time to time.

5.5.7 Soft and hard forks

Possible Bitcoin forks are illustrated in Figure 5.32. There is nothing profound about soft forks shown in Figure 5.32 (a). Such forks regularly occur on the top of any distributed database because miners are in constant competition with each other for the right to extend

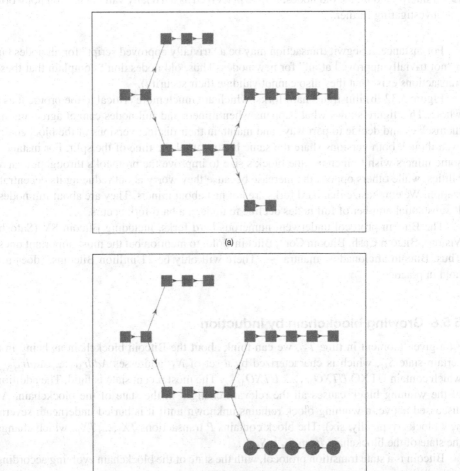

(a)

(b)

Figure 5.32 (a) Temporary forks are a fact of life in any distributed ledger including Bitcoin. They are routinely suppressed and the longest chain, which represents the truth, wins; (b) Yet, hard forks are possible, and, occur from time to time. Own graphics.

the valid Bitcoin chain. The longest-chain-wins rule is enforced automatically by miners' collective self-interest.

More generally, a soft fork is a retro-compatible change in the protocol. Segwit, which brought new capabilities, is such an example. For instance, it reinterpreted the meaning of some script operators, such that for updated nodes, they would have a new meaning as described in the spec of Segwit, but also in a way that nodes, which would not update, will still be able to process Segwit transactions even without knowing "what they do." The trick is often of the form of making a change that looks like:

(1) if interpreted on the updated nodes, it'll have the expected behavior of the update;
(2) if interpreted on the old nodes, it'll be perceived as a trivially valid operation not worth investigating further.

For instance, a Segwit transaction may be a "trivially approved script" for all nodes but a "not trivially approved at all" for new nodes. Thus, old nodes don't complain that these transactions exist (but they also cannot validate their security).

Figure 5.32 (b) illustrates hard forks, which are much more critical to the protocol as a whole. This figure shows what happens when miners and full nodes cannot agree among themselves and decide to part ways and maintain their distinct versions of the blockchain, even though both versions share the same history until the time of the split. For instance, some miners wish to increase the block's size to improve the protocol's throughput capabilities, while others oppose the increase because they worry about reducing its decentralization. We emphasize that hard forks are not just about miners. They are about full nodes. If substantial number of full nodes decline to update, a hard fork occurs.

The Bitcoin protocol underwent numerous hard forks, including Bitcoin SV (Satoshi Vision), Bitcoin Cash, Bitcoin Core, Bitcoin Gold, to mention but the most important ones. Thus, Bitcoin aficionados' mantra — "There will only be 21 million Bitcoins," does not hold in practice.

5.5.8 Growing blockchain by induction

At a given moment in time T_M, we can think about the Bitcoin blockchain as being in a certain state S_M, which is characterized by a set of N_M addresses $Addr_{M,1},..., Addr_{M,N_M}$, which contain UTXO $UTXO_{M,1},..., UTXO_{M,N_M}$. The most recent state is fluid. The addition of the winning block causes all the relevant changes in the state of the blockchain. As discussed above, a winning block remains unknown until it is buried underneath several new blocks (typically, six). The block contains P transactions $TX_1, ..., TX_P$, which change the state of the blockchain from S_M to S_{M+1}.

Bitcoin is a state transition protocol, with the state of the blockchain evolving according to the following rule:

$$APPLY\left(S_M, \{TX_1, ..., TX_P\}\right) \rightarrow S_{M+1} \text{ or } ERROR. \tag{5.37}$$

Full nodes check the validity of all transactions included in a block and the validity of the block itself to ensure that the blockchain grows correctly. Thus for every transaction, TX_p, the nodes need to check whether:

(1) TX is well-formed; otherwise, return ERROR.
(2) The sum of inputs is greater than the sum of the outputs; otherwise, return ERROR.
(3) TX is correctly signed; otherwise, return ERROR.

If there is an error, full nodes reject the block. We emphasize that similarly to conventional banking, the order of transactions impacts their validity. For instance, Alice has 10 BTC and wants to send 12 BTC to Bob as *TX*, while Charlie has 5 BTC and wants to send 3 BTC to Alice as *TX'*. The sequence *TX*, *TX'* results in the failure of *TX*, while the sequence *TX'*, *TX* results in its success.

Provided that all transactions in the block are valid, the nodes also need to check the block's header and verify whether:

(1) The previous block referenced by the current block exists and is valid; otherwise, return ERROR.
(2) The timestamp is current; otherwise, return ERROR.
(3) The Merkle root of all transactions is correct; otherwise, return ERROR.
(4) The PoW on the block is good; otherwise, return ERROR.

If there is an error, full nodes reject the block. The evolution of the state of blockchain caused by the addition of a valid block is more or less self-explanatory:

(1) Let S_M be the state of the ledger at the end of the previous block.
(2) Suppose that the block's transaction list has P transactions, $TX_p, p = 1, ..., P$.
(3) Define

$$S_M^p = APPLY(S_M^{p-1}, TX_p), \quad S_M^0 = S_M. \tag{5.38}$$

(4) If any application returns ERROR, exit, and return FALSE.
(5) If there are no errors, return TRUE, and register $S_{M+1} = S_M^P$ as the state at the end of this block.

Details are given in Figure 5.33.

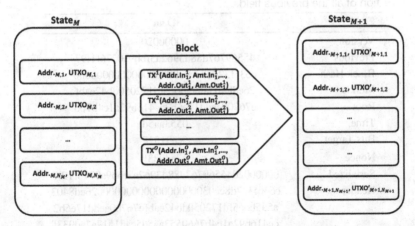

Figure 5.33 Bitcoin state transactions result in the state of Bitcoin blockchain evolving from S to S'. Own graphics.

Thus, we can think about the Bitcoin protocol as a vehicle for performing transitions in a Markov chain; see, e.g., Biais et al. (2019); Li et al. (2019).

5.6 Anatomy of Block 600,000

In this section, we summarize representative stats for Block 600,000. Recall the general format of a generic block header shown in Table 5.12. The specific header for Block 600,000 is presented in Tables 5.13 and 5.14 in human-readable, and parsed and serialized forms, respectively.

Table 5.13 600,000 block header in the human-readable form.

Field	Data
Version	0x20000000
Previous	0000000000000000000003ecd827f336c6
BlockHash	971f6f77a0b9fba362398dd867975645
Merkle	66b7c4a1926b41ceb2e617ddae0067e7
Root	bfea42db502017fde5b695a50384ed26
Time	19 Oct 2019, 00:04:21
Bits Target	1715a35c
Nonce	1,066,642,855

Table 5.14 600,000 block header in the parsed and serialized forms. Notice conversion from the big-endian (BE) to the little-endian format (LE). The serialized header is simply the concatenation of all the previous fields.

Field	Data
Version	00000020
Previous	45569767d88d3962a3fbb9a0776f1f97
Block Hash	c636f327d8ec03000000000000000000
Merkle	26ed8403a595b6e5fd172050db42eabf
Root	e76700aedd17e6b2ce416b92a1c4b766
Time	0553aa5d
Bits Target	5ca31517
Nonce	a7ad933f
Serialized	0000002045569767d88d3962a3fbb9a0776f1f97
Header	c636f327d8ec03000000000000000000026ed8403
	a595b6e5fd172050db42eabfe76700aedd17e6b2
	ce416b92a1c4b7660553aa5d5ca31517a7ad933f

Table 5.15 Block 600,000 header verification.

Serialized header
0000002045569767d88d3962a3fbb9a0776f1f97
c636f327d8ec0300000000000000000026ed8403
a595b6e5fd172050db42eabfe76700aedd17e6b2
ce416b92a1c4b7660553aa5d5ca31517a7ad933f
SHA256
dbef7e89e32cc45719faf452fafde765
dc694c02ea740c8b3163b0dd46010dc5
SHA256
915fcd96d1c84298a8fbfb9c13a9f7b4
760e9056683107000000000000000000
Byte Reversion
0000000000000000007316856900e76
b4f7a9139cfbfba89842c8d196cd5f91

We can represent this header in the serialized form and check that the byte-reversed double hash is clearly below the required threshold. Notice conversion from the big-endian to the little-endian format.[11]

Table 5.15 demonstrates that this block is self-consistent since the corresponding big-endian hash (Byte Reversion) is lower than the difficulty threshold given by Eq. (5.34) and Table 5.14.

The summary statistics for Block 600,000 are presented in Figure 5.34, while its place in the blockchain is shown in Figure 5.35.

5.7 Bitcoin pros and cons

There are heated debates between Bitcoin's supporters and its skeptics over its pros and cons; see, e.g., Bhutoria (2020).

Skeptics argue that:

(1) Bitcoin's volatility prevents it from being a store of value.
(2) Bitcoin is not suitable as a means of payment.
(3) Bitcoin's mining wastes energy on an immense scale.
(4) Bitcoin's usage facilitates illicit activity.
(5) Bitcoin has no backing.
(6) A competitor will replace Bitcoin.

---

[11]Mixing big-endian and little-endian formats in the Bitcoin protocol is likely a small design flaw. While standard hashes are big-endian, Bitcoin displays them in little-endian format because it considers hashes as little-endian integers rather than strings.

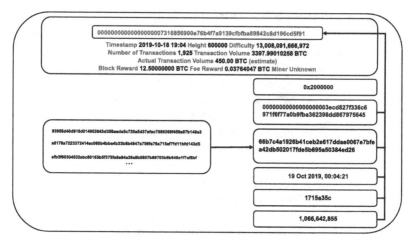

Figure 5.34 Details of Block 600,000. Own graphics.

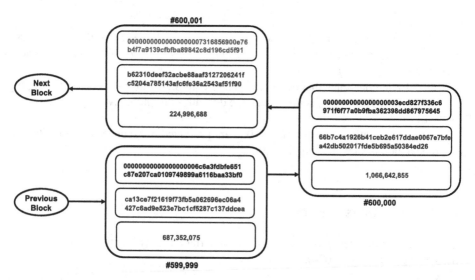

Figure 5.35 Block 600,000 as a blockchain member. Own graphics.

Supporters respond as follows:

(1) Bitcoin's volatility will decrease in time due to its greater adoption as an asset class and the development of Bitcoin's derivatives.

(2) Bitcoin deliberately optimizes its limited capacity and emphasizes the settlement of transactions outside of the conventional financial system.

(3) Bitcoin mining uses renewable energy; the energy consumption by the Bitcoin network is a legitimate use of natural resources.

(4) Bitcoin is a protocol, which is neutral and has properties valuable to good and bad actors alike, like the Internet; besides a share of the illicit Bitcoin transactions is low.

(5) Bitcoin has no associated cash flows, industrial utility, or legal standing; however, it is backed by its scarcity embedded in the Bitcoin protocol and efforts by miners maintaining consensus in the ecosystem.

(6) It is easy to fork Bitcoin's open-source software, but it is impossible to recreate its community and the associated network effects.

Not surprisingly, the truth is somewhere in-between:

(1) Because BTC price is prone to periodic bubble-bust episodes, see Section 10.3, it cannot be used as a conventional store of value. Due to the BTC scarcity and a large number of the so-called "HODLERs," not willing to part with their BTC under any circumstances, BTC is likely to retain some residual value, so investors with long time horizons can use BTC to store value.

(2) The Bitcoin protocol is not suitable for conventional payments for the simple reason that it processes about 3–6 TpS; see Section 5.5.4. Paying for coffee with BTCs is not an option. Still, one can use BTC for the immutable and final discharge of substantial obligations.

(3) Unquestionably, Bitcoin consumes an enormous amount of energy. According to the University of Cambridge estimates, in 2019, the Bitcoin protocol used an estimated 62 terawatt-hours (TWh) of electricity — more than such countries as Switzerland and the Czech Republic; about 0.28% of the total global electricity consumption. In 2019 the protocol processed only 120 million transactions; see Section 5.5.2. Such profligacy looks like sheer madness; however, given the protocol's very nature, this is the only way to maintain its integrity.

(4) Although criminals often use BTC for nefarious purposes, such as collecting ransoms, they use cash and, whenever possible, the banking system, on a much larger scale.

(5) BTCs have no intrinsic value. Accordingly, it is better to think of them as chips in a giant casino. There is a fixed number of BTCs and many potential players, which is enough to maintain their value in the long run. In addition to financial gains or losses, the players also receive entertainment from trading and owning their BTCs, a source of extra value. They should be compared and contrasted with conventional money, which, according to Aristotle, are derived from the law, and, hence, coercion rather than fun; see Section 12.2.4. There are some assets that can be stores-of-value for reasons other than their direct usefulness; see Chapter 10. Gold is a perfect example: it has survived since times immemorial despite its debatable value and better alternatives for most industrial applications. Fiat money has survived for 300 years despite being a depreciating asset; see Chapter 2.

(6) Nobody can be sure that Bitcoin will be here a hundred years from now. Bitcoin has shown one thing in the last 12 years; namely, its value does *not* lie in its "spec sheet." Although Ethereum goes further in many ways, Bitcoin is unequaled in many other ways, including its scarcity, adoption rate, robustness, and trust. Therefore, it is unlikely that Bitcoin will be superseded by "more robust and less wasteful technology" for a pretty long time. However, it is entirely possible, and even likely, that it will be superseded by more robust and less wasteful technology in the distant future. In this regard, an analogy with vinyl discs jumps to mind. Although there remain ardent aficionados of vinyl discs, the community has shrunk dramatically since their prime days.

5.8 Summary

After the publication of the seminal paper by Nakamoto (2008), Bitcoin took the world by storm. The fact that it had survived for more than a decade, and continues to thrive at present, speaks volumes. The Bitcoin protocol is a marvel of engineering, built in a very innovative way on top of excellent (if not the most modern) cryptographic primitives. Bitcoin cleverly combines cryptographic might with economics and game theory insights to make the protocol robust and mostly impervious to malicious interference.

However, the future of Bitcoin is uncertain. As shown in the chapter, it is not digital gold. Despite the original promises, it is highly centralized in its implementation, including mining and trading. Moreover, BTC ownership is exceedingly concentrated and, therefore, is profoundly undemocratic, or at least no more democratic than in the real world it wishes to improve. Besides, Bitcoin is not genuinely anonymous but rather pseudonymous. Despite the above, we feel that Bitcoin has a future as (highly volatile) store of value and possibly can find other exciting applications. For further discussion of BTC price dynamics and related topics, see Chapter 10.

We hope that this book will help the reader make her mind about Bitcoin's merits and weaknesses. In our mind, it is a great success at present, but, more likely than not, it will be eclipsed by other protocols in the future.

5.9 Exercises

1. Explain how one can use a good hash function to perform a proof-of-work (PoW) calculation?
2. Explain why two nonces are needed to find a block.
3. Derive the distribution of time it takes to mine a block successfully. Calculate mean and standard deviation for the time it takes for a miner whose share of hashing power is p to find a block. Assuming that a miner holds a proportion p of the total mining capacity, what is the probability that she may be first to find n blocks in a row? What are the implications in terms of censorship resistance and adverse chain reorganization?
4. Consider a secret miner owning more than 50% of the mining capacity and building an unpublished, parallel chain. Explain the possible threats.
5. Explain why miners may find it profitable to join forces with other miners — forming so-called mining pools. How does it impact the distribution of mining rewards?
6. Design an algorithm to estimate the number of bitcoins actually spent per block.

7. A vanity address is a bitcoin address containing a predefined series of characters, such as its owner's name. As an example, the following address

$$Address = \texttt{1GREATkHUc1ti5crqBX4wSF9hBjraEdmUk}$$

reveals the capitalized word GREAT, with the corresponding secret key

$$k_{WIF} = \frac{\texttt{5JSHCunVFzj38JNCHFL9P9JXQk}}{\texttt{732BzsW2ixRSiPNjuZ5L9W39w}} \; .$$

Download a vanity address generator, such as https://github.com/samr7/ vanitygen. Generate a keypair such that the corresponding address starts with 1EASY. Try to generate an address starting with 1HARDER. Why is it more challenging to do? Try to create an address beginning with 1IMPOSSIBLE. Why does it fail? Describe a possible vanity address generation algorithm in detail and propose a measure of the problem's difficulty. Explain how it relates to the PoW algorithm.

8. Choose the programming language of your choice and a bitcoin development library (e.g., https://github.com/MetacoSA/NBitcoin). Generate a keypair for the bitcoin testnet. The testnet blockchain is technically equivalent to the bitcoin blockchain but is only mined for testing and has almost no value. Identify a testnet faucet online (e.g., https://testnet-faucet.com/btc-testnet/) and send testnet coins to your address. Send testnet coins to your address a second time. Using a bitcoin testnet explorer online (e.g., https://blockstream.info/testnet/), search for your address as well as for the two incoming transactions. Develop a program to create a transaction out of your address programmatically based on the set of UTXOs identified on the testnet explorer. Experiment with multiple scriptPubKeys for the outputs. Build such a transaction, sign it appropriately, and broadcast it using any online service, such as https://live.blockcypher.com/btc-testnet/pushtx/. Use the bitcoin testnet explorer to document the transaction execution.

6 | Ethereum — A Distributed World Computer?

"What Ethereum intends to provide is a blockchain with a built-in fully fledged Turing-complete programming language that can be used to create 'contracts' that can be used to encode arbitrary state transition functions, allowing users to create any of the systems described above, as well as many others that we have not yet imagined, simply by writing up the logic in a few lines of code."

Ethereum white paper

6.1 Introduction

In this chapter, we discuss Ethereum — the second of the Big Three — its pros and cons, impressive accomplishments, and disturbing failures. Ethereum is a genuine attempt to address the limitations of Bitcoin scripts: The introduction of stateful and Turing-complete scripts in the form of smart contracts allows Ethereum to considerably augment the capabilities of previous generation DLTs and their possible applications. Whereas Bitcoin focuses on a single use case — that of regulating money flows — Ethereum offers a decentralized trusted compute platform capable of executing arbitrary code. ETH, Ethereum's native cryptocurrency, is not an end in itself: it merely acts as a token financing the execution of smart contracts by thousands of machines — contracts which may regulate the flows of entirely separate use cases.

While succeeding in rectifying some of the limitations of Bitcoin, Ethereum still suffers from some of its weaknesses:

(1) Limited privacy: Ethereum addresses are not genuinely anonymous and transaction can be traced by using the same methods as for Bitcoin; see, e.g., Chen et al. (2018); Klusman and Dijkhuizen (2018) among others.
(2) Bounded scalability: Ethereum improves the maximum transaction throughput as compared to Bitcoin but still is far from a capability to support widespread adoption.
(3) Lack of finality: Despite the faster block time leading to faster confirmations in average, Ethereum is still subject to chain reorganizations and cannot guarantee transaction finality.
(4) Volatile fees: Ethereum fees can vary dramatically during periods of high demand, making the network prohibitively expensive when it is most needed.

Ethereum's innovations come with new challenges, such as the maintenance of a fast growing blockchain preserving the state of every smart contract, or preponderent cybersecurity risks related to the complexity of smart contracts.[1]

ETH price dynamics and Google trends are presented in Figure 6.1.

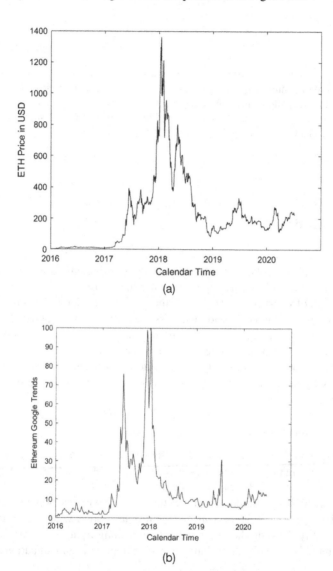

(a)

(b)

Figure 6.1 (a) The price of ETH; (b) The Google trend for Ethereum. Own graphics. Sources: (a) coinmarketcap.com, (b) Google.

---

[1] This situation is similar to what is known in risk management as a wrong-way risk.

This chapter is organized as follows. Section 6.2 outlines the main conceptual and technical differences between the Ethereum and Bitcoin protocols, while Section 6.3 covers finer points. In Section 6.4, we give a detailed description of the Ethereum ledger and its different accounts. Section 6.5 describes the famous Ethereum Virtual Machine. In Section 6.6, we discuss various types of transactions. Section 6.7 explains consensus building in Ethereum. Section 6.8 covers the concept of smart contracts, which are central to Ethereum as a whole. In Section 6.9, we present several practical applications, including ERC-20 tokens, initial coin offerings, stable coins, automated market makers, and distributed autonomous organizations. Our conclusions are summarized in Section 6.10.

6.2 Similarities and differences between Bitcoin and Ethereum

Ethereum is an ambitious extension of Bitcoin. The Ethereum white paper explains the purpose of the protocol as follows: "Satoshi Nakamoto's development of Bitcoin in 2009 has often been hailed as a radical development in money and currency, being the first example of a digital asset which simultaneously has no backing or intrinsic value and no centralized issuer or controller. However, another — arguably more important — part of the Bitcoin experiment is the underlying blockchain technology as a tool of distributed consensus, and attention is rapidly starting to shift to this other aspect of Bitcoin."

Not surprisingly, these two protocols have many similarities. While block sizes, mining times, mining rewards, and other particulars are quite different, Ethereum and Bitcoin are both public distributed ledgers relying on elliptic curve digital signature algorithm (ECDSA) and proof-of-work (PoW).[2]

However, there are conceptual differences. Bitcoin is a protocol supporting BTC as an alternative currency. Ethereum is a protocol running a distributed computing platform capable of storing and executing arbitrary code on a so-called EVM, or Ethereum Virtual Machine. While ETH is the second-largest cryptocurrency traded on exchanges, its primary purpose is to pay for the execution of this code by the network nodes, not to act as a mere currency per se. It is crucial to understand that this feature of Ethereum protocol conceals a self-defeating mechanism: As Ethereum adoption increases, leading to higher ETH demand and prices, utilization of Ethereum becomes prohibitively expensive and restricts its use.

This difference is fairly profound. In addition to external owned accounts (EOA), which operate just like Bitcoin accounts, Ethereum supports smart contracts (SCs), see also Section 3.7. In a nutshell, a basic smart contract can maintain a ledger for external tokens, and can even define applications unrelated to financial instruments. Thus, the real strength and distinguishing feature of Ethereum is to provide Consensus as a Service (CaaS) for any application that requires a high level of trust and auditability. In a way, an equivalent to BTC may trivially be reimplemented as a token contract on the Ethereum platform, including its monetary policy and UTXO management.

---

[2]Ethereum 2.0 leverages a PoS consensus algorithm.

Table 6.1 Similarities and differences between Bitcoin and Ethereum protocols.

	Bitcoin	Ethereum
Objective	An alternative currency	A distributed world computer
Native unit	BTC — a currency	ETH — a token
Storage	Wallet	Wallet
Traded on	Exchanges and peer-to-peer	Exchanges and peer-to-peer
Transaction style	UTXO-based	Account-based
Transaction purpose	Moving BTC	Moving ETH or executing smart contract
Implementation	Public blockchain	Public blockchain
Elliptic curve	secp256k1	secp256k1
Addresses	Secret-key controlled	Secret-key controlled and smart contracts
Ownership proof	ECDSA	ECDSA
Consensus algorithm	Proof-of-Work	Proof-of-Work
		Eventually Proof-of-Stake
Hash function	*SHA256*	*Ethash*
Consensus keepers	Miners	Miners
Transaction fees	Voluntary	Mandatory
Block size	2,000 TX	380 TX
Block mining time	approx. 10 min	approx. 15 sec
Supply mechanism	Mining	A combination of pre-mined supply and mining
Mining reward	6.25 BTC	2 ETH
Supply cap	21 million in total	18 million per year
Supply style	Deflationary	Inflationary
Launch date	January 2009	July 2015
Inventors	S. Nakamoto	V. Buterin, G. Wood, J. Lubin et al.

Commonalities and differences between Bitcoin and Ethereum are summarized in Table 6.1.

6.3 A more centralized protocol?

Ethereum is also perceived as being more centralized than Bitcoin, for good or bad reasons. In particular, all bitcoins in existence appear as a result of mining needed to operate the Bitcoin blockchain; even Satoshi Nakamoto earned his bitcoins through the mining process, earning block rewards for every block added to the chain. Ethereum however pre-mined the lion share of ETH in circulation (72 million out of 119 million) and distributed them to

Table 6.2 Some useful ETH units and their naming conventions. As of this writing (October 2020), average transaction size is 5 ETH, block mining reward is 2 ETH; cost of basic transaction is about 2,000 szabo; cost of gas is 100 Gwei.

Unit	Units Number per ETH	Usage
ETH (ether)	1	Transaction amounts and block rewards
finney	1,000	
szabo	1,000,000	Cost of a basic transaction
Gwei	1,000,000,000	Gas price
wei	1,000,000,000,000,000,000	The base indivisible unit

early contributors and investors to the project (see Table 6.2 on units and naming conventions). This fact is frequently cited by Bitcoin purists, who do not view Ethereum as a truly distributed protocol. On a related matter, Satoshi Nakamoto, the creator of Bitcoin, is as elusive as ever, while one of the founders of Ethereum, Vitalik Buterin, keeps a high profile and exerts a significant influence on the Ethereum ecosystem. A typical example of his power is the decision to execute a hard fork to counter the decentralized autonomous organization (DAO) hack; see Chapter 2 and Section 6.9.6. The importance, or lack thereof, of this additional source of centralization, remains to be seen.

Surprisingly, Ethereum's payment per-operation structure is archaic — this is how phone companies used to charge their customers on a per-minute basis for phone calls in the 20th century. By now, it became evident that subscription-based payments are the only way to pay for large-volume services. Recently, paying for Ethereum transactions became prohibitively expensive for all but speculative purposes. However, there is a good reason why Ethereum relies on the payment per-operation model. Ethereum payments provide anti-denial of service protection. By paying per canonical operation relative to its computational weight, one can ensure that no attacker can pollute the network by exploiting computationally expensive but financially cheap operations. In the future, other, more flexible alternatives may augment the existing per-operation payment model.

The Ethereum protocol development was funded by an online sale that took place between July and August 2014. The protocol went live on July 30, 2015 with the genesis supply of approximately 72,010,000 pre-mined ethers. As of September 15, 2020, the total supply of ether is about 112,600,000. Thus, about 64% of the total circulating supply is pre-mined — the fact often criticized by Bitcoin purists.

6.4 Ethereum account types

An Ethereum address can refer to two types of entities: (a) Externally Owned Accounts (EOAs) and (b) smart contracts (SCs). Although both EOAs and SCs have public addresses, only EOAs have associated secret keys. EOAs are comparable to Bitcoin addresses in principle but rely on a very different paradigm: whereas Bitcoin maintains the accounting of its system by tracking immutable UTXOs, Ethereum is more akin to traditional bank accounts: it stores an explicit balance for each address, mutated by transactions in and out. As a result, the state of an Ethereum account is more complicated than Bitcoin. An EOA contains the following fields:

(1) nonce, incremented for each outgoing transaction and acting an anti-replay protection;
(2) ETH balance;
(3) empty contract code;
(4) storage (which is initially empty by default).[3]

Unlike Bitcoin's UTXO paradigm, Ethereum's account model requires no picking and choosing UTXOs to satisfy various criteria, such as fee minimization, privacy optimization, etc. Thereby, the account model significantly simplifies the process of building transactions. However, the consequence of making accounts mutable is that one needs an anti-replay mechanism to ensure that a transaction signed by the wallet owner can never be re-executed a second time. This is trivially guaranteed under the UTXO model given that a UTXO no longer is "unspent" as soon as it is consumed for a transaction. However under the account model, the balance is mutated for each transaction but nothing inherently prevents a transaction from being replayed multiple times.

In order to prevent replays, Ethereum maintains a nonce as part of the account state on the blockchain. This nonce is incremented for every outgoing transaction and Ethereum guarantees that it only processes an outgoing transaction if the nonce it contains — and which is signed with the transaction — is precisely one more than that current nonce registered in the account on-chain. If a replay attack is attempted, the nonce has already been consumed, and the transaction can immediately be disregarded. Thus, the nonce as an outgoing transaction counter is critically important in deciding the transaction's validity.

To summarize, an EOA with no associated code can only send and receive ETH transactions and trigger contract code, and is controlled by a secret key.

The truly original feature of Ethereum is its support for SCs. A smart contract is an immutable piece of software. Once deployed on the blockchain, it maintains its own internal state and can receive requests from EOAs or other SCs. It can itself store ETH and trigger actions on other contracts based on its programmed behaviour. Smart contracts are guaranteed to execute as developed, with no theoretical means to alter their logic. In other words, an SC differs from an EOA because its behaviour and its ETH holdings are controlled by code, rather than by an external secret key.

In theory, the code of a smart contract is Turing-complete and can perform arbitrary complex operations. In particular, it can have a permanent state, it can operate arbitrary rules on this state and mutate it, and, if necessary, it can call other contracts. In practice, the high fees associated with the consumption of memory and the execution of the contract, as well as the archaic performance of the EVM, limit the reach of possible applications to small-size programs focused on security-critical code. For this restricted class of applications, Turing completeness is achieved precisely because such contracts rarely need to reach the limits of Ethereum.

---

[3]Eventually storage is populated with transaction's information.

We describe how Ethereum operates based on the white paper, Buterin (2013), yellow paper, Wood (2015), and a recent book, Antonopoulos and Wood (2018). As before, we distinguish between the Ethereum protocol and the native cryptocurrency ether (ETH).

6.5 The Ethereum Virtual Machine

The Ethereum Virtual Machine, or EVM is a sandboxed execution environment running on every Ethereum node. It is responsible for the execution of transactions and their related smart contracts and supports a Turing-complete set of operations. It runs in a segregated container guaranteeing isolation from the rest of the node logic and the network. In essence, it is the common standard all Ethereum nodes share for the execution of contracts and which makes the Ethereum "language" interpretable on each node of the network.

Importantly, we like to emphasize the distinction between Ethereum "distributed computing" and previous kinds of distributed computing networks that have existed for a long time. Originally, distributed computing meant horizontal scaling, i.e., the sharding of the computational effort across distributed nodes, each processing a tiny part of the overall problem; see, for instance, Folding@home[4]. For such networks, more nodes means increased performance; but the corollary is that trust in the result is low because there is no duplication of the work. There is no trivial means to identify data tampering or process corruptions on the distributed system. These distributed computations accelerate speed rather than increase trust.

Ethereum's goal is entirely different. It aims at having each node execute the same computation. Ethereum provides no horizontal scaling, at least in the current version of its network — arguably, it is slower than a single machine because of network latencies and the limitations of the EVM. However, such a redundancy enables one to achieve a high level of trust. By reaching a network-wide consensus on the execution of all smart contracts by all nodes, one can reach a high level of confidence in the output despite the hostile environment.

Like everything in life, reality does not have to be one of the two extremes. One obvious, albeit non-trivial, way of scaling distributed ledger technologies is to "shard" the execution of transactions and contracts, i.e., to position the DLT somehow between the two options mentioned above. Instead of all machines executing all computations, it is sufficient for a subset of nodes to duplicate each other so that the result is trustworthy.

Thus, the EVM is a massively replicated, redundant, distributed computer whose aim is to ensure proper execution of transactions and contracts at the cost of efficiency.

6.6 Transfers and smart contract calls

Ethereum is a state transition protocol, similar to the Bitcoin protocol, described by Eqs (5.37) and (5.38). One particularity however is that it relies on a single abstraction for both

---

[4]See https://foldingathome.org/.

simple transfers of ETH and advanced calls of smart contract methods. Even deploying a new contract on the Ethereum blockchain is achieved by a transaction.

Transfers contain the following information:

(1) nonce — for anti-replay;
(2) sender's address (implicit);
(3) recipient's address;
(4) signature's of the sender;
(5) amount of ETH to transfer;
(6) STARTGAS value, which determines the maximum number of computational steps the originator of the transaction is willing to pay for;
(7) GASPRICE value representing the actual fee per computational step the sender is willing/able to pay;
(8) DATA field to specify an optional smart contract method to call and its parameters.

The last two steps are crucial anti-denial of service measures and represent a unique feature of the Ethereum protocol. Gas pays for miners' services. Simply put, gas is a unit measuring the computational effort needed to perform operations and store data. Every successfully executed operation in Ethereum, be it an elementary ETH transfer or a sophisticated SC, requires gas. Miners are paid an amount in ETH equal to the total amount of gas needed for executing a request times the gas price agreed upon by the transaction's initiator.

A typical Ethereum transaction affecting only Alice's and Bob's addresses is shown in Table 6.3.

The evolution of the Ethereum system caused by such a transaction is shown in Figure 6.2.

The amount of gas needed to support a simple fund transfer is 21,000 gas. If Alice indicates her willingness to pay STARTGAS $\geq$21,000 gas; miners will process her transaction, provided, of course, that Alice has the requisite funds. If she only agrees to pay STARTGAS <21,000 gas for the transaction, her gas will be spent, but the transaction will not be completed and will not change the state of the Ethereum ecosystem. On the other hand, excess gas equal to the difference between STARTGAS and actual gas consumption will revert to Alice. Therefore, there is a built-in incentive to show a relatively high value of START-GAS. However, if it becomes too high, the miners will not process Alice's transaction to avoid clogging their blocks.

Table 6.3 Moving ETH between Alice and Bob.

	Scenario$_1$	Scenario$_2$
State$_n^{i-1}$	{nonce m_n^{i-1}, Alice: 50, Bob: 20}	{nonce m_n^{i-1}, Alice: 50, Bob: 20}
TX	{send 10 ETH from Alice to Bob}	{send 60 ETH from Alice to Bob}
Result	Success	Failure
State$_n^i$	{nonce m_n^i, Alice: 40, Bob: 30}	{nonce m_n^{i-1}, Alice: 50, Bob: 20}

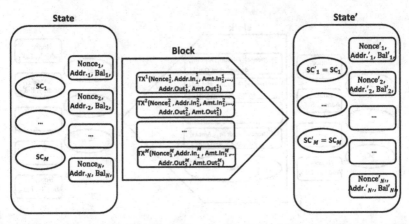

Figure 6.2 Ethereum state transactions result in the state of Ethereum blockchain evolving from S to S'. In this figure, we assume that no SC transitions are initiated. Own graphics.

The price of gas, GASPRICE, is a very different story. If Alice indicates her price per unit of gas lower than the market, the miners may not process her transaction. If, on the other hand, she agrees to pay the price higher than the market, the miners will process her transaction in priority, consuming the needed amount of gas (such as 21,000 for the standard transaction) at Alice's (rather than market) price.

In summary, one can include a reasonably high STARTGAS number and expect to get the excess back, but indicating a low GASPRICE will result in transaction not being executed at all, thus causing unnecessary irreversible losses. However, like everything else in the Ethereum protocol, the situation is highly nuanced — showing a very high STARTGAS number will make the transaction unattractive to the miners since the blocks they form has a total gas limit, which currently stands at 10 million gas. Thus, a standard transaction showing 100,000 STARTGAS will prevent four other transactions (and their associated fees) from being included in the block. Driven by their enlighted self-interest, miners will reject it. A simple analogy is in place. Putting sufficient fuel in an airplane is a must for it to reach the intended destination. However, putting in too much will make the plane so heavy that it will not take off in the first place.

A more general transaction triggering the execution of a smart contract is shown in Figure 6.3. It requires more operations than a simple transfer from Alice to Bob; how much more depends on the complexity of the underlying SC.

The Bitcoin protocol is very explicit regarding miners' renumeration, which consists of mandatory block rewards and voluntary (but highly advisable) transaction fees, see also Sections 3.6 and 5.5. In contrast, the monetary policy of the Ethereum protocol is deliberately opaque. With his characteristic flair, Vitalik Buterin put his thoughts as follows: "I think we've been consistent on the issuance question. The issuance is whatever it needs to be to ensure reasonable level of security." Figure 6.4 shows that Ethereum follows an inflationary monetary policy, which, in itself, is not necessarily a mortal sin.

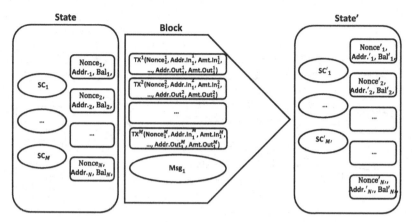

Figure 6.3 Ethereum state transactions result in the state of Ethereum blockchain evolving from S to S'. In this figure, we assume that a SC transition is initiated. Own graphics.

In contrast to Bitcoin, the transaction fee is mandatory; it can be calculated as follows:

$$\text{transaction fee [ETH]} = \text{gas price} \left[\frac{\text{ETH}}{\text{gas}} \right] \times \text{gas used [gas]}. \tag{6.1}$$

Occasionally, Ethereum suffers from uncomfortably high transaction costs, making it unsuitable for high-scale projects or retail applications. In its current form, the platform would not, for instance, be capable of hosting a retail payment token without overflowing the blockchain and suffering from surging and highly volatile transaction costs. In 2020, Ethereum's cumulative transaction fees dominated those of Bitcoin, as illustrated in Figure 6.5.

6.7 Consensus

As in Bitcoin, Ethereum miners, supporting consensus in the Ethereum protocol, tend to coalesce in mining pools. These pools are colossal so that Ethereum's mining is not sufficiently decentralized as claimed; see Figure 6.6.

Given that Ethereum's block generation rate is approximately 40 times higher than in Bitcoin (a block every 15 seconds vs. every 10 minutes), the former protocol routinely creates many more orphan blocks than the latter one. Recall that an orphan block is a valid block, not included in the blockchain, in favor of a different block mined around the same time. While the Bitcoin protocol discards orphan blocks so that miners' computational efforts are wasted and viewed as the cost of doing business, see Chapter 5, the Ethereum protocol cannot afford such extravagance.

Instead, Ethereum elevates orphan blocks to the status of uncles and rewards them with partial mining rewards. The rationale behind this approach is threefold. First, frequent emergence of uncle blocks is a natural outcome of shorter block times comparable with the network lag, hence the protocol needs to adapt and find a meaningful remedy. Next,

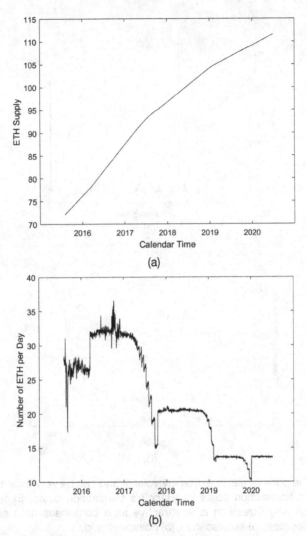

Figure 6.4 ETH supply growth chart: (a) the total number of ETH, in millions; (b) the number of ETH mined per day, in thousands. Own graphics. Source: etherscan.io.

rewarding uncles increases the network's security by incentivizing extra work even if the blockchain itself is not using them directly. Finally, rewarding uncles makes the whole system a bit fairer and decreases its otherwise significant centralization. Instead of allocating most rewards to large mining pools controlling a large portion of hashing power, it leaves some small rewards for individual miners.

The high-level operation of the Ethereum ecosystem is detailed in Figures 6.7–6.9. In Figure 6.7, we show the size of Ethereum blocks. In contrast to Bitcoin, there are two limitations: the block's size in bytes and the cumulative amount of gas for all the block's transactions (currently, 10 million gas). Given that even the most basic transaction moving

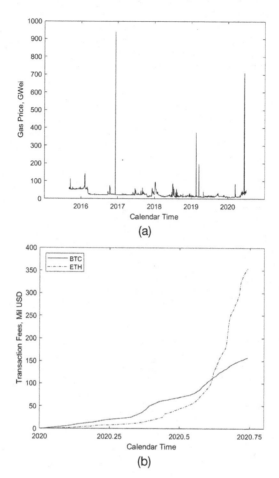

Figure 6.5 (a) The price of gas is extremely volatile and can be very expensive; (b) Cumulative Ethereum's transaction costs vs. Bitcoin's transaction costs. Expensive gas is a significant reason why Ethereum is unattractive as a consensus as a service provider. Own graphics. Sources: (a) etherscan.io, (b) coinmetrics.io.

ETH from one account to another requires 21,000 gas, this limit implies that an Ethereum block can contain no more than 380 transactions, compared with 1,500–2,000 transactions per Bitcoin block. Figure 6.7 (a) shows the size of an Ethereum block in bytes, while the complementary Figure 6.7 (b) shows the gas limit. Due to this limit's low level, the often-made claim of EVM being Turing-complete seems to be a gross exaggeration.

Figure 6.8 shows the mining time for an Ethereum block. This time is more or less regular but, on occasion, spikes up. Besides, this figure demonstrates the number of blocks produced per day is relatively steady, with sharp drops mirroring spikes in the mining time.

The complementary Figure 6.9 presents the number of uncle blocks generated per day. While the number of blocks included in the blockchain is stable, the uncle block count exhibits substantial variability.

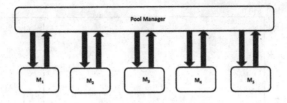

(a) A typical mining pool.

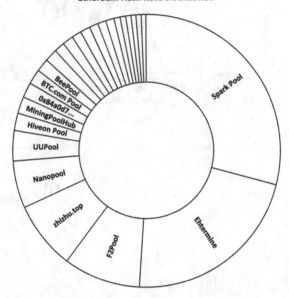

(b) The hash rate distribution among mining pools.

Figure 6.6 (a) Miners organize themselves in gigantic pools governed by managers; (b) The number of pools responsible for the vast majority of hash rate is very small. Comparison with Figure 5.29 shows that Ethereum is even more centralized than Bitcoin. Own graphics. Source: (b) etherscan.io.

6.8 Smart contracts

6.8.1 Definitions

An SC is a computer code that executes deterministic commands upon receiving legally relevant inputs; see Section 3.7. In theory, SCs do not rely on trusted intermediaries, making them cheaper to manage than conventional legal agreements. Nick Szabo, a well-known computer scientist, was the first to introduce SCs in the mid-1990s; see Szabo (1997). He described them as follows: "[A] set of promises, specified in digital form, including protocols within which the parties perform on these promises." In contrast to Ethereum SCs, SCs envisaged by Szabo are stateless.

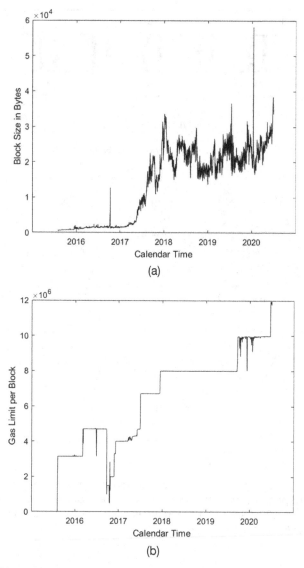

Figure 6.7 (a) Ethereum block size is much smaller than the Bitcoin block size; (b) More importantly, the amount of gas per block is limited as well, so that the oft-repeated claim that Ethereum is a Turing-complete computer makes no sense. Own graphics. Source: etherscan.io.

It is customary to view vending machines as the oldest technology relying on SCs for its operation. The argument goes as follows: a customer deposits a dollar in a vending machine and receives a can of coke in return, or if the coke is not available, she gets her dollar back. Already at this basic level, one can anticipate numerous problems, the most obvious being

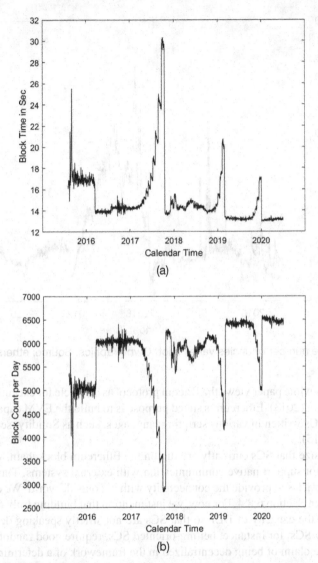

Figure 6.8 While Ethereum blocks are much smaller than Bitcoin blocks, they are processed much faster. (a) Ethereum produces blocks at a relatively stable speed. (b) As a result, it produces a relatively stable number of blocks per day. The dips in Figure (b) correspond to the spikes in Figure (a). Own graphics. Source: etherscan.io.

what to do if the machine is defective and swallows the dollar. Besides, at the risk of being pedantic, multiple mundane actions, such as using a key to open a lock, or pressing a button to call an elevator, can be viewed as basic SCs. We emphasize that 99.999% of the time, the door opens, and the elevator comes. It is 0.001% when the lock breaks or the elevator gets stuck between floors when human intervention is badly needed.

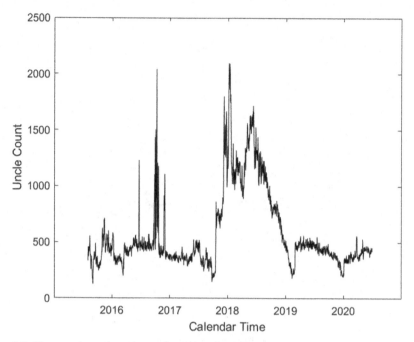

Figure 6.9 The number of uncles varies a lot. Own graphics. Source: etherscan.io.

Ethereum's white paper views the Bitcoin protocol as a vehicle for executing elementary SCs; see Buterin (2013). Ethereum's stated purpose is to build the EVM capable of running very general SCs written in various scripting languages, such as Solidity; see Reitwiessner and Wood (2015).

We emphasize that SCs currently are run on the Ethereum blockchain, which, by construction, cannot support native communication with external systems. Thus, SCs require the so-called oracles to provide the connectivity with the outside world. We discuss oracles in some detail in Section 6.8.4. For now, we just mention that building truly distributed oracles is beyond the capacity of DLT so that SCs are not strictly speaking decentralized. In addition, many SCs, for instance, betting-oriented SCs, require good randomness sources, challenging the claim of being decentralized in the framework of a deterministic EVM.

6.8.2 Example — A forward contract

Recall that a forward financial contract is an agreement between two parties, Alice and Bob. We say that Alice owns (is long) the forward contract if she agrees to purchase from Bob, who sells (is short) the forward contract, an agreed amount of a commodity or currency for a fixed price at a predetermined future time. Such contracts are essential building blocks of the financial infrastructure and can be used for both hedging and speculative purposes. A detailed description of conventional forward contracts is presented in Lipton (2001), among others. Since forward contracts are inherently risky obligations, both parties should be prepared to suffer losses.

Let us implement a forward contract as an SC, restricting ourselves to the Ethereum's native token, ETH. As before, a forward contract requires two parties, Alice and Bob. Assume that both of them have some ETH but hold opposite views of its future dollar price. Alice thinks that ETH price will go down temporarily and wishes to accumulate more ethers before they rally back. They want to structure a contract so that Alice makes ETH if its price drops in the short run and (naturally) loses money otherwise. Since a forward contract is a zero-sum game, Bob has a mirror image payout. Alice and Bob can design the corresponding contract along the following lines:

(1) Alice deposits 100 ETHs.
(2) Bob deposits 100 ETHs.
(3) At the contract's inception, the USD value of 100 ETHs is X USD.
(4) At the contract's maturity, say in one month, 100 ETHs are worth Y USD.
(5) The SC sends X dollars worth of ether to Alice and the rest (if anything is left) to Bob.

At the inception of the contract and its maturity, the price of ETH is calculated by querying an agreed-upon data feed contract.

Let us discuss what happens step-by-step. Denote relative price change by $Z = Y/X$. If the ETH price in 30 days goes down, $Y < X, Z < 1$, Alice gains more ETHs compared to the buy-and-hold strategy, and if it goes up, $Y > X, Z > 1$, she loses ETHs. However, she cannot gain more than 100 ETHs in addition to her initial stake of 100 ETHs, reflecting the total amount of collateral and corresponding to $Z \leq 1/2$. Obviously, for $Z = 1$, Alice gets back 100 ETHs, regaining her original position. As ETH continues to rally and $Z > 1$, she receives less and less ETHs that are always worth X dollars. In the limit, she can lose her entire stake of 100 ETHs. Alice's and Bob's payouts are shown in Figure 6.10.

The beauty of the situation is that the counterparty risk is absent in the SC, since the contract is funded in advance. Of course, such funding is neither possible nor entirely desirable in real life, so that the perceived advantage is illusory. Moreover, the whole premise of financial transactions is to reduce collateral and increase leverage to magnify a potential gain. Besides, if the price of ETH drops by a factor of two or more, the entire collateral of 200 ether can be worth less than X dollars. It is essential to understand that this contract requires an oracle extrinsic to Ethereum in the form of a centralized, or, at least, semi-centralized, feed. Thus, this contract is not entirely decentralized. Recently, ChainLink made some progress toward solving the above-mentioned critical issue; see Section 6.8.4.

6.8.3 Advantages and disadvantages of smart contracts

SCs have several desirable properties. They are autonomous and can be entered between two parties directly without using the services of intermediaries. SCs are trustworthy, as long as the corresponding code itself is trustworthy and free of bugs, the agreed terms are enforced automatically. SCs are reliable since their execution is guaranteed by the entirety of miners supporting the Ethereum protocol's integrity at large. Besides, SCs are reasonably efficient. Because most contracts are either highly standardized or relatively simple, it is

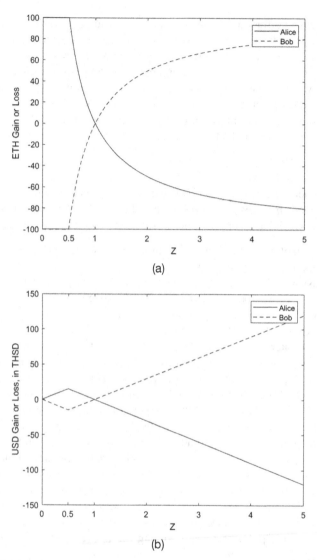

Figure 6.10 Alice's and Bob's payoffs: (a) in ETH; (b) in USD. To be concrete, we assume that today's ETH price is 300 USD/ETH, so that $X = 30,000$. Own graphics.

easier and cheaper to implement them as a code than as a legal document. Finally, SCs are accurate since a properly implemented code can be made very precise.

Theoretical advantages of SCs are summarized in Figure 6.11.

Not surprisingly, SCs have several drawbacks, some of them severe. They are not smart enough and require all potential financial obligations to be fully collateralized ahead of time, thus consuming collateral voraciously. Complicated SCs consume a lot of gas and are very expensive to run. Potentially, they can be too big to fit in a block. SCs are very rigid

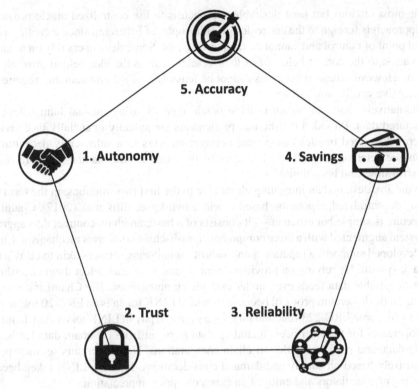

Figure 6.11 Potential advantages of smart contracts. Own graphics.

because making changes is not possible once they are deployed.[5] Occasionally, rigidity is perceived as a desirable feature. However, it can quickly turn into its opposite if further changes in the deployed contract are required. Since SCs are visible to all, anyone, who can find a bug that cannot be fixed in real-time, will exploit it for profit. Rigidity does not help in this regard.[6] SCs, by construction, know nothing about the outside world or randomness. As such, they need external oracles, which are described next. These oracles are expensive and not sufficiently decentralized.

6.8.4 Oracles

As mentioned in Section 6.8.1, the deployment of meaningful SCs is often impossible without a good oracle. Yet, such oracles are hard to build, so, not surprisingly, there is a short supply of possible technologies, which severely limits the usefulness of the existing SCs; see Beniiche (2020).

---

[5]This rigidity is Ethereum-specific. Some protocols do allow upgrading of SCs.

[6]Such an attack happened during the infamous June 17, 2016 DAO attack.

The most obvious but least desirable approach is to use centralized oracle providers. This approach is foreign to the entire design philosophy of Ethereum since it creates a centralized point of control and cannot be tamper-resistant. Some developers rely on notarization to attest to the correct behavior of their oracles. This is the idea behind provable.xyz protocol. However, these attestations cannot be feasibly verified on-chain and require further recursive verification.

Alternatively, one can obtain reliable oracle data by using manual human input of unstructured data. Provided that human participants are suitably financially incentivized, such crowd-sourced oracles can produce correct answers in a sufficiently decentralized fashion; see Peterson et al. (2015). However, by their very nature, manual-input oracles are expensive, slow, and have limited scope.

ChainLink designed an intriguing alternative to the first two solutions, in the form of a partially decentralized, reputation-based oracle network; see Ellis et al. (2017). ChainLink architecture is simple but efficient — it consists of a basic on-chain contract data aggregation system augmented with a more comprehensive off-chain consensus mechanism. Chain-Link developed supporting reputation and security monitoring services akin to eBay's user feedback system. By relying on providers' reputations, users can select them accordingly and enjoy reliable data feeds even under extreme circumstances. The ChainLink network depends on the Ethereum protocol because its native LINK token is an ERC-20 token, augmented with some ERC-223 functionality. Users have to pay in LINK tokens to ChainLink Node operators for their services, including data retrieval from off-chain data feeds, formatting data, and performing the off-chain computations. Node operators set their prices competitively based on supply and demand considerations. In 2020, LINK token became the darling of speculators and enjoyed an enormous price appreciation.

6.9 Applications of Ethereum smart contracts

6.9.1 Potential use cases

Initially, SCs took the world by storm. Later, with experience, it became clear that their ability to perform complex tasks is relatively limited.

SCs have several applications. First and foremost, financial applications, including financial derivatives, hedging instruments, saving accounts, and other similar use cases. Next, semi-financial applications, pursuing non-financial purposes, such as DAOs, but having a financial component. Finally, with various degrees of success, SCs have been used for online voting and communal decision making.

6.9.2 Tokens and token standards

As mentioned earlier, one could view Ethereum protocol as a provider of CaaS. Thus, it is natural to use the Ethereum protocol to maintain external ledgers. In some respects, these ledgers are similar to Bitcoin being public and cryptographically secured. However, they are different from the Bitcoin ledger in the following crucial way — their consensus is externally provided and does not require any mining. Thus, one could characterize these ledgers as being centralized-decentralized. The corresponding tokens are pre-mined and distributed from a single source. Therefore, maintaining the ledger's internal consistency by making sure that simple accounting rules are satisfied, is the only requirement.

Initially, developers proposed several competing approaches to the most efficient maintenance of token-bearing ledgers, but, eventually, best practices emerged in the form of several standards, such as the ERC-20 standard for fungible tokens and the ERC-721 standard for nonfungible ones.

The fungibility of ERC-20 tokens means that individual units are identical and interchangeable. Accordingly, any two tokens represent the same value.

The ERC-20-compliant token has to have three optional functions and six mandatory ones. Optional functions include a token name, token symbol, and decimal. The meaning of first two functions is self-explanatory. The third one is an integer between 0 and 18, which shows the token's smallest unit. If Decimal is 0, the token itself is the smallest unit; if Decimal is 18, 10^{-18} is the indivisible base unit, similar to wei in Ethereum. In addition to optional, every ERC-20 contract has six mandatory public functions, whose meaning is self-explanatory:

(1) totalSupply(), returns (uint256 totalSupply); totalSupply gets the total token supply.
(2) balanceOf(address _owner), returns (uint256 balance); balanceOf gets the account balance.
(3) transfer(address _to, uint256 _value), returns (bool success); transfer sends _value tokens to address _to.
(4) transferFrom(address _from, address _to, uint256 _value), returns (bool success); transferFrom sends _value amount of tokens from address _from to address _to.
(5) approve(address _spender, uint256 _value), returns (bool success); approve allows spender to withdraw funds multiple times from the account up to the predetermined value amount. When called again, this function overwrites the current allowance with another value.
(6) allowance(address _owner, address _spender), returns (uint256 remaining); allowance returns the amount which address _spender is still allowed to withdraw from address _owner.

The ERC-20 token standard, being the first of a kind, suffers from several drawbacks. The ERC-223 token standard, superseding ERC-20, gained considerable traction to address these shortcomings. The ERC-223 standard is safer than its predecessor since it doesn't allow token transfers to contracts that cannot process token receiving and handling. Still, many existing tokens rely on the ERC-20 standard.

The ERC-721 standard allows one to create non-fungible tokens (NFTs) on Ethereum. Every ERC-721 token is unique and, allowing imagination to run wild, can be viewed as one-of-a-kind collectible. Since ERC-721 tokens are unique, they can represent ownership over assets, such as title deeds, art, and digital collectibles. Of course, establishing real assets ownership is not just (or even mostly) a technological problem and requires building a robust legal foundation first.

For a brief period in 2017 and 2018, ERC-721-based CryptoKitties gained considerable popularity. CryptoKitties is an online game allowing users to collect and breed virtual cats with unique genomes defining their appearance and traits. Since each kitty is an Ethereum ERC-721 token, all of them are one-of-a-kind. By construction, a CryptoKitty is cryptographically secure. Hence, it cannot be replicated, taken away from the owner, or destroyed. A simple question of who needs them in the first place has not crossed the enthusiasts' mind.

By the end of 2018, cumulative sales of CryptoKitties exceeded US$27 million, including a CryptoKitty, sold for US$140,000 on May 12, 2018.[7] The CryptoKitties-related traffic was so high that it clogged and slowed down the entire Ethereum blockchain in late 2017 and dramatically increased gas prices. This fact shows that Ethereum is far from being robust with respect to its spare capacity.

At the beginning of 2021, NFTs again caught the interest of the general public and experts alike. An extensive market for NFTs' trading appeared practically overnight. Up to now, the most striking example is the sale of digital artwork by Beeple, called EVERYDAYS: THE FIRST 5000 DAYS, for US$69.3 million in 2021 at an auction organized by a British auction house Christie's founded in 1766. The full potential of NFTs remains to be seen, but they open new possibilities of trading in digital objects, such as JPG files. One of the problems with such files is their infinite reproducibility. By signing such a file digitally and using Ethereum as a CaaS provider to track its transition from one owner to the next, one can artificially create digital scarcity for the signed file. The best way of understanding the difference between the signed and JPG and its unsigned version is by the analogy between a signed and unsigned baseball paraphernalia. While physically identical, a signed ball or another item is orders of magnitude more expensive than the unsigned one. Thus, one can think about an NFT as an autographed collectible object. The problem of fakes, which afflicts the market for collectibles, is much less of an issue for NFTs because they are perfectly traceable from the moment they are created, provided, of course, that the original signer is legitimate. However, NFTs are not universally welcomed by all — the main point of contention being the fact that they are not environmentally friendly. Perhaps this problem can be resolved if and when Ethereum switches to a PoS consensus.

6.9.3 Crowdfunding via initial coin offerings

An initial coin offering (ICO) is a type of funding relying on issuing Ethereum-based ERC-20 tokens, or, occasionally, tokens based on other protocols. Typically, entrepreneurs sell ICOs on Ethereum, but private placements are not uncommon. As a rule, in an ICO, "tokens" or "coins" are sold to speculators or investors in exchange for Bitcoin or Ethereum, or, seldom, fiat. These tokens promise monetary rewards when the ICO's funding goals are met or during the project's life. In principle, an ICO can be an attractive capital source for startup companies because they do not dilute ownership. On the pros side, ICOs allow startups to avoid intermediaries, such as angel investors and venture capitalists. On the cons side, in most cases, ICOs fall in legal limbo concerning existing regulations. Not surprisingly, ICOs have been prone to scams, get rich quick schemes, and securities law violations. Authorities summarily ban them in several countries, such as China and South Korea. The question of whether or not tokens sold via ICOs are securities is very nuanced.

---

[7] While reasonable people can disagree about the bubble-like behavior of the cryptocurrency prices, no one can argue that Cryptokitties prices have not experienced a bubble.

One cannot answer it without studying the actual offering in detail; see, e.g., Blandin et al. (2019); Walch (2015).

A typical ICO half-life is short; fewer than half of all ICOs survive longer than four months. For example, almost half of ICOs sold in 2017 failed by February 2018. Despite this abysmal record of failure, they remained popular longer than they should have been — in 2017, ICOs raised US$7 billion, while in 2018, they brought in US$12 billion, of which US$4.2 billion went toward EOS protocol, and US$1.2 billion to Telegram, the latter project eventually failed. In fairness, in 2019, the ICO market was down 97% compared with 2018 and is small at present.

6.9.4 Stablecoins

Ethereum-based stablecoins are typically ERC-20 or ERC-223 tokens pegged to an asset or a basket of assets, viewed as stable in a conventional economic sense. The Ethereum protocol is often regarded as a natural medium for stablecoins, while the Bitcoin protocol is considered ill-suited for anything other than recording BTC transactions. However, the biggest by far and, arguably, the most controversial of stablecoins is Tether, a Bitcoin Omni Layer token.

Given their paramount importance for the future real-life applications of DLT, we dedicate Chapter 8 to their detailed analysis. Here we mention that there are six types of stablecoins:

(1) fully collateralized by individual fiat currencies;
(2) fully collateralized by baskets of fiat currencies;
(3) partially collateralized by fiat;
(4) overcollateralized with ETH;
(5) algorithmically stabilized;
(6) asset-backed.

Unfortunately, the structure of the Ethereum protocol is such that none of these stablecoins is genuinely decentralized. The reason is easy to comprehend — there are no direct connections between the Ethereum ecosystem and the real economy, except that miners have to pay for their hardware and electricity in fiat. Thus, merely asking the question: "What is the fiat price of a token?" automatically destroys decentralization. Conceptually, this is similar to the uncertainty principle in quantum mechanics, which states that one cannot assign exact simultaneous values to a physical system's position and its momentum.

Even without the arguments mentioned above, fiat or asset collateralized stablecoins are centralized simply because the corresponding collateral must be held centrally outside of the Ethereum blockchain. For ETH-backed and algorithmically stabilized coins, centralization comes from the fact that an oracle is needed to know ETH's market price determined on the exchange.

6.9.5 Automated market makers

Let us design an SC, also called an automated market maker (AMM), capable of making markets between two tokens TN_1, TN_2. At present, AMMs are an essential integral part of the burgeoning decentralized finance (DeFi) ecosystem. They allow anyone to become a market maker by delivering TN_1 and TN_2 simultaneously in the right proportion to the collateral pool. Subsequently, anyone can remove one of the tokens from the pool by simultaneously posting the other token to the pool.

The predetermined rules defining the actual exchange rate can vary. This section considers three possible choices: the constant sum, the constant product, and a mixture of the two. Several papers cover this material; see, e.g., Angeris et al. (2019); Egorov (2019); Schär (2020); Zhang et al. (2018).

Assume that initial prices of these tokens are equal to each other, and consider an automated market maker defined by the following constant sum rule:

$$X + Y = \Sigma_0,$$

$$X_0 = Y_0 = N, \quad \Sigma_0 = 2N. \tag{6.2}$$

Here X, Y are the numbers of TN_1, TN_2, respectively. Eq. (6.2) yields:

$$Y = \Sigma_0 - X, \quad \left|\frac{dY}{dX}\right| = 1. \tag{6.3}$$

The first equation shows that the pool becomes exhausted when $X = \Sigma_0$. Increasing X from N to $2N$ is a rational thing for an arbitrageur to do when TN_2 becomes more expensive than TN_1. At the same time, the marginal price of TN_2 expressed in terms of TN_1, given by Eq. (6.3), is constant and equal to 1.

Such a market maker is acceptable when TN_1, TN_2 represent stablecoins, so that their prices weakly fluctuate around their equilibrium values. In the absence of transaction fees, it is still rational to exhaust the pool even if the deviation from equilibrium is minuscule. However, in the more realistic situation when transaction fees are nonzero, the corresponding deviation has to be above a threshold.

Now we consider a more interesting (and practically important) constant product rule:

$$XY = \Pi_0,$$

$$X_0 = Y_0 = N, \quad \Pi_0 = N^2. \tag{6.4}$$

It is clear that

$$Y = \frac{\Pi_0}{X},$$

$$\left|\frac{dY}{dX}\right| = \frac{\Pi_0}{X^2}. \tag{6.5}$$

Thus, regardless of the arbitrageur's actions, the pool will never be exhausted and shall be able to exist indefinitely. The price of TN_2 expressed in terms of TN_1 is no longer constant, of course. It increases (decreases) when X decreases (increases).

It is natural to look for a rule more general than the constant sum and constant product rule. The constant sum and constant product rules, given by Eqs (6.2) and (6.4), can be written as follows:

$$\left(\frac{\Sigma}{\Sigma_0} - 1\right) = 0,$$

$$\left(\frac{\Pi_0}{\Pi} - 1\right) = 0,$$

(6.6)

where $\Sigma = X + Y$, $\Pi = XY$ are the current sum and product, respectively. We can combine these rules as follows:

$$\left(\frac{\Pi_0}{\Pi} - 1\right) - \alpha\left(\frac{\Sigma}{\Sigma_0} - 1\right) = 0,$$

(6.7)

where $\alpha > 0$ is an adaptive parameter that characterizes the transition from a constant product to a constant sum rule. We put Π in the denominator to avoid the possibility of exhausting the entire pool, by making sure that:

$$Y(X) \underset{X \to 0}{\to} \infty, \quad X(Y) \underset{Y \to 0}{\to} \infty.$$

(6.8)

The constant sum, constant product, and mixed rule curves, as well as relative prices of TN_2, expressed in terms of TN_1, and impermanent losses are shown in Figure 6.12.

This figure shows that all of the above rules have the so-called negative convexity, which means that a market maker experiences a loss whenever the tokens' relative price deviates from its equilibrium value. The hope is that this price is mean-reverting and will go back to the equilibrium value sooner or later. That's why the loss is called impermanent. Obviously, if the price never reverts to the mean, then this loss becomes permanent. The market maker is compensated for her losses by collecting bid-ask spread. Further discussion of this important topic is given in Chapter 10.

6.9.6 Decentralized autonomous organizations

Given the nature of Ethereum as a CaaS provider, it is natural to explore some nontrivial possibilities for using this outside consensus for building new business models. One such case is to create a decentralized autonomous organization (DAO), occasionally called a decentralized autonomous corporation (DAC). A DAO is governed by rules encoded in smart contracts, controlled by its members engaged in peer-to-peer collaboration, and does not have a centralized government. All corporate activities such as financial transactions are recorded on the Ethereum blockchain, or, for that matter, any other blockchain which maintains the smart contract framework. This way, the costs of preserving consensus are offset by the fact that a DAO does not have central management.

The idea of a DAO is audacious. Not surprisingly, there are numerous issues one needs to address before it reaches the mainstream. The main one is conceptual — real-life organizations or corporations are way too complex to be described by a simple set of predetermined rules. This statement is evident because these rules are encoded in smart contracts, which are notoriously fickle to program and not powerful enough, given the gas limits per block and other issues. The very idea of a DAO reminds one of "... an extreme reductionistic, mechanistic approach to biology that champions decomposition of biological systems

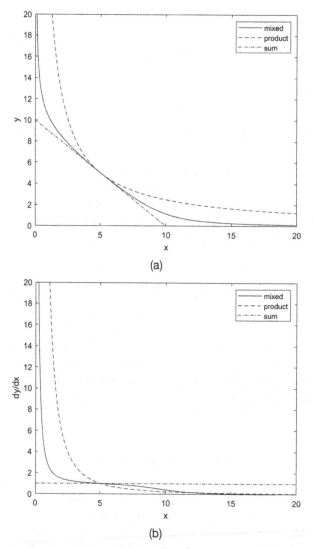

Figure 6.12 (a) Y as a function of X for three different types of AMMs: the constant sum, product, and mixed rule; (b) Relative price of TN_2 expressed in terms of TN_1 equal to $|dY/dX|$ for three different types of AMMs. Own graphics.

into their molecular constituents and emphasis on such constituents in explanations of biological phenomenon"; see Bechtel (2007). It is clear to anyone who spends more than a month or two working for a corporation that its decomposition in a series of simple operations described by a smart contract is practically impossible.

On a more practical note, the legal status of a DAO is unclear at best. As is often the case, ensuring democratic participation in the decision-making process is hard since DAO's members can lose interest in its operations. In fairness, conventional corporations suffer from shareholder fatigue as well, so for one shareholder activist, there are thousands of passive shareholders routinely voting with the management. But the most vexing issue is to ensure DAO's security via reliable smart contracts.

The DAO, already mentioned in Chapter 1, is the best-known example of a (failed) DAO. The DAO was launched with grand fanfare in June 2016, with a US$150 million war chest raised via crowdfunding. The purpose was to run a venture capital (VC) firm democratically, without venture capitalists. Unknown attackers immediately hacked the corresponding smart contract and transferred US$50 million worth of Ethereum to themselves. Vitalik Buterin and his entourage decided that they cannot tolerate this hack. They initiated an Ethereum hard fork, which restored the Ethereum blockchain to its state before the hack.[8]

Additional details are given in Barinov et al. (2019); Hsieh et al. (2018); Jentzsch (2016).

6.10 Summary

This chapter covers some of the most relevant aspects of the Ethereum protocol. In many respects, it is an extension of the seminal Bitcoin protocol discussed in Chapters 3 and 5, augmented with some additional striking capabilities. Because of that, Ethereum is more potent than its predecessor, but, at the same time, due to its overarching ambitions, it has many more weaknesses.

On the pro side, Ethereum can do what Bitcoin does best, namely, moving the native token — ETH — from one address to the next. Moreover, because Ethereum has a transaction counting nonce concept, there is no need to spend the entire amount stored at a particular address every time the owner wants to pay in ETH. Besides, as was mentioned earlier, one can view Ethereum as a CaaS provider. The actual consensus provision relies on smart contracts. Thus, Ethereum is well-suited for building tokens on top of its underlying blockchain.

On the con side, Ethereum is frequently expensive to use and, on occasion, unreliable. Given that smart contracts' verification requires a lot of effort from the miners, Ethereum must rely on an elaborate framework allowing users to pay for miners' services. Ethereum users pay to miners for each operation they perform in gas; the price of gas continually fluctuates and reflects how busy the network is at any given moment. This fee structure is Ethereum's Achilles' heel — it is simultaneously rigid and, very often, prohibitively expensive. The network capacity is relatively small, so the Ethereum blockchain is frequently slow and unreliable. The idea of the Ethereum blockchain being a Turing complete world computer is not valid in practice. Not surprisingly, many developers of practical real-life applications move their activities to the Polkadot protocol and some others, which are more

---

[8]Two conclusions, related to this sad saga, jump to mind: (a) strictly speaking, it is not possible to hack a smart contract because the code-is-law principle; (b) the Ethereum blockchain is not an immutable source of truth and can be manipulated as necessary.

powerful and easier to use. In a recent interview with Camila Russo from the Defiant Podcast, FTX exchange co-founder and CEO Sam Bankman-Fried put it succinctly: "At some point, it's just clear that we weren't getting around that. Either we're going to build it on Ethereum or we are going to build something that we thought was going to be really exciting, but not both."

In summary, Ethereum is a bold attempt at improving Bitcoin and bringing it to the next level. However, in its present form, it is rather difficult and expensive to use. Time will tell if switching to the proof-of-stake consensus can be a game-changer.

Additional details can be found in Antonopoulos and Wood (2018); Bistarelli et al. (2019); Buterin (2013); Wood (2015).

6.11 Exercises

1. An ERC-20 smart contract is a fungible token contract specification. Search for colored coin protocols on the Bitcoin ledger and compare them to the ERC-20 specification. Explain how an ERC-20 transfer is implemented within an Ethereum transaction and provide a detailed description of how the transaction's data field is formed. Describe how another smart contract may trigger an ERC-20 token transfer. Deploy your own ERC-20 token and execute a transfer programmatically.

2. Study the application of smart contracts to non-fungible tokens. Review the ERC-721 and ERC-1155 and explain how they relate to and differ from ERC-20 tokens.

3. Consider an application of smart contracts relying on an external state, such as the value of an off-chain asset, the weather in a particular location, and the result of soccer competition. Explain why it is nontrivial to define a smart contract relying on an external state without creating a single-point-of-failure. Search for the ChainLink protocol and explain how they propose to solve the problem of a decentralized oracle.

4. In the context of decentralized finance, peer-to-peer exchanges are key. Search for existing decentralized exchange (DEX) protocols and explain their differences. In particular, study the Uniswap protocol. Describe in detail the advantages and disadvantages of the Automated Market Maker (AMM) model and how it avoids the dependency on an oracle.

7 | Ripple — A Simple Solution to a Complex Problem?

7.1 Introduction

This chapter describes the Ripple protocol — the last one of the big three (Bitcoin, Ethereum, Ripple) — its pros, cons, and actual prospects. Many authors, including the present ones, consider Ripple as a non-crypto crypto since it relies on preminted XRP tokens, has no blocks, and does not use proof-of-work (PoW). Instead, each participant in the Ripple network chooses a set of validators from a Unique Nodes List (UNL), whom she trusts and expects to behave honestly. She only follows consensus decisions reached by these validators. Typically, a list of validators is provided to the new participants upon joining the platform. The protocol uses some cryptographic primitives such as digital signature algorithm (DSA) and is public, but other than that is relatively pedestrian in its implementation. Given this simplicity, we can afford to be brief.

The chapter is organized as follows. We discuss the main features of the Ripple protocol including commonalities and differences between Ripple and Bitcoin in Section 7.2. Potential applications of Ripple, including their pros and cons are presented in Section 7.3. Conclusions are drawn in Section 7.4.

7.2 Ripple protocol

Ripple is a money transfer protocol, while ripple, or XRP, is the underlying native currency. Ripple is entirely different from Bitcoin and Ethereum. Compared to the other two of the Big Three, it is a much more modest edifice by every measure. For starters, XRPs are pre-mined, so Ripple is not decentralized by its very nature, although, as we have seen before, for very different reasons, neither are Bitcoin and Ethereum. However, the degree of decentralization does matter! An objective observer concludes that Bitcoin and Ethereum are trying hard to stay decentralized, while Ripple is not even trying.

We base our exposition on various Ripple materials; see Chase and MacBrough (2018); Moreno-Sanchez et al. (2018); Schwartz et al. (2014).

XRP price dynamics and Ripple Google trends are presented in Figure 7.1, which shows that XRP is exceptionally volatile and, at some point, was grossly overvalued. The interest in Ripple perfectly correlates with the market price of XRP.

By itself, it is not particularly surprising since XRP has no intrinsic value and is pre-mined. Thus the monetary policy lies entirely in the hands of the Ripple Foundation, who can, and often does, release XRP at will.

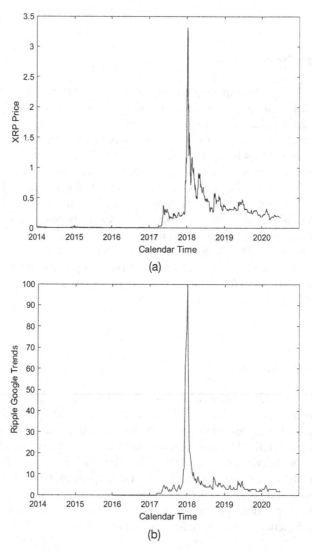

(a)

(b)

Figure 7.1 (a) The price of XRP; (b) The Google trend for Ripple. Own graphics. Sources: (a) coinmarketcap.com, (b) Google.

The Ripple protocol is designed to maintain a consistent ledger by using validating nodes. The ledger evolves in a way, which is capable of accommodating new transactions consistently. The ledger's evolution is shown in Figure 7.2.

Figure 7.3 presents the Ripple ecosystem, which consists of validating nodes, tracking nodes, and clients.

The main components of the Ripple ecosystem are validating servers, maintaining the consensus, tracking servers, recording transactions, and clients, who use the system to initiate new transactions. Any server can assemble recent transactions' batch and propose ledger updates, which validators must agree upon. Any party, including banks, market makers, payment service providers, and universities, can own a server. Validators check account balances instantly so that irreversible payments, which do not allow chargebacks, are executed within a few seconds. The objective is to maintain a coherent and conflict-free ledger, which does not permit double-spending attacks. We distinguish between the last-closed ledger, which is unique, and several open ledgers, competing to become the next closed ledger. Each server maintains a UNL, which consists of nodes it trusts. The list is received upon joining the Ripple ecosystem and expanded as part of doing business. Every ledger has two identifiers — the ledger index, which numbers ledgers incrementally, and the ledger hashes, which serve as digital fingerprints of the ledger's contents.

Ripple transactions and their effects are similar to transactions in the Bitcoin and Ethereum protocols. A single transaction's lifecycle starts with it being created and signed by the account owner, who then broadcasts it to the network. If badly formed, the transaction fails immediately. However, for a correctly formed transaction, there are two paths: it may provisionally succeed but later fail. Alternatively, it may provisionally fail but then succeed.

The ledger changes are due to user-level transactions, including payments, changes to account settings or trust lines, and offers to trade. Every transaction cryptographically signed by an account owner authorizes one or more changes to the ledger; there are no other mechanisms to introduce changes in the ledger. Every transaction included in a ledger destroys a certain (small) amount of XRP. This amount represents a transaction cost and has a deflationary impact on the Ripple ecosystem. There is a simple but essential difference between candidate transactions proposed for inclusion in the ledger and validated transactions included in the ledger. Only the latter transactions are immutable.

Initially, successful transactions can subsequently fail, or vice versa, depending on the order of transactions in the round. This is a familiar problem touched upon in Chapters 2 and 5. Depending on the order of transactions arrival, the account balance at a particular address might or might not be sufficient to process them.

The consensus is achieved by voting of validators. It is an example of a Byzantine Fault-Tolerant (BFT) consensus, considered in further detail in Chapter 11. During the consensus stage, new valid transactions are included in the ledger via the following process. Initially, similarly to Bitcoin, validating nodes pick newly broadcasted transactions from the memory pool, order them as they see fit, and propose a new ledger. It is worth noting that no blocks are needed in the Ethereum protocol since it does not use the PoW.

Further, validating and tracking nodes broadcast the proposed ledger, which contains the list of new transactions, to the network. Every consensus round attempts to validate a new ledger and change the last-closed ledger's state consistently — a successful round results

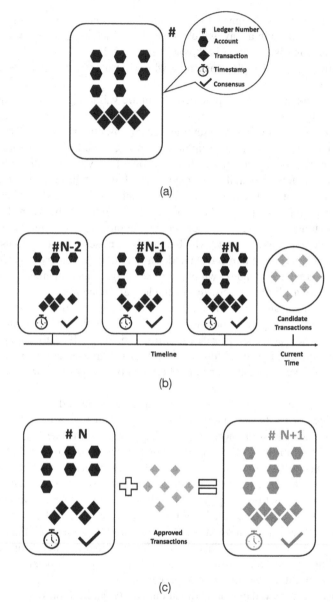

Figure 7.2 The Ripple ledger evolves in time to accommodate new transactions. This evolution is controlled by validators; (a) The validated ledger at a particular moment in time; (b) New transactions are proposed; (c) The ledger absorbs legitimate transactions and transitions to the next state. Own graphics. Source: Ripple.

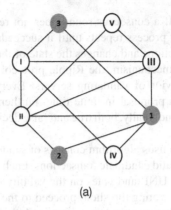

(a)

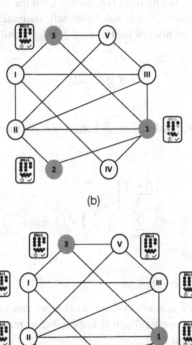

(b)

(c)

Figure 7.3 (a) Ripple ecosystem consists of validating nodes represented by Arabic numerals 1, 2, 3, and tracking nodes represented by Roman numerals I–V; (b) Validating nodes vote for the proposed states of the ledger; (c) Once consensus is reached, both validating and tracking nodes record the new state of the ledger. Own graphics. Source: Ripple.

in a newly accepted ledger. If a consensus round does not reach the minimum threshold of votes and fails, the voting process repeats until it succeeds. The final validated ledger includes the approved transactions and changes the state of the ledger appropriately.

As any other consensus mechanism, the Ripple protocol is based on three plausible assumptions about the behavior of validating servers: Every non-faulty server decides whether to accept or reject a proposal in finite time. Furthermore, all non-faulty servers arrive at the same decision. And finally, both true and false decision regarding a given transaction is possible.

The Ripple protocol consensus algorithm consists of sequential rounds: Every proposing server compiles a list of valid candidate transactions. Each server includes all candidate transactions coming from its UNL and votes on the validity of these transactions. Transactions passing the minimum voting threshold proceed to the next round. The final round requires an 80% agreement. Voting rounds continue until the 80% threshold is reached.

It is straightforward to show that N validators will maintain correctness of the Ripple protocol, provided that the number of faulty ones, f, is bounded above as follows:

$$f \leq \frac{N-1}{5}. \tag{7.1}$$

Alternatively, if the probability of malicious behavior is p_c, then the probability of correctness p^* has the form:

$$p^* = \sum_{i=0}^{\left[\frac{N-1}{5}\right]} \binom{N}{i} p_c^i (1 - p_c)^{N-i}, \tag{7.2}$$

where $\binom{N}{i}$ is the binomial coefficient.

As with many other BFT consensus frameworks, we cannot guarantee that consensus will eventually be reached. Specifically, it is impossible to reach consensus if servers split into cliques, as shown in Figure 7.4.

Commonalities and differences between Ripple and Bitcoin are summarized in Table 7.1. Below we provide further explanations.

7.3 Applications

Ripple's stated objective is to make the movement of money seamless, emphasizing the cross-border activities. As such, the main competition for Ripple comes from the traditional correspondent banking model and its ultimate foundation layer — SWIFT. One can think of Ripple as a modernized and computerized version of Hawala described in Section 2.5.3. Ripple tries to find chains of banks willing to extend credit lines to each other, exceeding

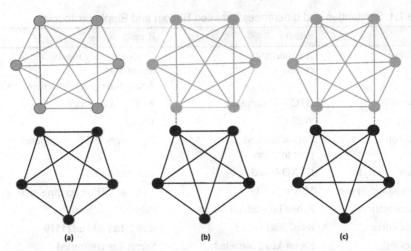

Figure 7.4 (a, b) If validators are split into cliques, consensus becomes impossible. (c) When the connectivity of validating nodes is sufficiently robust, consensus can always be reached. Own graphics. Source: Ripple.

the amount of money one wishes to move across the system. As such, Ripple payments are not payments in the conventional sense of the word. After Alice sends her money to Bob, there is a chain of reverse transactions, which need to be closed at some point — the same issue that afflicts Hawala protocol; see Section 2.5.3.

In general, having multiple payment routes is an advantage. However, even with a plethora of roads, some are better to avoid. The reader will do well to recall the following conversation between Napoleon and the Russian general Balashev from *War and Peace* by Leo Tolstoy: "Napoleon ... naïvely asked Balashev through what towns the direct road from there to Moscow passed. Balashev, who was on the alert all through the dinner, replied that just as 'all roads lead to Rome', so all roads lead to Moscow: there were many roads, and 'among them the road through Poltava, which Charles XII chose'." [1]

Under normal circumstances, credit extended by one bank to another is not an issue. However, in the case of default, it might quickly become a severe problem indeed. For the uninitiated, who think that this is just a hypothetical possibility, I have two words — the Herstatt risk; see Galati (2002). This risk is essentially a cross-currency settlement risk due to asynchronous money flowing in and out of the system. For instance, a bank can receive money as a part of the transaction and default before honoring its money transfer side. This situation happened on June 26, 1974, when the troubled German bank Herstatt received payments in Deutsche Marks from several counterparties and was supposed to initiate USD-denominated reciprocal payments. However, Herstatt ceased its operations before sending payments due to time zone differences between Cologne and New York. As a result, the counterparty banks did not get their dollars. Although this type of risk is

---

[1] The general was alluding to the decisive victory of Peter the Great over Charles XII of Sweden in the battle of Poltava on July 8, 1709. According to Tolstoy, the irony of this answer escaped Napoleon.

Table 7.1 Similarities and differences between Bitcoin and Ripple protocols.

	Bitcoin	Ripple
Objective	An alternative currency	Cross-currency messaging technology A real-time gross settlement system
Native unit	BTC — a currency	XRP — a currency
Storage	Wallet	Wallet
Traded on	Exchanges and peer-to-peer	Exchanges and peer-to-peer
Transaction style	UTXO-based	Account-based
Transaction purpose	Moving BTC	Moving XRP or sending message
Implementation	Public blockchain	Public blockchain
Elliptic curve	secp256k1	secp256k1 and ed25519
Addresses	Secret-key controlled	Secret-key controlled
Ownership proof	ECDSA	ECDSA and EdDSA
Consensus algorithm	Proof-of-Work	Low-latency Byzantine agreement protocol
Hash function	$SHA256$	N/A
Consensus keepers	Miners	Validators
Transaction fees	Voluntary	Mandatory and voluntary
Block size	2,000 TX	N/A
Ledger update time	approx. 10 min	approx. 3–5 sec
Supply mechanism	Mining	Pre-mined supply and mining
Mining reward	6.25 BTC	N/A
Supply cap	21 million in total	100 billion in total Deflationary in theory, inflationary in practice
Supply style	Deflationary	Inflationary in practice
Launch date	January 2009	July 2012
Inventors	S. Nakamoto	J. McCaleb, A. Britto, D. Schwartz, and R. Fugger

significantly reduced due to the Continuous Linked Settlement's prevalence for fiat transactions, the Ripple system functions outside of it and is still vulnerable.

The way value moves in the Ripple ecosystem is shown in Figure 7.5.

It is worth noting that if the chain of banks willing to deal with each other cannot be found, or, more likely, is uneconomical, Alice can choose a different route. She can convert her dollars into XRP on an exchange (exposing herself to the credit risk of the exchange), send Bob XRPs, and let him deal with his exchange to convert them into Euros (the credit risk again).

The Ripple way of moving value should be compared and contrasted with the conventional correspondent banking framework discussed in Section 2.5.1; see especially

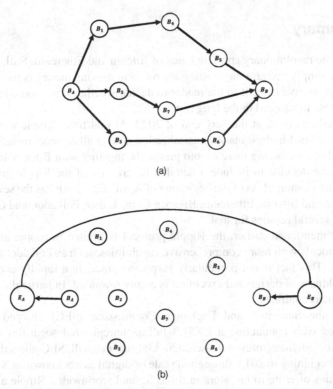

(a)

(b)

Figure 7.5 Value can be moved across the Ripple network from Alice to Bob in two ways. (a) Several chains of mutually trusting banks can be used, such as $B_A- > B_1- > B_4- > B_5- > B_B$, $B_A- > B_2- > B_7- > B_B$, and $B_A- > B_3- > B_6- > B_B$; (b) If such chains cannot be found, or are too expensive to use, the first fiat currency is converted into XRP on E_1, XPR is moved across the network to E_2, where it is converted into the second fiat currency. Own graphics.

Figures 2.15 and 2.16. The sheer costs of establishing a correspondent banking relationship prevent smaller banks from having many partners. As a result, their choices in moving value are both limited and expensive. Ripple, if it were able to reduce the Herstatt risk, would be a welcome addition to, or, even, a replacement of the existing payment rails.

Ripple envisions several possible participants in RippleNet. It wants to attract banks looking to streamline payments for their corporate and retail customers and banks interested in processing payments for and providing liquidity to other banks. The other important segment is formed by payment service providers wishing to improve liquidity provision, platform businesses sending high volume, low-value payments to a global base of suppliers, and, finally, organizations moving money across the global supply chain.

While making some inroads in these market segments, transaction volume on the Ripple network pales compared to the existing banking network volume. It is likely to stay this way for a long time, predominantly because the Ripple protocol cannot solve the all-important Herstatt risk problem.

7.4 Summary

Ripple lacks the revolutionary spirit and zeal of Bitcoin and Ethereum. Still, it might gain some traction simply because the existing approach to moving money is too frustrating to many participants. We leave it to the reader to decide whether it is good enough a *raison d'être* for Ripple to succeed in the long run.

We finished this book at the very end of 2020. At that time, Ripple was doing well, commanded the third-highest market capitalization among all cryptocurrencies, and could be considered one of the big three crypto protocols, together with Bitcoin and Ethereum. Accordingly, we decided to include a detailed description of the Ripple protocol in the book. Times are changing! As of the beginning of April 2021, XRP has the seventh-highest market cap, behind Bitcoin, Ethereum, Binance Coin, Tether, Polkadot, and Cardano.

There are several reasons for that.

Firstly, as mentioned earlier, the Ripple protocol is limited in scope, allowing more advanced protocols with more comprehensive capabilities, such as Polkadot and Cardano, to overtake it. This fact is not particularly surprising since, in a rapidly evolving crypto world, the boldness of design and execution is amply rewarded. In particular, Ripple does not support smart contracts.

Secondly, the Securities and Exchange Commission (SEC) charged Ripple and two executives with conducting a US$1.3 billion unregistered securities offering; see https://www.sec.gov/news/press-release/2020-338. In a nutshell, SEC alleged "that Ripple raised funds, beginning in 2013, through the sale of digital assets known as XRP in unregistered securities offering to investors in the U.S. and worldwide. Ripple also allegedly distributed billions of XRP in exchange for non-cash consideration, such as labor and market-making services."

Time will tell if Ripple could defend itself against SEC allegations, put its act together and regain its place among the big three, for instance, by positioning itself as a "neutral bridge currency" facilitating exchanges between digitized fiat currencies.

7.5 Exercises

1. Recently, Ripple published a white paper, https://ripple.com/wp-content/uploads/2021/01/cbdc-whitepaper-2020.pdf, proposing to use XRP as a "neutral bridge currency" facilitating exchanges between digitized fiat currencies. Describe the pros and cons of their proposal.

Central Bank Digital Currencies and Stablecoins

8.1 Introduction

In many countries, including the US, cash and debit cards, which combine both digital and physical aspects of cash, play an important role in the economy. The Fed 2015 Diary of Consumer Payment Choice by the FRBSF (2015), observed: "Cash continues to be the most frequently used consumer payment instrument. Cash is widely used in a variety of circumstances. Cash dominates small-value transactions. The average value of cash holdings has grown."

Similarly, Yves Mersch, Member of the Executive Board of the European Central Bank (ECB), in a speech given on February 14, 2018, noted: "... printed euro banknotes will retain their place and their role in society as legal tender for a very long time to come. There is no viable alternative to euro cash. There is good reason to believe that banknotes do not only have to take the form of printed paper, cotton, or polymer. However, printed banknotes will remain our core business. And if there is public demand for digital central bank money, this should only be a technical variant of cash" (ECB, 2018).

Hence, building a digital analog of cash addresses well-defined societal needs. The often-repeated statement that credit cards and other digital forms of payment already play the role of digital cash is misguided since the nature of such payments is very different from a simple act of cash payment, which is anonymous and extinguishing all obligations.

On the one hand, we observe a heated arms race between different central banks researching, experimenting, or building central bank digital currencies (CBDCs). On the other hand, numerous enterprises, from huge conglomerates to small private companies, develop their own stablecoins, which are variations on the same theme. In this chapter, we discuss some of these efforts and put them in a proper context.

One can envision two types of CBDCs: (a) wholesale CBDCs designed to circulate only among selected financial institutions; (b) retail CBDCs intended for everyday usage by the whole population, much like physical cash at present. Wholesale CBDCs are very technical — they can dramatically change to the better domestic and cross-border payments discussed in Sections 2.4 and 2.5, but are not going to affect the way ordinary people live their lives. At the same time, retail CBDCs, if correctly implemented and deployed, can have a profound impact on society as a whole.

The reason behind CBDCs' and stablecoins' importance is easy to discern — they might be the only way of bringing advantages of distributed ledger technology (DLT) to the real economy. Besides, there is a geopolitical aspect of these efforts. The government, which manages to digitize its currency in the best possible way, has a shot on making it a reserve currency of the future.

In principle, it might become possible to open a checking account at the central bank directly or indirectly, with retail banks becoming obsolete. In practice, it might be easier to open an account at a Narrow Bank (NB); see the discussion below.

On the one hand, the absence of physical cash would alleviate societal ills, such as crime, drug trafficking, and the like, or, at least, make them more challenging to conduct. It would facilitate commerce and help the unbanked become participants in the digital economy through smartphones, thus improving society. On the other hand, if cash is relegated to the dustbin of history, central bankers will have full control over interest rates and could set them as negative as they like.

Suppose central banks want to impose meaningfully negative rates (in Switzerland, it is in the tune of -0.75% at present). In that case, they have to issue digital cash, or, at least, abolish large-denomination notes. Otherwise, as is mentioned in Chapter 2, physical lower bound (PLB) would prevent them from achieving their goals. In theory, negative interest rates are a useful tool if one wants to fight the enormous pile of excess cash reserves kept by commercial banks at central banks; see Figure 2.7. However, the potency of this tool is hard to assess at the moment. If one wishes to fight anemic economic growth, she needs to stimulate demand rather than chase cash holdings, which will dwindle on their own. It is worth mentioning that attempts of central banks in the eurozone, Japan, and Switzerland to conduct the negative interest rate policy didn't significantly increase economic activity in their own countries.

This chapter is organized as follows. In Section 8.2, we discuss an essential concept of an NB; assets of such a bank are central bank cash and short-term government obligations. Hence, such a bank is immune to market forces and cannot default, except for operational reasons. Section 8.3 deals with CBDC. We consider two possible approaches to issuing CBDC: either emittance by the central bank itself or affiliated NBs. We argue that narrow banks are central to any realistic scheme for issuing CBDC. In Section 8.4, we introduce stablecoins, which are private versions CBDCs. We distinguish four categories of stablecoins: (a) stablecoins fully collateralized with fiat; (b) stablecoins partially collateralized with fiat; (c) stablecoins overcollateralized with the underlying cryptocurrencies; (d) dynamically stabilized coins. We argue that only fully collateralized stablecoins are worthy of their name. In Section 8.5, we discuss digital trade coins and articulate their potential as modern versions of the Spanish reals and Austrian thalers described in Chapter 2. We draw our conclusions in Section 8.6.

8.2 Narrow banks

8.2.1 A brief history

We start with a brief discussion of narrow banking history, as given by Lipton et al. (2018a).

The transition from the High Middle Ages to the Renaissance and the early modern period was primarily due to technical and financial innovations, especially fractional reserve banking. However, from the very beginning, fractional reserve banking firms were prone to collapse. For instance, in Florence, the great banking houses, such as Bardi, Peruzzi, and Medici (to mention but a few), all failed. The natural reaction to this state of affairs resulted in the emergence of the narrow banking idea, pursued with vigor and enthusiasm by visionaries, financial reformers, and, from time to time, authorities for hundreds of years; see, e.g., Pennacchi (2012); Dittmer (2015); Roberds and Velde (2014); http://narrowbanking.org/, and references therein.

On several occasions, the actual transformation of fractional-reserve banks into NB has been tried. In 1361, Venice's Senate prohibited lending out depositors' money, effectively making Venetian banks narrow. However, banks systematically circumvented this prohibition, with failures following in due course. A case in point is the Pisano & Tiepolo bank, which failed in 1584, underwent conversion into a state bank, and defaulted again in 1619. The Bank of Amsterdam, chartered as an NB in 1609, started to lend its reserves in secret, failed in 1791, and was taken over by the city; see, e.g., Frost et al. (2020).

In the 19th century, banks, acting in self-interest, reduced their lending activities, thus becoming much more narrow than they were in the early modern period, or are today. Commercial banks followed the real bills doctrine and lent predominantly short-term. Their loans, with maturities of two to three months, provided short-term working capital and trade credit. These loans were collateralized by the borrower's wealth or the goods in transit; see Bodenhorn (2000); Pennacchi (2012).

The situation changed in the 20th century when, encouraged by the Federal Reserve Bank's creation in 1913, commercial banks abandoned the real bills doctrine. To expand their business, banks started to lend for much longer maturities, established revolving lines of credit for some of their borrowers, thus overemphasizing their maturity transformation ability, sometimes at the expense of prudence. The consequences became apparent during the Great Depression of 1929, with numerous catastrophic bank failures, thus causing the idea of an NB to come to the fore yet again.

In the U.K., the Nobel Prize winner Frederick Soddy was an ardent partisan and popularizer of this idea; see Soddy (1933). In the U.S., several influential Chicago economists came with the so-called Chicago Plan for the abolition of fractional reserve banks; see Knight et al. (1933); Fisher (1935); Hart (1935); Douglas et al. (1939); Phillips (1996). The core tenets of the plan can be summarized as follows:

(1) The government should own Federal Reserve Banks outright.
(2) The government should suspend the gold standard.
(3) The assets of all member banks should be liquidated, and all existing banks dissolved.
(4) New NBs accepting only demand deposits subject to 100% reserve requirement in cash and deposits with the Fed should be created.
(5) Investment trusts handling saving deposits should be created.

These proposals are clearly not lacking the revolutionary spirit. Although, a practical conversion of fractional-reserve banks into NBs never happened due to enormous political pressure from the former, the idea was never forgotten. It gained a new lease of life during and after the Savings and Loan (S&L) crisis in the 1980s and 1990s; see, e.g., Friedman (1960); Tobin (1985); Gorton and Pennacchi (1993). The NB idea came to the fore again during and after the Global Financial Crisis (GFC); see, e.g., Kay (2010); Kotlikoff (2010); Phillips and Roselli (2011); Beneš and Kumhof (2012); Pennacchi (2012); Admati and Hellwig (2014); Cochrane (2014); Dittmer (2015); Lainà (2018); Lipton et al. (2018a).

The recent technological advances, including the development of blockchains, clearly help develop the NB project further; see Lipton et. al (2016); Dembo et al. (2018); Lipton et al. (2018a).

8.2.2 Possible designs of a narrow bank

An NB design is such that it cannot default except due to operational failures, which can never be entirely eliminated but only minimized by using state-of-the-art technology, thus providing the safest payment system. By construction, the NB's asset mix includes solely marketable low-risk securities and central bank cash in the amount exceeding its deposit base. An NB does not need deposit insurance, thus eliminating its perverse effects on the system as a whole, not least the associated moral hazards. Deposits in an NB are as close to actual currency as possible.

Depending on specific details, NBs keep all depositors' money as liquid cash deposited with a central bank or liquid government bonds, becoming thereby necessary stabilizers and facilitators of the overall global financial system. There is an acute need for such stabilizers after the GFC, which resulted in a massive concentration of assets in a handful of systemically important banks and, in the process, made the overall financial system even less robust and stable than it was before the crisis. In addition to steadying the whole banking sector, NBs naturally perform the record-keeping function in the economy while refraining from issuing loans.

Since a narrow bank is as good as its assets, one needs to pay particular attention to the choice of these assets. There are several approaches for designing an NB; see, e.g., Pennacchi (2012); Dembo et al. (2018); Lipton et al. (2018a):

(1) 100% Reserve Bank: Assets — central bank reserves and currency. Liabilities — demandable deposits and shareholder equity. Depending on the circumstances, these deposits can either be non-interest-bearing, interest-paying, or interest-charging. The latter setup might be necessary if the interest rate paid by the central bank is negative. The bank finances itself by a combination of deposits (debt) and shareholders' equity.

(2) Treasury money market mutual fund: Assets — Treasury bills or repurchase agreements collateralized by Treasury bills. Liabilities — demandable equity shares having a proportional claim on the assets. The bank is financed solely by equity.

(3) Prime money market mutual fund: Assets — short-term Federal agency securities, short-term bank certificates of deposits, bankers' acceptances, highly rated commercial paper, and repurchase agreements backed by low-risk collateral. Liabilities — demandable equity shares having a proportional claim on the assets. As before, the bank is financed solely by equity.

In Figure 8.1, we show the balance sheet of a representative NB.

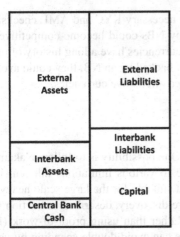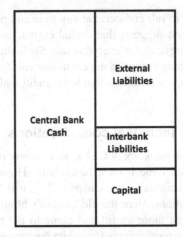

Figure 8.1 Balance sheets of a fractional reserve bank (left) and a narrow bank (right). The difference speaks for itself. Own graphics.

A detailed design for a modern NB, including its international dimensions is given by Dembo et al. (2018).

8.3 CBDC

8.3.1 Central banks as issuers of CBDC

Experience shows that central banks routinely open accounts for licensed banks and selected non-banking financial firms, such as central clearing counterparties. However, they do not allow a broader range of corporate or individual participants (particularly if they wish to be anonymous) to have an account with them. There are several reasons for that. For instance, central banks are too small to solve the Know Your Customer (KYC) and Anti-Money Laundering (AML) problem independently; they don't want to face potential political complications in agreeing or denying opening an account.

8.3.2 Narrow banks as issuers of CBDC

Keeping a one-to-one parity between the fiat currency and digital tokens is possible only if the equal amount of fiat currency is held in escrow; see Section 8.4.2 below. However, the collateral put in a fractional-reserve bank cannot be safe at all times. Since bank depositors are junior unsecured creditors of a bank, their deposits will be in danger in case of default. Even if the recovery of a portion of these deposits is possible, the money will not be available until resolving all the bankruptcy issues, which is a very lengthy process.

Thus, an NB is a natural emitter of CBDC. A properly constructed NB issuing its digital currency offers a much more palatable and practical solution than CBDC by allowing access to full collateral at any time and providing necessary KYC and AML checks. This approach suggests that digital currencies issued by NBs could become competitive with fiat currencies for everyday use. Such alternative currencies have a long history of use for smoothing bumps in the economic environment. If backed by an NB, they could avoid the bankruptcy problems that have traditionally plagued alternative currencies.

8.3.3 Technical considerations

One can think about CBDC in a variety of ways. One possibility is to follow Nakamoto's idea and issue it on a blockchain. However, due to various limitations inherent in the blockchain itself, see Chapters 5–7, it is likely not suitable for the large scale heavy-duty applications. Here the old Chaum's blind signature discovery, described in Section 4.5.6, might be more useful and come to the rescue. Rather than using proof-of-work (PoW) or other mechanisms for achieving consensus, one can avoid double spending by using a single-usage unique number per unit of currency issued by the central bank itself. Recall that paper banknotes have unique numbers already; see Figure 2.3. This number can be verified in a decentralized fashion by a set of validators (or notaries), each knowing only part of the overall number. In this way, validators can use secure multi-party computations to check the numbers' database and address the double spend problem in a Byzantine fault-tolerant way.

Major technical specifications for CBDC have to amalgamate the original Okamoto and Ohta ideas and some additional considerations articulated by the Bank of England; see Okamoto and Ohta (1991); Bank of England (2020). We summarize these requirements in Table 8.1.

Table 8.1 Key technical specifications for CBDC.

Requirement	Summary
Anonymous or pseudonymous	Transactable without revealing parties' identities
Secure	Secure against cyber and physical attacks
Divisible	Subdivisable as needed
Future-proof	Capable of upgrading without impacting operations
High-throughput capable	Capable of 20,000 transactions per second
Innovation-ready	Capable of handling innovative features and services
Instantaneous and final	Near-instantaneous, real-time, with settlement finality
Interoperable	Interoperable with traditional payment systems and other CBDCs
Non-reproducible	Cannot be copied and reused
Usable online	Securely usable online
Usable offline	Securely usable offline
Resilient	99.9999% operational availability 24/7/365
Transferrable	Transferrable to others

8.4 Stablecoins

8.4.1 Background

Since none of the existing blockchain ecosystems have banks worthy of the name, they have to rely on algorithmic monetary policies for cryptocurrency creation. Regardless of their creators' statements, in essence, they resurrect, in one way or the other, old ideas of central planning, without fully recognizing the consequences. More important and counterintuitive is the fact that newly emerging DLT-based crypto ecosystems suffer from the same drawbacks as the current banking system because they continue commingling payments and monetary creation (however primitive and ill-advised).

Bitcoin, Ethereum, and Ripple have tried to play the role of money but have ended up as a new speculative asset class.[1] Despite their impressive technical achievements, existing cryptos cannot replace fiat money for at least four reasons. First, they are not stable enough, even for short transaction times, which prevents their usage for conventional commerce. Recent experience shows that all decentralized crypto coins exhibit extreme volatility and thus are unsuitable for commercial applications; see Figures 5.1, 6.1, and 7.1. Second, crypocurrencies are operationally inconvenient for an average economic agent, which is a prerequisite for their wide adoption as a means of payment. Realistically speaking, for such an agent, the only way to obtain cryptocurriencies is to buy them on a centralized exchange, which is both costly, inconvenient, and adds rather than removes an extra intermediary. Third, as already mentioned earlier, most crypto protocols' built-in monetary policies are simplistic and inflexible. Some observers consider this as an advantage, see Chapter 3. However, historical experience suggests otherwise and shows that having a fixed supply of money results in numerous economic distortions, some of which are hard to anticipate in advance. Fourth, as explained in Chapters 5–7, cryptocurrencies are not truly decentralized while pretending to be so.

Not surprisingly, the status quo in the cryptocurrency land is less than satisfactory. As a result, developers and entrepreneurs directed much attention to designing stable crypto coins, which can play the role of tokenized cash and be viewed as private versions of CBDCs. Stablecoins come in several flavors. Below, we give a brief discussion of the most visible types.

An adequately designed stablecoin can serve as the cornerstone of a fully digital financial ecosystem, facilitating fast payments on a commercial scale with minimal reliance on the existing banking system, yet doing so in a regulatory-compliant fashion. According to Sila (2018) and Hardjono et al. (2019a), such a coin addresses the gaping hole in the open-access Transmission Control Protocol (TCP) and the Internet Protocol (IP), also known as TCP/IP, namely, the lack of specifications for money and identity.

It is easy to see why stablecoins are necessary within the cryptocurrency universe. The vast majority of transactions occur on exchanges not fully integrated with the existing banking system. Accordingly, having a stable token to quickly and cheaply get in and out of

---

[1]An aphorism by Viktor Chernomyrdin (1938–2010), a former Prime Minister of Russia, jumps to mind. "We wanted it as good as possible, but it turned out as always."

cryptocurrencies is necessary. Thus, entrepreneurs created such tokens to link the crypto universe and the existing banking system.

We argue that rather than designing monetary policy independently, which is a challenging task in both practice and theory, efficient applications of DLT should initially aim to tokenize existing stable financial instruments. These instruments could be denominated in fiat currencies, for instance, the US dollars. Later on, if these efforts are successful, one should tokenize more diverse asset pools. Such pools can counterbalance fiat currencies and be very useful for cross-border transfers.

In our estimation, and contrary to numerous unfounded claims, it is not currently possible to build a genuinely decentralized stablecoin. This observation is partly because the mere determination of a crypto coin's price in terms of a fiat currency requires an oracle and automatically breaks decentralization. Such oracles are prone to manipulation and, more importantly, at odds with the coin's claimed decentralized nature. Ideally, one should be able to directly observe this price from the blockchain in a distributed fashion. Unfortunately, at present, it is not quite possible for both conceptual and technical reasons. Hence, any potentially successful stablecoin has to combine centralized and decentralized features. The degree of centralization vs. decentralization can vary, as explained below.

8.4.2 Stablecoins fully collateralized with fiat

Nicolas Copernicus, who, in addition to his mathematical and astronomical studies, consulted the Prussian Diet on monetary policy issues, articulated his requirements for sound money as follows; see Taylor (1955): "1. It must not be changed in value except after ripe deliberation by the government authorities ... 2. One single place must be chosen for the minting of the money, which must be minted in the name of the entire country and not in the name of a single city. ... 3. When the new currency is issued, the old currency must be demonetized and withdrawn from circulation. ... 4. It is essential to have an inviolable and unchangeable rule to mint only 20 marks and no more from a pound of silver, deducting only the quantity of silver necessary to cover the expenses of coinage. ... 5. Too great a quantity of money must not be issued. ... 6. All the different kinds of coins should be issued at the same time. ..."

It is truly astonishing to realize to what extent some of Copernicus's laws are valid today. For instance, the designers of the proposed cryptocurrency Libra miss the third postulate entirely. They do not understand and refuse to accept that Libra can create enormous inflationary pressures, especially in the emerging markets that it is supposed to help in the first place. However, such a cavalier approach to creating new currencies is quite typical for most technologically innovative but financially naive stratagems.

Custodial stablecoins fully collateralized with fiat are relatively centralized. A single party performs the creation and destruction of such coins. However, once a coin is created, it can freely move on the corresponding blockchain before being eventually destroyed. Given this semi-centralized design, custodial coins are particularly prone to regulatory influences, and, therefore, must be regulatory compliant to survive. Several coins of this nature, including Tether (USDT), TrueUSD (TUSD), USD Coin (USDC), and Sila, already exist; several others are currently on the drawing board. We briefly consider a representative example below.

Tether is one of the earliest and best known custodial coins, issued on the Bitcoin blockchain through the Omni Layer Protocol; see Tether (2016). As its name suggests, USDT is supposed to be linked (tethered) to the underlying fiat currency, such as USD, since every tether in circulation is collateralized with a dollar held in a dedicated bank account.[2] While simple in theory, in practice, the situation is exceptionally opaque, mostly because nobody knows if the corresponding collateral exists, where it is held, and who administers it. Given the above, Tether is not regulatory compliant. Besides, as per Tether's legal disclaimer: "Beginning on January 1, 2018, Tether Tokens will no longer be issued to the U.S. Persons." Not surprisingly, all of the above issues resulted in USDT recently breaking the peg to the USD. Tether is geared toward crypto exchanges rather than for regular commerce.

Despite its glaring shortcomings, Tether's imposing feature emanates from its very nature as an essential instrument for speculators. Namely, USDT has an exceptionally high velocity. Although its capitalization is about 60 times smaller than Bitcoin, the dollar value of BTC and USDT traded per day is broadly comparable. For instance, according to coinmarketcap.com, on September 26, 2020, BTC capitalization and daily volume were US$198 billion and US$52.5 billion, respectively, while for USDT, they were US$15.3 billion and US$37.0 billion. Thus, the velocity of USDT is much higher than the velocity of either BTC or USD. Many observers believe that Tether is actively used to manipulate Bitcoin prices; see, e.g., Griffin and Shams (2019); Leising et al. (2018).

The Utility Settlement Coin (USC), developed by a consortium of banks called Fnality and a fintech startup called Clearmatics is a promising candidate for use in the interbank market.[3] Initially, USC can be an internal token for a consortium of participating banks. These coins have to be fully collateralized by these banks' electronic cash balances held by the Central Bank; see Lipton et al. (2018a). Eventually, these coins can circulate among a larger group of participants. However, in this case, the issuance of USCs has to be outsourced to an NB.

By construction, all stablecoins are centralized to a greater or lesser degree, so their issuers represent a single point of failure and an irresistible attraction for both regulators and hackers (for different reasons, obviously). Operationally, for some coins, there is no transparency regarding the underlying reserves, making it difficult for users to trust them. Typically, collateralized coins rely on third parties, most notably banks, to keep the corresponding collateral and execute transactions on their behalf, making their actual costs excessive and forcing them to impose high transaction fees on users. Ironically, reliance on existing legacy banks brings to the forefront the issue of stability of the banking system itself, which an alternative crypto ecosystem is supposed to rectify in the first place.

Unless very carefully structured, fiat-backed coins suffer from the high costs of carrying collateral, which leads to low profitability, which is a significant danger in its own right. For example, creators of custodial stablecoins that are reliant on bank balance sheets to hold reserves have to pay a high price to effectively rent those balance sheets (and their capital buffers). This fact vastly reduces custodial coins' cost-effectiveness and requires a much broader customer base to reach a profitable scale.

---

[2] Recall that tether means a rope or chain with which an animal is tied to restrict its movement.

[3] Alexander Lipton is an Advisory Board member at Clearmatics.

An NB is a natural candidate for issuing a stable fiat-backed cryptocurrency, which can be viewed as a private digital version of cash. A client can deliver fiat currency into an NB either from their existing account with the NB itself or via a wire transfer from another bank. In the former case, the bank already knows the client and does not need to do due diligence. In the latter case, the NB performs the necessary KYC steps. In return, the client receives digital tokens, while the NB immediately deposits the funds with the central bank. The client can use their tokens to transfer value on a distributed ledger; see Figure 8.2.

The token circulates on a distributed ledger. The consensus is maintained either by PoW, proof-of-stake or by third-party notaries. When the current owner decides to convert her token back into fiat currency, she returns the token to an NB, which either credits the existing account or wires the funds to an account in a different bank. In the former case, the bank knows the client already. In the latter case, the NB performs the necessary KYC steps.

Following the flow diagrams of Figures 8.2 (a), (b), initially, the NB issues a digital token to $Client_1$ in exchange for fiat currency. The currency comes from the $Client_1$ account at the NB. Next, $Client_1$ passes this token to $User_1$ in exchange for goods and services. The act of ownership transfer is recorded on a distributed ledger maintained by miners or notaries. In the course of doing business and transacting, the NB provides AML checks. Eventually, $User_1$ passes the token to $Participant_2$ in exchange for goods and services. At time T_4, $User_2$ gives the token to $Client_2$ in exchange for goods and services. And finally, $Client_2$ passes the token to the NB in exchange for fiat currency, which is deposited in the $Client_2$ account at the NB.

The tokens retain their value in a narrow band around par by arbitrage. When the token price falls significantly below par, its owner can immediately give it back to the bank and receive the full value. When the token price increases considerably above par, the pent-up demand causes an NB to issue more tokens into the ledger by using its capital or attracting outside investors, thus reducing the price. Issuing stablecoins by an NB is the only way of satisfying the third postulate of Copernicus.

8.4.3 Coins partially collateralized with fiat

Saga is a relatively rare example of cryptocurrencies having a time-varying degree of collateralization. By design, it starts its life as fully collateralized and eventually becoming fully free-floating; see Saga (2018). In our opinion, such coins cannot be stable in the long run, regardless of the theoretical arguments put forward by their backers. History has been brutal to such schemes. For instance, when governments start to manipulate their coinage's gold content, their coins' value plummets precipitously. The story of the so-called assignats, used during the French Revolution, and eventually became worthless, illustrates this point with extreme clarity; see, e.g., Goetzmann (2017).

In the 1970s, inflation triggered by various causes, including the US dropping the gold peg led to sharp dollar devaluation. Paul Volcker eventually brought it under control and stabilized the US dollar by using all the tools available to the Fed.

It is clear that Saga does not have such tools at its disposal and hence is very prone to a death spiral. Generally speaking, its approach is similar to someone building a sturdy table with four legs, and then when everyone is happily sitting around the table, starting to remove legs one by one until the table collapses.

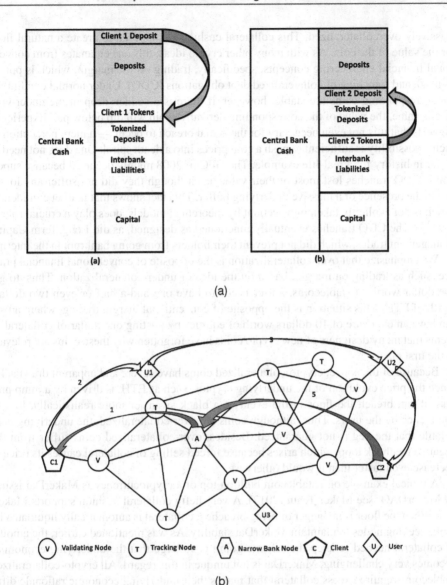

Figure 8.2 Stablecoin issuance by a narrow bank. Such a coin is always covered by reserves, and does not put inflationary pressures on the economy. (a) Bank's ledger transitions. (b) Blockchain transitions. Own graphics.

8.4.4 Coins (over)collateralized with crypto

An alternative approach to creating stablecoins is to use unstable native crypto coins, specifically ETH, as collateral. By providing ETH as collateral to a smart contract, an agent can get stablecoins in return. These coins, representing a sliver of the total locked value, are massively overcollateralized. This collateral cushion is supposed to create a natural floor for the value of the coin. As with many other crypto ideas, this one emanates from conventional financial engineering concepts, specifically, trading on the margin, which is put on its head, and tranching of collateralized debt obligations (CDO). Under normal conditions, the corresponding coins are stable; however, if there is a sudden drop in the underlying asset's value, the value of the corresponding derivative coin will dip below par. Experience suggests that it is not even necessary for the actual breach to occur — a mere perception of such a possibility is sufficient to push a coin prices into a death spiral. One does not need to go far in history for a real-life example. The GFC of 2008 mainly occurred because super-senior CDO tranches lost most of their value, even though they did not suffer any losses due to the presence of a massive underlying buffer. This fact shows that market confidence, which is not explicitly taken into account by theoretical models, does play a crucial role in real life. The CDO tranches eventually functioned as designed, as did Long-Term Capital Management trades, which did not prevent their holders from going bankrupt in the interim.

We emphasize that over-collateralization is the opposite of conventional financial practice, such as trading on margin, based on the idea of under-collateralization. Thus, to get one dollar worth of stablecoins, a user needs to have one-and-a-half or even two dollars worth of ETH. This situation is the opposite of conventional margin trading, where a user can have an exposure of 10 dollars worth of equities by posting one dollar of collateral. It seems that many designers of new crypto coins have forgotten why these coins are relevant in the first place.

Because of the above, crypto-collateralized coins have severe and apparent drawbacks. Since the price evolution of the underlying cryptos, such as ETH, is driven by a jump process, it can breach the floor in the event of a black swan, or more realistically, loss of confidence by the users. Thus, although coins are over-collateralized, the underlying is so volatile that the peg is not guaranteed. Besides, over-collateralized coins suffer from the negative feedback loop, which arises because forced selling of collateral causes its price to decrease even faster than it would otherwise.

A typical example of a stablecoin built on top of a cryptocurrency is MakerDai issued by MakerDAO; see Maker Team (2017). A very hefty collateral cushion supports Maker-Dai. When the floor is in danger of being breached, collateral is automatically liquidated by profit-seeking nodes to maintain MakerDai stability. As was mentioned earlier, the amount of collateral needed to keep the scheme going is so large that the underlying economics becomes very challenging. MakerDai is not unique in this regard. All crypto-collateralized stablecoins require excess collateral that makes their underlying economic rationale difficult to fathom.

8.4.5 Dynamically stabilized coins

While the idea of dynamic stabilization of a coin contradicts common sense and historical experience, it has recently gripped the investor community's imagination and hence is worth investigating in some detail. There are several obvious issues with algorithmically stabilized coins.

Monetary policies underpinning dynamically stabilized coins emanate from obsolete and outdated economic considerations. Either by necessity or by design, these policies are vague and either unproven or proven by pre-crypto experiences not to work. Centralized mechanisms, such as a centralized reserve, have to be maintained to ensure price stability in any case. Historically, all the live projects have suffered drops in value, just breaking their stated *raison d'être*. Besides, so far, no one has been able to design trusted decentralized (price) oracles.

Basis is a representative example of such a coin; see Al-Naji et al. (2017). The algorithm, denuded of amenities, works as follows. If the value of the coin is going up (a relatively straightforward case), then, not surprisingly, new coins are issued and distributed amongst the holders on a pro-rata basis. If the coin's value is going down (a more complicated case), then bond-like instruments are issued in exchange for coins, which are burned. As a result, the number of coins in circulation shrinks, and, in theory, their price increases. This method is the essence of the seigniorage concept; see Sams (2015).

When formulated as a cryptocurrency stabilization scheme, this algorithm sounds meaningful. However, rephrased in more familiar terms, for instance, as a company trying to keep its stock price constant, the algorithm shows its true colors. If the stock price increases, additional shares are issued, thus pushing the stock price down. When the stock price is falling, the same theory prescribes the company to sell bonds and use the corresponding proceeds to buy its stock, thus pushing its price up. This stabilization scheme represents yet another purely theoretical construct that cannot and does not work in practice and will collapse when bonds come due — not to mention that selling bonds in the middle of a market panic might be an insurmountable obstacle.

Algorithms of this type are not new. For example, Baron Munchausen is famous for using one when pulling himself and his horse out of a mire by his hair. More recently, academic economists entertained similar ideas. They proposed a mechanism for fixing gold price by a government in terms of its fiat without keeping any gold in reserve; see Black (1987). Even with all the coercive instruments at their disposal, no government could ever achieve such a feat. The probability of a crypto algorithm that lacks such tools succeeding is even lower. Without going too far back in history, the recent crisis of the Argentinian peso was accelerated by the increase of interest rates to 60%, which, according to the above theory, is supposed to arrest the currency freefall. A similar price dynamics for the Venezuelan government bonds comes to mind of anyone passingly familiar with financial markets.

As an aside, the Quantity Theory of Money (TQM), which is a foundational concept underpinning Basis has been discredited for decades and does not pass scientific analysis, not to mention common sense; see Lipton (2016b). The only saving grace is that a similar unflattering observation is valid for most traditional macroeconomic theories; see Lipton (2016a). As was his habit, John Maynard Keynes put it in a characteristically pithy fashion: "Practical men, who believe themselves to be quite exempt from any intellectual influence, are usually the slaves of some defunct economist. Madmen in authority, who hear voices in the air, are distilling their frenzy from some academic scribbler of a few years back."

8.5 Digital trade coins collateralized with assets

8.5.1 General considerations

USC and stablecoins are helpful from a technical perspective, but they do not solve monetary policy issues. To address these issues in earnest, we need to build a counterweight for fiat currencies by designing digital trade coins (DTCs) collateralized with real assets.

In a stable economic and monetary environment, it makes sense to collateralize crypto coins with fiat. It is not a viable option in some parts of the world since the local fiat is not stable enough. In these instances, coins collateralized with real assets come to the rescue. Recently, several such coins have been proposed by competing startups, including Digital Trade Coin (DTC) (of which one of the present authors, AL, is a co-inventor), Oilcoin, Sweatcoin, Tiberiuscoin, Libra, and several others, see, e.g., Lipton et al. (2018a,b); Tiberiuscoin (2017).[4] These coins have significant utility value for supply-chain financing and well-defined trading environments, such as oil trading or cross-border trade financing. Besides, several authors convincingly argue that even in the presence of stable fiat, asset-backed coins have considerable advantages by limiting the freedom of central banks to manipulate their currencies, see, e.g., Lipton et al. (2018b).

While undeniably useful, fiat currencies are not well suited to 21st-century commerce, especially trade and supply chain finance, because of high handling and exchange costs. Besides, the USD's world reserve currency status causes severe trade imbalances that can trigger trade wars and exacerbate international frictions. Several financial authorities have suggested that it makes sense to complement fiat currencies with a supranational currency, similar to Spanish Peso de Ocho, or Austrian Thalers, introduced and described in Chapter 2 to alleviate these ills. Such supranational currency is impervious to adverse actions by a single central bank or other parties.

Recently, Zhou Xiaochuan, Governor of the People's Bank of China, advocated the need for such currency in Zhou (2009) as follows:

I. The outbreak of the crisis and its spillover to the entire world reflect the inherent vulnerabilities and systemic risks in the existing international monetary system.

II. The desirable goal of reforming the international monetary system, therefore, is to create an international reserve currency that is disconnected from individual nations and is able to remain stable in the long run, thus removing the inherent deficiencies caused by using credit-based national currencies.

III. The reform should be guided by a grand vision and begin with specific deliverables. It should be a gradual process that yields win-win results for all.

The idea of anchoring the value of a paper currency in baskets of real assets is old. For example, gold and silver, as well as bi-metallic collateral, have been used for centuries. Two approaches are common in discussing such currencies: (a) a redeemable currency backed by a basket of commodities; (b) a tabular-standard currency indexed to a basket of

---

[4]Initially, Libra (if it ever takes off the ground) intends to use fiat as collateral; subsequently, it plans to use a basket of currencies, making Libra fall into the DTC category.

commodities. For example, F. Graham, inspired by the Great Depression's events, proposed an automatic countercyclical policy requiring 100% backing of bank deposits with commodities and goods; see Graham (1940). B. Graham suggested backing the USD with a commodity basket of 60% commodities and 40% gold.; see Graham (1933). Hayek extended proposals by both Grahams to establishing a universal basket of commodities, which every country would use to back its currency; see Hayek (2009). Keynes proposed the bancor, an international currency defined as a gold weight supposed to be a multilateral transaction currency; see Keynes (1943). Kaldor developed a new commodity standard, which he also called bancor, a commodity reserve currency; see Kaldor (2007).

Creating an independent international reserve currency will reform the global monetary system and transform it into a system that remains stable in the long run and does not suffer from the inherent deficiencies of credit-based fiat national currencies. In recent years, an MIT team (of which one of the authors, AL, is a member) has designed an asset-backed digital currency called DTC well-suited to operate as a medium of trade and exchange for smaller nations and supranational organizations. Today, developing a digital supranational currency backed by diverse and widely held assets is possible for the first time. Such a digital currency could combine historical currencies' best features, including the finality of settlement, partial anonymity, and usability on the web. DTC is mostly immune to direct central banks' policies controlling the world's reserve currencies. However, while mitigated, indirect influence of the US dollar will remain due to pricing most commodities in dollar terms. Consequently, it has enormous potential to improve the stability and competitiveness of trading and natural resource producing economies. DTC is a much-needed counterpoint for today's fiat currencies and a way forward toward ensuring financial stability and inclusion; see Lipton et al. (2018a,b).

8.5.2 How to build a DTC

Here we briefly describe how to build a DTC. The foundational layer of a DTC consists of real assets, such as oil, metals, crops, mooring rights, and other similar commodities and services of value:

(1) The sponsors bring their commodities to the pool's administrator, who issues DTC in a one-to-one ratio.
(2) The administrator sells DTCs to the public and deposits the corresponding fiat currencies with the affiliated NB.
(3) The administrator passes the proceeds through to sponsors.
(4) As a result, the administrator possesses real assets. The sponsors end up with fiat currency. The general public owns DTCs that are always convertible into fiat at the current market price.

We illustrate the way generic blockchains operate in Figure 8.3, while the structure of wholesale and retail DTCs is presented in Figures 8.4–8.5, respectively.

The price P_{DTC} of DTC will be close to (but not strictly at) the market price of the corresponding asset pool, P_M. Indeed, if P_{DTC} falls significantly below P_M, it becomes rational for economic agents to put DTC back to the administrator. In turn, the administrator has to sell a fraction of the pool's assets for cash and pass it to these agents.

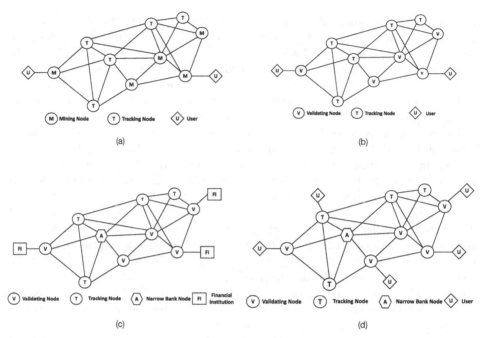

Figure 8.3 Four examples of distributed ledgers: (a) Bitcoin; (b) Ripple; (c) wholesale CBDC; (d) retail CBDC. Own graphics. Source: Lipton et al. (2018b).

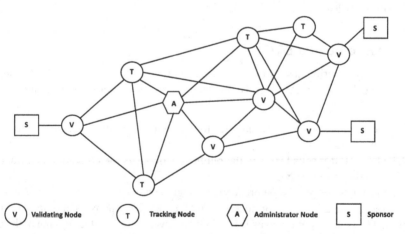

Figure 8.4 Design for a wholesale DTC used only by sponsors. Own graphics. Source: Lipton et al. (2018b).

Suppose P_{DTC} increases significantly above P_M. In that case, sponsors supply additional assets to the administrator. In return, the administrator issues additional DTC and passes them to sponsors, who sell them for cash, thus pushing the price down. This mechanism

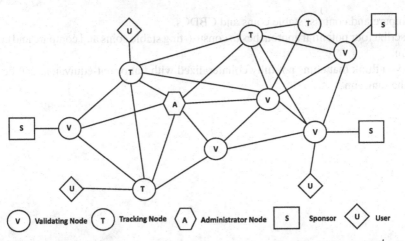

Figure 8.5 Design of a retail DTC used by sponsors and users alike. Own graphics. Source: Lipton et al. (2018b).

ensures that $|P_{DTC} - P_M|/P_M \ll 1$, which is a big advantage, especially because conventional cryptocurrencies habitually exhibit extreme volatility.

8.6 Summary

Among many potentially useful applications of distributed ledgers in finance, CBDCs and stablecoins are very promising avenue. If physical cash disappears, it is possible to imagine a future where everyone has access to central bank cash, albeit indirectly. In this scenario, retail banks may bifurcate into NBs and investment pools. Asset-backed cryptocurrencies can serve as a much-needed counterpoint for fiat currencies and a very potent geopolitical tool.

The time has come for central banks and other regulators to open competition between slow-moving incumbent financial institutions and nimber newcomers, especially once fully exploiting new technologies, including distributed ledgers.

Additional insights are given by Adrian and Griffoli (2019); Bordo and Levin (2017); Borgonovo et al. (2018); Danezis and Meiklejohn (2015); Duffie (2019); ECB (2012, 2015); Griffoli et al. (2018); Grym (2018); Kahn et al. (2019); Lipton and Pentland (2018); Lipton et al. (2020); Oh and Nguyen (2018); Senner and Sornette (2019); Raskin and Yermack (2018); Taskinsoy (2019); Williamson (2019).

8.7 Exercises

1. Explain the motivation for issuing CBDCs. Is DLT the best approach to minting CBDCs? What do you think about using Chaum's technique instead?
2. Explain the pros and cons of stable coins.

3. Compare and contrast stable coins and CBDCs.
4. Describe four potential approaches to constructing stable coins and compare and contrast them.
5. Do you think that coins not fully collateralized with fiat or fiat-equivalent can be stable in the long run?

9 | Wallets and Key Management

9.1 Introduction

There was a time when robbing assets necessarily involved physical force, gunpowder, and death. The Hollywood industry became expert at depicting all sorts of historical or fictional narratives around the clash of noble defenders of property rights and faithless criminals looking to acquire assets by any coercive means. From highwaymen being heroes of violence to impressive bank hold-ups, these depictions have the merit of emphasizing one painful truth about security: that of its imperfection.

One of the first learnings in the field of security is that a security framework should never pretend to absoluteness. Rather, it is defined as a set of counter-measures against specific threat vectors. A well-designed framework models attacks that it aims to control but also classes of risks that it knowingly does not address. Security is also a dynamic field: it is an ever evolving arms race between attackers, who will use ingenuity to come up with new ways of breaking defense mechanisms, and security engineers reinforcing the defense and widening the protection surface.

A simple look at the history of bank robberies provides an interesting material illustration to this principle. Bank vaults, which were designed to secure money, valuables, and documents against theft and destruction, have been notoriously fragile to all sorts of attacks. After thousands of years perfecting locking mechanisms — from the oldest Egyptian lock designs, to Roman warded locks impeding picking, to the Chinese combination locks — one could have expected vaults to finally have reached an ultimate level of protection. However practice revealed that the preferred attack of bank robbers during the 19th-century Gold Rush was not to break the lock that engineers and metallurgists had spent so much time perfecting; rather, the preferred attack vector was to break into the bank and directly extract the whole safe out a window to a distant location to force it open, therefore entirely bypassing the lock itself — an attack vector which had not been planned appropriately.

This new form of attack naturally led banks to invent new protections, such as heavier and larger safes that would become much harder to steal altogether. To that, robbers responded by blasting off the door with explosives on premise, which consequently led banks to further improve the resilience of their vaults as well as use harder materials and better thought architecture and locks. This made vaults almost impossible to extract out of the bank or to blast. But it also revealed a new weak point in the whole security model: the human risk. It became more effective to threaten or kidnap the bank manager and coerce him into sharing the key or revealing the lock combination than to physically force the vault open.

The Digital Revolution initiated in the late 1950s and amplified with the global connectivity provided by the Internet brought an entirely new playground for criminals and redefined security best practices as well as the universe of threat vectors and their defense mechanisms. It became obvious to criminals that risking their life to rob would be favorably exchanged for the seemingly risk free, yet often rewarding, extraction or tampering of digital data. Cyber criminals did not disappoint: in 2017, more than half of chief executive officers (CEOs) expect cybersecurity and data breaches to threaten stakeholder trust in their industries over the next five years.[1] The average per-record-compromised cost of data breaches reaches new all-time highs every year, crossing the US$4 million bar in 2020 with a 60% increase over the last decade; see Ponemon (2017).

The unavoidable automation trend strongly suggests that persistence and size of negative cyber shocks will further grow in the future. The increasing diversity of perpetrators, from state and non-state actors, cyberterrorists, and hacktivists, also shows how financial gain no longer is the only motivation. Intuitively, the financial sector is an obvious high risk industry. Information security risk management defines risk as a combination of consequences and likelihood (ISO, 2011), such that

$$\text{Risk} = f(\text{Threat}, \text{Vulnerability}, \text{Consequences}).$$

For a financial institution, threat level is high due to cybercrime, hacktivism, proxy organisation, and surveillance of communication; dependence on highly interconnected network implies high vulnerabilities; and the dematerialized activity means that cyberattacks lead to big consequences; Kopp et al. (2017). Distributed ledger technologies (DLTs) and cryptocurrency service providers arguably are even more critical targets. The aggregate value of the cryptocurrency market being in the hundreds of billions at the time of writing, and its networks being fully available publicly, it is one of the biggest honey pots of the Internet. Its security foundations should be solid enough to sustain constant attacks on its infrastructure, its accounting ledger, and its users — attacks designed with the specific intent to steal funds, disrupt the service availability, or generate panic as a market manipulation technique or as an end-goal in itself. Although the Bitcoin network and the other major cryptocurrencies have been generally resilient to protocol-level attacks, there have been many extraordinary thefts since 2009, including some of the most prominent exchanges such as Mt Gox, Bitstamp, Bitfinex, or Binance (see Table 9.1).

As introduced in Chapters 3, 5, and 6, distributed ledgers rely on asymmetric cryptography to represent and secure ownership. In the simplest case, a wallet is a direct mapping to a public key, whose secret key is held confidential by the owner of the wallet. The owner

---

[1] See Risk in Review 2017 Study, PricewaterhouseCoopers.

Table 9.1 A selection of exchanges hacks between 2013 and 2019.

Institution	Date	# Estimated losses (USD mn)
Inputs.io	Oct. 2013	1.3
GBL	Oct. 2013	5
Bitcoin Internet Payment Services	Nov. 2013	1
MT Gox	Jan. 2014	470
BitPay	Dec. 2014	1.9
EgoPay	Dec. 2014	1.1
Bitstamp	Jan. 2015	5.3
Bitfinex	May 2015	0.3
Gatecoin	May 2016	2
DAO Smart Contract	Jun. 2016	50
Bitfinex	Aug. 2016	72.2
CoinDash	Jul. 2017	7
Tether	Nov. 2017	31
NiceHash	Dec. 2017	64
Coincheck	Jan. 2018	534
BitGrail	Feb. 2018	170
Coinsecure	Apr. 2018	33
Cryptopia	Jan. 2019	16
Coinbene	Mar. 2019	105
Binance	May 2019	40
Upbit	Nov. 2019	49

Sources: ORX News, Financial Times.

can at any time broadcast an instruction signed with his secret key to be verified by the network. In particular, a transaction instruction can move funds from one account to another. Key management is therefore critical: without the key, an instruction cannot be signed and therefore neither can it be approved by the network; the corollary is that if the key is stolen, any instruction may be instantly authorized by the thief. Loss, destruction, or theft of the secret keys is therefore a critical risk for the owner, who must define appropriate controls for the whole wallet architecture and the key management.

In this chapter, we emphasize that best practices may strongly vary based on the particular use case: an individual using cryptocurrencies for day-to-day payments, an investor storing reserves for the long term, an institutional custody provider, or a broker, they all might have very different requirements — therefore various implementations. The following sections aim to cover these points by presenting general practices and architectures and their associated threats. Section 9.2 presents the architecture of a cryptocurrency wallet. Section 9.3 introduces the notion of key generation and entropy. Section 9.4 reviews the main key derivation standards for the generation of new secrets. Section 9.5 presents the most common approach for key backup and recovery of retail wallets. Section 9.6

relates two well-known cryptocurrency exchange hacks. Section 9.7 introduces some of the best practices for institutional digital asset custody providers. Conclusions are presented in Section 9.8.

9.2 General wallet architecture

9.2.1 Necessary components

The architecture of a wallet generally involves the combination of the four below components (also see Figure 9.1):

- a network node
- an indexer
- a wallet application
- a keystore

Historically, these components were all merged into a single application. Bitcoin Core, the Bitcoin full node originally developed by Satoshi Nakamoto, provides indexing and wallet features and also takes care of the key management. Today, nodes are mainly used to communicate with the peer-to-peer network and validate consensus rules. They offer APIs to request the state of the blockchain such as the content of a block or current fee levels. Indexing, wallet, and key management services someimes provided by network nodes are most of the time ignored and outsourced to external, specialized components.

9.2.2 Network node

As detailed in Section 5.4, a network node tracks the contemporaneous state of the blockchain and maintains a local copy of part or the whole ledger; it also acts as a relay for instruction and transaction broadcasts. A node generally exposes an API for third-party modules to build additional functionality on top of its core services. Optionally, one may rely on a fully-validating node, or full node, to locally validate the full set of consensus rules and filter out inconsistent requests emerging from the network. In particular, a full node

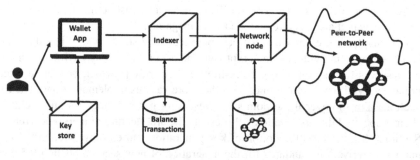

Figure 9.1 Generic architecture of a cryptocurrency wallet. Own graphics.

enforces that the data format of blocks and transactions is valid, that all transaction signatures are valid, that accounting rules within blocks are valid, that there does not exist double spending instances in a single blockchain, and that block rewards fit the reward schedule, do not over or under inflate. In case of validation failure, a full node will disregard the request (e.g., a transaction or a block) entirely.

Because it is not meant to provide advanced querying services, a node generally limits its data processing to the parsing of native ledger primitives, such as the blocks of a blockchain; it does not aim to reorganize information to facilitate business queries. For instance, a node can efficiently answer the question "what is the content of block 12344?" but generally cannot efficiently answer the question of "what is the transaction history of the wallet formed by this given set of addresses?". The reason for this is that a blockchain does not provide a constant access time view over the transaction history of an address or wallet; rather, it requires a block-by-block, linear scan of the blockchain to build a consolidated response, which is a process that full nodes do not natively implement for scalability reasons in general. Nothing technically prevents a node to embed more data processing features however, and some software actually do: Bitcoin Core is such example as it provides basic walleting features as a standard service.

9.2.3 Blockchain indexer

A blockchain indexer, or simply *indexer*, dynamically filters and restructures information extracted from a node in order to build a context-specific data set adapted to querying (see Figure 9.2). An indexer tracks the content of specific wallets and addresses and maintains a consolidated view over the respective balances, transactions, and other relevant part of the state such that queries do not require a linear scan of the underlying ledger; it also maintains the information required for transaction building, such as the UTXO set for Bitcoin-like protocols, or the latest nonce for Ethereum-like coins, see Sections 5.3 and 6.4, respectively. Although not security critical in the sense that an indexer cannot spend funds without access to secret key material, an indexer, jointly with its database, is a crucial component of a wallet service that, if compromised, can create security weaknesses in the whole service: a malfunctioning or compromised indexer may tamper with transactional data with the intent to counterfeit accounting entries — for instance to fake an incoming transaction within the wallet or hide a double spend attempt. It is therefore not unusual that a wallet relies on multiple independent indexers to cross-validate information, or that it falls back to direct verification of the information within the blockchain by requesting a node from its view of the blockchain.

Bitcoin, for instance, encourages a *Simplified Payment Verification* (SPV) introduced in Section 5.4 to verify blockchain inclusion of transactions, as described in section 8 of Satoshi Nakamoto's seminal paper: SPV allows a wallet to verify the inclusion of a transaction within the longest chain — and therefore its validity — without maintaining a full node or a complete copy of the blockchain. Instead, it keeps a copy of the block headers of the longest chain only, which are available by querying network nodes until it is apparent that the longest chain has been obtained. A proof of inclusion is constructed by demonstrating inclusion of the transaction in the Merkle tree of a block within the longest chain, as in Figure 4.21.

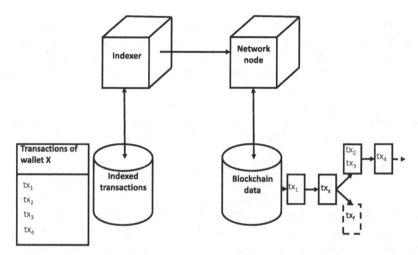

Figure 9.2 Indexing of blockchain transactions. The transactions of a particular wallet are spread across multiple blocks, mixed with transactions of unrelated wallets, and with irrelevant forked transactions. An indexer listens to transactions and reorganizes them into a curated list. Own graphics.

9.2.4 Wallet application

A wallet application is the central point of interaction between the wallet owner, the key material required for transaction approval, and an indexer proxying for a node. It provides a user interface through a webpage, a desktop application or a mobile application, or an API to be consumed programatically. A wallet exposes the balance, transaction history, and transaction status to its user and provides the tools for transaction creation, authorization and broadcast. A wallet is responsible for the applications of operational best practices and for the analysis and treatment of exception cases. In particular it must answer to the following points:

(1) what workflows to enforce for a transaction approval process
(2) what level of fees to recommend for a given transaction
(3) when to consider a transaction as confirmed in the sense that reversal has a very low probability
(4) how to respond to the non inclusion of a broadcasted transaction in the blockchain or an orphan transaction due to a fork
(5) when to create new addresses to include in the wallet (for compatible protocols)

9.2.5 Keystore

A wallet relies on a *keystore* to persist secret keys. A keystore may be as simple as a file system or a database, but additional protections are generally deployed to protect the key material at rest and during use. Protections include encrypting the secret keys with other keys, splitting secret keys into multiple distributed shards, or purpose-built hardware with electronic protections.

A hardware security module, or *HSM*, is a device offering cryptographic features in a trusted hardware environment. It guarantees data confidentiality and integrity in presence of both cyber threats and physical threats to the system. It is the core security component of a public key infrastructure, which manages roles and procedures needed to use and store digital certificates and provide public-key encryption. It typically provides services for the crypto-secure generation and safeguarding of key material, the computation of digital signatures, data encryption, the storage of passwords and credentials, and it occasionally offers a secure execution environment to define conditional access to the HSM services. Although HSMs have historically been used mainly in critical business contexts, they have become more common also on the retail market in the late 2000s with various portable devices such as YubiKeys and cryptocurrency hardware wallets such as Ledger Nano or Trezor. Even recent mobile phones now embed security chips which provide a hardware secure environment for the management of the lifecycle of keys and secrets in general.

9.3 Wallet genesis

9.3.1 Key space

A wallet corresponds to a set of secret keys and respective addresses; addresses are used as pseudo-anonymous account identifiers on the distributed ledger. The whole security model of distributed ledgers relies on the capacity to prove ownership with digital signatures. Therefore it is essential that secret keys are generated from a good enough entropy source so that they cannot be guessed or brute forced: randomness being the only defense mechanism, key collision should also be statistically improbable, even when the computational cost of key generation is low.

Let us consider the example of a 32-bit key offering 4,294,967,296 combinations. Such keys might look good enough given that guessing the key of a specific user has a probability of $\frac{1}{2^{32}}$, which is about 30 times less probable than winning the jackpot at EuroMillions lottery. However, as compared to the lottery where choosing one combination may cost several Euros, generating cryptographic keys is inexpensive and may easily be computationally scaled to millions of attempts per second at negligible cost. A full iteration through all possible combinations may thus be achieved in a short instant. In practice there are multiple users, each possibly managing many keys, therefore dramatically increasing the probability of a collision as reminded by the Birthday paradox in Section 4.9.

Most cryptocurrencies rely on keys of 256 bits, therefore have approximately 10^{77} combinations. In order to illustrate the magnitude of this number, we shall loosely compare it to the number of atoms in the universe, estimated to 10^{80}. Generating a secret key is akin to selecting a specific atom in the universe and guessing someone else's key amounts to guessing the right atom in the universe. The improbability of a successful collision is such that it is deemed computationally infeasible and the risk can be ignored in practice.

There is no need to test for pre-existence of a key after generation as there should never be two generation events leading to the same outcome within the bounds of human existence.

9.3.2 Brain wallet

We emphasize that a 256 bits integer does not necessarily embed 256 bits of entropy. In order to illustrate this point, we present what is probably the most terrible, yet exciting idea that Bitcoin entrepreneurs came up with: *brain wallets*. A brain wallet is a wallet whose keys are deterministically derived from a single human-readable and memorizable (therefore "brain") passphrase. The passphrase may be anything from a single word, a long sentence, or a series of uncorrelated symbols. The advantage of this approach is obvious: one may simply remember the passphrase in order to access his wallet from any location in the world, with no need for backup or physical security devices. The wallet is a direct mapping from some secret information deep inside the memory of your brain! Figure 9.3 illustrates the principle.

The wallet key 6663C658 D7F16F48 7D63C156 DCBC28D1 2878FDCC 7A097E9F 8D53091F 4B498152, expressed in hexadecimal form, looks like a fully random 256-bit integer. In fact, a simple binomial test is unable to reject the null hypothesis of Bernoulli draws for the bits at a 5% level and it is expected that other tests would fail at identifying any divergence from i.i.d. random Bernoulli draws. The hash function does its job properly in taking a source and compressing it into a truly random-looking integer.

However, the seed of the hashing algorithm has dangerously low entropy. Information Theory's father C. E. Shannon shows in Shannon (1951) that the entropy of printed longer than 100-letter sequences of an English text should be bounded between 0.6 and 1.3 bits per letter. His analysis relies on a human prediction approach and ignores punctuation and character case. Other studies further estimate entropy per character between 1.25 and 1.77; see, e.g., Cover and King (1978); Teahan and Cleary (1996); Kontoyiannis (1997). Blindly applying these results to our 17 characters-long seed "Bitcoin is secure"— and ignoring the aggravating factor that it contains two of the most widely used words in the cryptocurrency community, "Bitcoin" and "secure"— we are able to confidently conclude that entropy is lower than 32 bits.

Brain seed   **Wallet key**

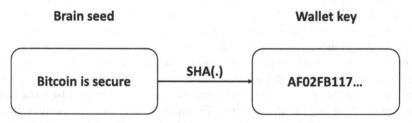

Figure 9.3: Brain wallet key generation. Own graphics.

The defect in the architecture of brain wallets led attackers to build simple brute-force tools to break brain seeds. Rather than iterating through the 256-bit integer space, attackers would focus on the English language, apply dictionary attacks and scale out the system with parallelism to test billions of combinations. Researchers of the study titled "The Bitcoin Brain Drain: A Short Paper on the Use and Abuse of Bitcoin Brain Wallets" by Vasek et al. (2016) analyzed 300 billion passwords based on a wide range of word lists and identified 884 brain wallets worth over 1,000 BTC between September 2011 and August 2015. Because brain wallets used no salt and passed plaintext passwords through a single hash iteration, the authors could trivially implement a large-scale measurement of the use of brain wallets in Bitcoin. They reveal the existence of an active attacker community that rapidly steals funds from vulnerable brain wallets in nearly all cases they identify. "Many brain wallets are drained within minutes, and while those storing larger values are emptied faster, nearly all wallets are drained within 24 hours," claim the authors. Table 9.2 illustrates the number of wallets broken by the authors by password data source, as well as the associated BTC value. One can see that even without investing effort in brute-forcing complex passwords, the simple use of public dictionaries and word databases compromises a substantial number of seemingly secure wallets.

Table 9.2 Wallets compromised by dictionary attacks for different data sources. Some wallets still hold funds at the time of the attack.

Dictionary Source	# Wallets	# Non-Empty	Total BTC
Urban Dictionary	296	3	561.95
Two Words	13	3	0.79
Eng/Slang Urban Dict.	63	14	0.90
Eng. Wikipedia	250	0	38
WikiQuotes	35	0	60.96
Phrases	283	0	578.69
xkcd	90	3	97.66
Lyrics	329	4	230.45
Blockchain.info tags	112	0	577.93
Rootkit	123	2	4.50
MySpace	59	0	1.14
RockYou	415	3	113.82
LinkedIn	213	0	10.11
LEET MRL	3	0	0.01
Prince MRL	295	4	88.93
CrackStation	640	3	396.09
Naxxatoe	388	0	41.56
Skull Security	414	3	71.73
Uniqpass	490	3	134.95

9.3.3 The need for entropy

This attack emphasizes the importance of generating keys that are truly random in that they hold enough entropy and are not vulnerable to prediction or bias. Most key generation systems rely on a process called a cryptographically secure Pseudo Random Number Generator (or *PRNG*) for the creation of keys. A PRNG is a deterministic process producing outputs that are computationally indistinguishable from true randomness. It is qualified as Cryptographically Secure (CSPRNG) when additional security requirements are verified, including testing of conditional independence between the output bits, and demonstrating that the leak of the PRNG state does not compromise past generations. Equipped with a CSPRNG and with some initial entropy, an application may create keys on demand in a deterministic way and be confident that every key in the sequence is strong.

However, the weak point still lies in the capacity of the system to properly collect the initial entropy within the application to feed the first state of the CSPRNG. Computers are inherently bad at operating random tasks due to their deterministic nature: with full knowledge of the state of the system at a given time, such as contents of the disk, RAM and CPU registers, one can replicate the machine behavior and compute any future state. This shows that a computer ultimately cannot generate true randomness through the consideration of its own state only: it must rely on sources of noise. Some implementations extract entropy from ad hoc internal sources such as fan noise, disc interrupts, and temperatures, or external inputs such as mouse movements and keyboard pressures, which contain some inherent level of randomness. Alternatively, website *Random.org* is one example of a platform offering true random numbers as a service. The platform aggregates randomness from atmospheric noise, which, the company claims, for many purposes is better than the pseudo-random number algorithms typically used in computer programs.

9.3.4 A concrete collision attack

In 2015, the largest BTC wallet provider *blockchain.info* was subjected to a grave security flaw. Blockchain.info security model relies on the generation of a wallet key on the mobile device or computer of the wallet owner directly. The key is immediately encrypted with a passphrase that is only known to the wallet owner. The secret keys, encrypted with the user passphrase, are stored on the servers of Blockchain.info but can only be decrypted with the user passphrase — which makes Blockchain.info resilient to many forms of attacks and makes the possible consequences of a data breach less catastrophic.

The key generation itself was implemented with two sources of entropy at the time: one, the local entropy as provided by the user device; the other, entropy provided by Random.org. As multiple hacks led to the suspicion that Blockchain.info had a security flaw, a security audit was triggered and did reveal critical flaws in the implementation of the service. The most serious of the security issues was related to the use of unencrypted HTTP connections for the requests to Random.org services. Random.org had updated its APIs in January 2015 to require the use of the encrypted, HTTPS protocol. The lack of update of the Blockchain.info wallet application led to the generation of a secret key corresponding to the address 1Bn9ReEocMG1WEW1qYjuDrdFzEFFDCq43F irrespective of the address defined by the user. The owner(s) of this address received more than 47 BTC at the time because of the flaw, worth approximately US$500,000 in 2020. Worse, in some cases the Android application failed to capture the local device source of entropy and ignored the failure entirely, therefore fully relying on the entropy provided by Random.org exclusively. In such cases, the secret key and address were deterministically derived from an external source of entropy, which may be observed at rest or in transit or tampered with (e.g., to reduce entropy undetected). "It only seems to affect a tiny number of devices," Johns Hopkins University professor Matt Green told Ars. "On those devices it's catastrophic unless you've patched."

9.3.5 True random number generators

Specialized, cryptographically secure hardware random number generators, also called true random number generators, have become standard components in modern security modules in order to generate large amounts of entropy on demand. These devices generate random numbers from a physical process, rather than following an algorithm such as a CSPRNG. They are normally based on microscopic phenomena that generate statistically random noise signals, such as thermal noise, photoelectric effect, and other quantum effects. A well-known implementation relies on a beam splitter mechanism (Figure 9.4): by shooting photons through a semi-transparent mirror, one can measure reflection and transmission occurrences, respectively and obtain a stream of truly random, independent binary outcomes. Such components have become standard in security hardware modules today and are considered best practice.

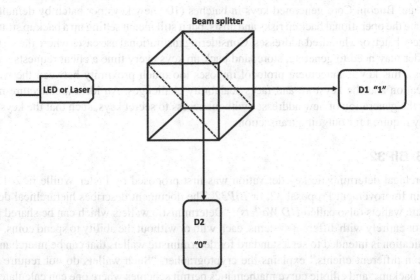

Figure 9.4 Beam splitter-based true random number generation. Own graphics.

9.4 Deterministic key derivation

9.4.1 The need for multiple keys

A single secret key is generally not enough to properly manage cryptocurrencies. The first reason is purely mathematical: a valid secret key for a given protocol — for instance the secp 256k1 elliptic curve of Bitcoin — might not be a valid secret under another cryptography algorithm. One must therefore assume that at least one key per distributed ledger protocol is required. Second, using the same key in multiple contexts is a direct information leak impacting privacy: because of the public nature of permissionless distributed ledgers, the correlation between multiple accounts relying on the same key is evident to an external observer, which can naturally conclude that the several accounts are controlled by a unique entity. Third, one may need multiple accounts on a single ledger either for privacy reasons, e.g. to allocate the assets onto multiple apparently segregated addresses, or for operational reasons. Bitcoin protocol even promotes the multi-key paradigm, or *no address reuse*, to a standard best practice: addresses and their corresponding secret keys should always be used only once. As a consequence, privacy is improved by making clustering analysis harder for an external observer with no knowledge about a wallet address set.

9.4.2 The original Bitcoin Core approach

The reference Bitcoin client, Bitcoin Core, originally dealt with the plurality of keys by generating new random keys on demand. For every new address in a wallet, a secret key would be generated from some source of entropy and stored in the filesystem. Although straightforward on paper, this approach comes with multiple operational frictions. First, it becomes a critical requirement to back the key material up regularly, as new keys are generated for each transaction and need immunization against risks of loss. To alleviate the issue, Bitcoin Core generated keys in batches (100 new keys per batch by default) to alleviate the operational backup risks and costs, but it still meant setting up a backup strategy for every batch of a hundred addresses. Considering institutional use cases where the service provider may need to generate, store, and back up keys every time a client requests a new address, this key management protocol imposes too much proximity between the wallet application and the keystore and faces scalability challenges. An ideal architecture must allow the generation of new addresses without access to secret keys, such that the keystore is only required for outgoing transactions.

9.4.3 BIP32

Hierarchical deterministic key derivation was first proposed by Pieter Wuille in 2013 as Bitcoin Improvement Proposal 32, or *BIP32*. The document describes hierarchical deterministic wallets (also called *HD Wallets*): "Deterministic wallets which can be shared partially or entirely with different systems, each with or without the ability to spend coins. The specification is intended to set a standard for deterministic wallets that can be interchanged between different clients," explains the cryptographer. "Such wallets do not require frequent backups, and elliptic curve mathematics permit schemes where one can calculate the

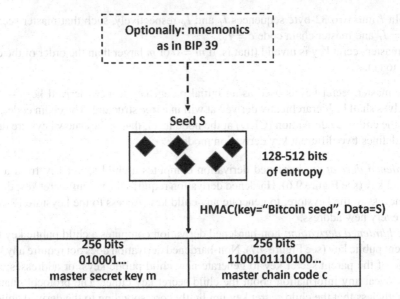

Figure 9.5 Derivation of the BIP32 master extended key from a seed. Own graphics.

public keys without revealing the secret keys. This permits for example a webshop business to let its web server generate fresh addresses for each order or for each customer, without giving the webserver access to the corresponding secret keys (which are required for spending the received funds)."

BIP32 defines a protocol to generate a virtually unlimited number of keys based on a single secret also called seed, akin to a PRNG (see Figure 9.5). Because of the deterministic relation between the seed and any derived key, a backup of the seed is sufficient to backup the entire set of keys; this removes the risks associated with the recurrent backing up of new keys. BIP32 also defines a capability to derive addresses independently of the seed or any secret key material, yet with the guarantee that secret keys corresponding to such addresses can be reconstructed independently when accessing the keystore. This is a powerful feature to separate the wallet application from the keystore and guarantee that the keystore is only accessed when strictly necessary: generating new addresses for incoming transactions can be done with no risk of key leakage.

Reusing Pieter Wuille's notation, the algorithm starts as follows:

(1) Generate a random seed sequence S of a chosen length between 128 and 512 bits, and back it up; optionally, the seed may be derived from a mnemonic code as presented in Section 9.5.2.

(2) Calculate $I = HMAC(Key = $ "BitcoinSeed", $Data = S)$, where the underlying hash function of the message authentication function is SHA-512. Here HMAC stands for hash-based message authentication code.

(3) Split I into two 32-byte sequences I_L and I_R respectively, such that master secret key $m = I_L$ and master chain code $c = I_R$.

(4) If master secret key is invalid (that is, if $m = 0$ or m larger than the order of the curve), go to (1).

The master secret key is used as an initial parent key to new derived keys — which themselves shall be hierarchically derived at will in a tree structure. The chain code characterizes the child key derivation (CKD) at the node level, that is, how new keys are derived. BIP32 defines two different key derivation modes:

- *hardened derivation*: hardened derivation computes a child secret key from a parent secret key (see Figure 9.6). Hardened derivation requires the parent secret key, therefore an access to the keystore, to generate any child key. Access to the keystore is required for every new address.
- *non-hardened derivation*: non-hardened derivation computes a child public key from a parent public key (see Figure 9.7). Non-hardened derivation does not require any knowledge of the parent secret key to generate new child public keys, or addresses; it does not reveal any information about the child secret key either. The protocol guarantees nonetheless that the child secret key implicitly corresponding to the derived child public key exists, and that it may be independently calculated from the parent secret key available in the keystore.

Both derivation schemes are complementary and may be used jointly to compute a key that has been partly derived with a hardened scheme, partly derived with a non-hardened scheme. Non-hardened derivation was proposed precisely to allow the creation of new addresses with no access to secret keys, so that the keystore may be isolated for security reasons.

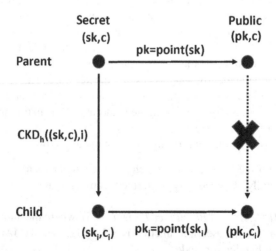

Figure 9.6 Hardened child key derivation with BIP32. Own graphics.

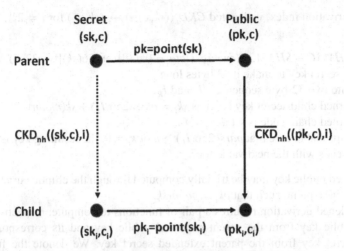

Figure 9.7 Non-hardened child key derivation with BIP32. Own graphics.

Continuing Peter Wuille's specification, we assume the public key cryptography used in Bitcoin below, namely the elliptic curve digital signature algorithm (ECDSA), described in Section 4.7, with curve secp256k1, but note that the scheme may be extended to other curves and protocols. Variables below are either integers modulo the order of the curve (referred to as n), coordinates of points on the curve, or byte sequences. Addition (+) of two coordinate pair is defined as application of the EC group operation; concatenation (||) is the operation of appending one byte sequence to another. As standard conversion functions, we define the below primitives:

(1) *point(p)*: returns the coordinate pair resulting from elliptic curve point multiplication (repeated application of the EC group operation) of the secp256k1 base point with the integer p.

(2) *ser32(i)*: serializes a 32-bit unsigned integer i as a 4-byte sequence, most significant byte first.

(3) *ser256(p)*: serializes the integer p as a 32-byte sequence, most significant byte first.

(4) *serP(P)*: serializes the coordinate pair $P = (x, y)$ as a byte sequence using SEC1's compressed form: $(0x02 or 0x03)||ser256(x)$, where the header byte depends on the parity of the omitted y coordinate.

(5) *parse256(p)*: interprets a 32-byte sequence as a 256-bit number, most significant byte first.

In order to prevent the key derivation scheme to only depend on the parent key itself, BIP32 introduces the notion of extended secret key (sk, c) and extended public key (pk, c), where sk is a secret key, pk is a public key, and c is an additional 256 bits of entropy called *chain code*. The child key derivation $CKD_{mode}((k, c), i) \rightarrow (k_i, c_i)$ takes an extended key and generates a new, child extended key with index i, where i is a 32 bit integer defining the derivation branch *and the derivation mode*. We set *mode* to h when hardened derivation is used, and *mode* to nh when non-hardened derivation is used. The hardened derivation function, which computes a child-extended secret key based on a parent-extended secret key and a derivation index, is denoted $CKD_h((sk, c), i) \rightarrow (sk_i, c_i)$ for $i = 2^{31}, \ldots, 2^{32} - 1$, such that:

(1) Let $I = HMAC - SHA512(Key = c, Data = 0x00||ser256(sk)||ser32(i))$, where 0x00 pads the secret key to make it 33 bytes long.
(2) Split I into two 32-byte sequences, I_L and I_R.
(3) The returned child secret key is set as $sk_i = parse256(I_L) + sk(modn)$.
(4) The returned chain code is set as $c_i = I_R$.
(5) In the improbable case that $parse256(I_L) \geq n$ or $k_i = 0$, the resulting key is invalid and one proceeds with the next value for i.

The derived public key may be trivially computed through the elliptic curve multiplication with the base point, such that $pk_i = point(sk_i)$.

Non-hardened derivation requires a pair of functions to compute, respectively, a child extended public key from the parent extended public key and its corresponding child extended secret key from the parent extended secret key. We denote the functions as follows:

$$\left\{ \begin{array}{l} \overline{CKD}_{nh}((pk, c), i) \rightarrow (pk_i, c_i) \\ CKD_{nh}((sk, c), i) \rightarrow (sk_i, c_i) \end{array} \right\} \tag{9.1}$$

for $i = 0, \ldots, 2^{31} - 1$, where CKD_{nh} guarantees that $pk_i = point(sk_i)$. In other words, $\overline{CKD}_{nh}((pk, c), i)$ is able to derive a child extended public key (pk_i, c_i) from a parent extended public key(pk, c) with no knowledge of the parent secret key sk — and yet it guarantees that there exists a function $CKD_{nh}((sk, c), i)$ that computes the corresponding child extended secret key (sk_i, c_i) given the parent extended secret key (sk, c).

$\overline{CKD}_{nh}((pk, c), i)$ is defined as follows:

(1) Let $I = HMAC - SHA512(Key = c, Data = serP(pk)||ser32(i))$.
(2) Split I into two 32-byte sequences, I_L and I_R.
(3) The returned child public key is defined as $pk_i = point(parse256(I_L)) + pk$.
(4) The returned chain code is defined as $c_i = I_R$.
(5) In the improbable case that $parse256(I_L) \geq n$ or pk_i is the point at infinity, the resulting key is invalid, and one proceeds with the next value for i.

The corresponding function for secret key derivation $CKD_{nh}((sk, c), i)$ is defined as follows:

(1) Let $I = HMAC - SHA512(Key = c, Data = serP(point(sk))||ser32(i))$.
(2) Split I into two 32-byte sequences, I_L and I_R.
(3) The returned child secret key is set as $sk_i = parse256(I_L) + sk(modn)$.
(4) The returned chain code is set as $c_i = I_R$.
(5) In the improbable case that $parse256(I_L) \geq n$ or $k_i = 0$, the resulting key is invalid and one proceeds with the next value for i.

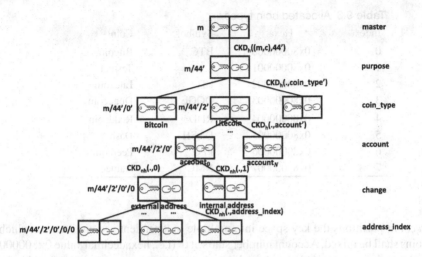

Figure 9.8 Example hierarchy as based on BIP44. Own graphics.

9.4.4 BIP44

Armed with the above child key derivation functions and other BIP32 primitives, Bitcoin Improvement Proposal 44 (BIP44) by Marek Palatinus and Pavol Rusnak is the main attempt at standardizing a logical hierarchy for deterministic wallets (see Figure 9.8). It aims to support not only Bitcoin, but any cryptocurrency relying on ECDSA with curve secp256k1 and support multiple accounts, multiple chains per account, and millions of addresses per chain. BIP39 defines the following five levels of derivation:

m/purpose'/coin type'/account'/change/address index,

where the/-separated path is executed by hierarchical applications of the child key derivation function. Apostrophes indicate that a hardened derivation function CKD_h is used and absence of apostrophes refers to non-hardened derivation function $\overline{CKD}_{nh}$ and CKD_{nh}. Because hardened derivation branches have their left-most bit set to 1, notation i' for $i' = 0, \ldots, 2^{31} - 1$ is to be interpreted as derivation branch $i = i' + 2^{31}$ is used in CKD_h. We describe the use of each derivation level below:

(1) *purpose* is set to the constant 44' (i.e., hexadecimal value 0x8000002C) and simply indicates that the derivation scheme relies on BIP44 specification. Alternative specifications may change the purpose to explicitly indicate a divergence from the BIP44 tree structure.

(2) *coin type* maps to a particular cryptocurrency network. This allows a single master key to be used for an unlimited number of independent coins with no address reuse, and therefore no privacy issues. Table 9.3 shows an extract of the allocated coin types, which is regularly updated depending on new coin registrations. We can see in particular the derivation path for Bitcoin set to 0', Bitcoin testnet 1' and Litecoin 2'.

Table 9.3 Allocated coin types.

Index	Hexa	Symbol	Coin
0	0x80000000	BTC	Bitcoin
1	0x80000001		Testnet
2	0x80000002	LTC	Litecoin
3	0x80000003	DOGE	Dogecoin
4	0x80000004	RDD	Reddcoin
5	0x80000005	DASH	Dash
6	0x80000006	PPC	Peercoin
7	0x80000007	NMC	Namecoin

(3) *account* partitions the key space into multiple independent accounts across which no coins shall be mixed. Account number starts at $0'$ (i.e., hexadecimal value 0x80000000) and may be incremented as required in order to generate new segregated address sets. Notably, account is the last hardened derivation level; keys derived within an account are computed using non-hardened derivation.

(4) *change* is set either to 0 for addresses which are shared externally, e.g. for receiving payments, and therefore are visible outside of the wallet, or to 1 for internal addresses which are not shared externally. The main use case for internal addresses is the generation of change addresses as in Bitcoin transactions; we remind however that some cryptocurrency networks (e.g., Ethereum) only support one key per account and have no notion of change address, therefore derivation 1 is only used for Bitcoin-like protocols.

(5) *address index* selects a specific address within the account. Addresses are numbered from index 0 and incremented as required. For cryptocurrency networks only supporting one address per account, address index is set to the constant 0.

9.4.5 Extensions to BIP32

Although BIP44 already supports the many cryptocurrencies using ECDSA on curve secp256k1, we note that some cryptocurrencies have opted for alternative digital signature algorithms. Table 9.4 illustrates the digital signature algorithms and curves used by some of the main cryptocurrencies in terms of market capitalization. It is apparent that except for IOTA, which relies on custom cryptography, two additional elliptic curve standards have gained relevance in the cryptocurrency market, respectively secp256r1 and ed25519; see Bernstein et al. (2012). Consequently, there exist multiple extensions to BIP32 and BIP44 aiming to provide a unique hierarchical key derivation framework supporting multiple curves. We refer to SLIP-0010[2] and Khovratovich and Law (2017) as two possible extensions.

---

[2]Universal secret key derivation from the master secret key, by Jochen Hoenicke and Pavol Rusnak.

Table 9.4 Elliptic curves used by different protocols.

Currency	DSA
Bitcoin	secp256k1
Litecoin	secp256k1
Ethereum	secp256k1
Ripple	secp256k1 & ed25519
Bitcoin Cash	secp256k1
Cardano	secp256r1
Stellar	ed25519
IOTA	Winternitz one-time signature
NEO	secp256r1
Dash	secp256k1
Zcash	secp256k1
NEM	ed25519

9.5 From mnemonic to seed

9.5.1 Export and human-readability

Hierarchical deterministic key derivation reduces the complexity of key management down to a single secret seed, from which billions of keys can be computed following strict specification. Although the seed may be generated and stored within secure computing devices during its complete lifetime, it is common practice to export it in a human-readable format for back up. Secret keys being large integers, they are hard to manipulate manually: human brain was never wired for memorizing, processing, and communicating long unintelligible strings of symbols. The operational pain and risk of error in dealing with 128 to 512 bit long numbers by hand would simply be too high to tolerate. A first approach would be to convert the binary representation of the seed into a hexadecimal string to reduce the number of characters by a factor of four, down to 32 for 128 bits of entropy. Other encodings with larger alphabets can further decrease the number of symbols: Base64 for instance is regularly used for serialization of large numbers in Internet communications, and Base58 is the Bitcoin standard for human-readable encoding of big integers.

9.5.2 BIP39

Bitcoin Improvement Proposal 39 (BIP39) by Market Palatinus et al. specifies a method to transform a mnemonic code into a binary seed. The code is defined by a sequence of human-readable words generated from a natural language dictionary, such that each word is unambiguous, easy-to-spell, and embeds enough entropy that a short sequence may be used as a seed. In a way, one can loosely consider BIP39 as the secure version of a brain wallet: it transforms a sentence into a secret, and guarantees that the sentence is random

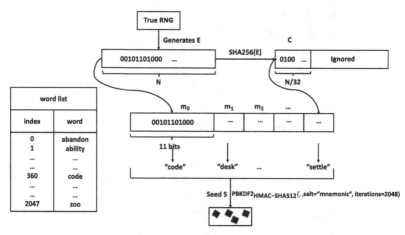

Figure 9.9 RNG to mnemonic to seed following BIP39 specification. Own graphics.

enough that it cannot be brute-forced. Quoting the authors, "BIP39 is meant to be a way to transport computer-generated randomness with a human readable transcription." It makes manipulation and storage of a large random number easier and less prone to errors while guaranteeing enough entropy for the generation of ECDSA secret keys, in particular in the context of a hierarchical deterministic key derivation algorithm such as BIP32. Figure 9.9 illustrates the combination of BIP39 and BIP32 such that the master extended key is derived from a seed, itself computed from a mnemonic code: the mnemonic code is sufficient to recover the complete hierarchy of keys and therefore to demonstrate ownership of any of the derived accounts.

We describe the algorithms below:

(1) An initial entropy E of N bits is generated, where $N \in \{128, 160, 192, 224, 256\}$.

(2) A checksum C is set to the first $N/32$ bits of $SHA256(E)$.

(3) The concatenation $E||C$ of the initial entropy and the checksum is split into subsequences of 11 bits, respectively $m_0, m_1, \ldots, m_{\frac{N+N/32}{11}}$.

(4) Integers (m_i), each encoding a number from 0 to 2047, serve as a sequence of indices into a wordlist to construct the mnemonic code.

(5) The 512-bit seed S is the output of the Password-Based Key Derivation Function 2 (PBKDF2) function with mnemonic as input, and salt set to constant string "mnemonic" (optionally augmented with secret passphrase). The iteration count is set to 2048 and HMAC-SHA512 is used as the pseudo-random function.

The wordlist, or dictionary, proposed by the authors of BIP39 is a careful selection of natural language words satisfying the following characteristics:

- *smart selection of words*: the wordlist is created in such way that first four letters unambiguously identify a complete word

- *similar words avoided*: word pairs like "build" and "built," "woman" and "women," or "quick" and "quickly" are disallowed to minimize the risk of errors and facilitate memorization
- *sorted wordlists*: the word list is sorted to allow efficient lookup of the code words, such as with binary search

We list below the first 30 words of BIP39 English wordlist and note that alternative versions exist in Japanese, Korean, Spanish, Chinese, French, Italian, and Czech. One can immediately verify that, by design, there does not exist a pair of words starting with the same sequence of four letters, and that words with too much lexical proximity have been excluded. We refer to BIP39 specification for the full wordlist.

abandon, ability, able, about, above, absent, absorb, abstract, absurd, abuse, access, accident, account, accuse, achieve, acid, acoustic, acquire, across, act, action, actor, actress, actual, adapt, add, addict, address, adjust, admit.

In practice, one still needs to generate the initial entropy E. Although this can be achieved free of electronic devices — for instance by flipping a coin repeatedly — one generally relies on some form of hardware module to bootstrap BIP39. The module will gather entropy, execute BIP39 algorithm, store the seed in secure memory, and output the mnemonic code for export. The wallet owner may then copy the mnemonic code on a piece of paper or any resilient substrate and rely on it for recovery in case of hardware failure or loss. Figure 9.10 illustrates a custom-designed, highly resilient tool for persisting a BIP39 mnemonic code created by company CryptoSteel: the creators claim that the tool is stainless, fireproof, shockproof, and waterproof and therefore make it the ideal solution for long-term seed storage. By combining the qualities of BIP32, BIP39, and a robust mnemonic storage tool, one has been able to create the secure version of a brain wallet.

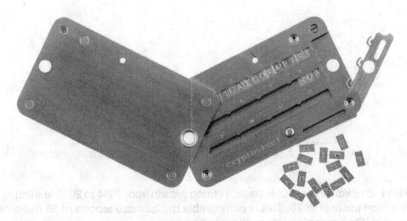

Figure 9.10 Hard back up of the BIP39 seed using specialized, robust store.

9.6 The relativeness of security

9.6.1 Custodial wallet services

Bitcoin was originally designed to favor an independent economy excluding banking involvement. *Be your own bank and don't trust no bank!* has been the strongest convergence argument of the Bitcoin community, which made its mandate of encouraging self-management of secret keys and therefore removal of counterparty risks related to custodial wallet providers. Although respectful of the original Bitcoin ideology, simple market observation reveals that many investors actually do not see it that way: centralized custodians such as Coinbase, Bitstamp, or Gemini have seen their user base and assets under management grow massively over the years, to a point that it becomes reasonable to wonder whether there aren't more users relying on centralized custody services than there are independent ones. In 2020, Coinbase, one of the largest cryptocurrency custodians on the market, claims 35 million users relying on their services. Figure 9.11 illustrates the growth of Blockchain.com user base over the second half of the decade, whose model arguably is in between a custodial and a non-custodial wallet provider. As institutional capital flows into cryptocurrencies, it is reasonable to assume that the growth of custodial wallet services will accelerate even more in the future.

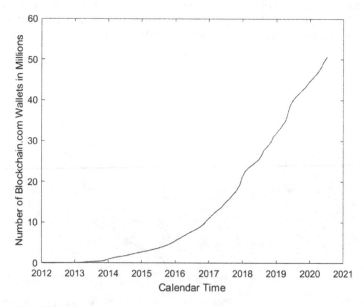

Figure 9.11 Blockchain.com user base: a strong growth from 2014 to 2019 leading to more than 30 million users in 2020. This is comparable to Coinbase reports of 35 million active users in 2020. Own graphics. Source: Blockchain.com.

This trend largely reflects the complexity of cryptocurrency self-custody. Storing cryptocurrencies requires special expertise in security and possibly cryptography, a particular care, and special infrastructure. Being your own bank does not only mean taking back power and control: it means bringing back all the responsibility and liability with it — something which is often ignored by the cryptocurrency community! Some people may be able to achieve it, but most will see value in externalizing this risk to a specialized body and paying for the service.

As a general principle security cannot be optimized without consideration for other requirements, particularly regarding the availability of the funds. Intuitively, a highly guarded asset will be harder to access; alternatively, an easily accessible, ready-to-use asset may suffer from more attack vectors. We take the illustration of cash, which may be stored in a wallet held in the pocket of its owner, or may be deposited in a more secure location such as a vault. Getting cash from the wallet is immediate and inexpensive and makes it ideal for frequent payments that require immediate processing. This flexibility comes at the cost of security: a wallet generally has no physical security or access controls; risks of loss or destruction are also tangible. Getting cash from a bank vault, however, guarantees a physically secure storage and enforces strict identity verification for access. It is evident that many important risks associated with the limited security provided by a wallet are solved with a bank vault, which makes it better suited for large savings. This security comes at the cost of the availability of funds: access to cash requires an in-person visit to the vault and is therefore slow, possibly expensive due to travel costs and processing fees — and notably may open new threat vectors for instance during transport. A natural strategy is therefore to rely on the combination of both a bank vault for long-term savings and on a wallet for smaller reserves and day-to-day payments.

Rather than optimizing security as an absolute, which would always favor a bank vault over a personal wallet, one must consider the availability requirement and use case as a preliminary condition. It is a general security principle that security and availability are antithetic: one will aim to minimize the attack surface conditional on a specific availability requirement, rather than as an isolated objective.

9.6.2 A practical example

We illustrate a possible approach with the example of a BIP39 paper wallet. The wallet owner takes a fair coin and flips it 256 times in a row, while writing down every occurrence of heads as a 0, and every occurrence of tails as a 1. Assuming the conditions of the experiment are ideal — in particular the coin flip exercise accurately follows a sequence of independent Bernoulli events of probability $1/2$ — one is now armed with 256 bits of entropy. By following BIP39 step by step, the wallet owner is able to extract the corresponding mnemonic code with the objective to write it down on a piece of paper. The paper is stored in a physically secure location and can be recovered at any time to extract secret keys and sign transactions.

Although this process may look secure given its limited exposure to a computer or the Internet and the absence of trust in third-party software or hardware — a remote attacker cannot hack a piece of paper after all — it is particularly painful to operate, it ignores important threat vectors, and it fails to satisfy basic requirements inherent to wallet use cases. First, the mnemonic code is stored on a simple piece of paper, in a single location. It implies that in case of disaster such as a fire, flooding, or earthquake, the mnemonic code might no longer be recoverable or even accessible. It also means that there is a unique attack point onto which the efforts of an attacker may be concentrated. This central point of failure is a regular reason for losses when considering some of the public cryptocurrency hacks.

Second, the mnemonic code, and therefore the key material, being physically secured on a static substrate, every transaction authorization requires the wallet owner to fetch the mnemonic and import it into a computing device to calculate the signatures for a specific transaction. In fact, the mere computation of the mnemonic code from the entropy source, or the subsequent computation of an address, which is required to receive coins in the first place, cannot realistically be performed in absence of a computer given the thousands of integer operations required. The dependence on computing devices opens multiple threats related to the operational risks during import and export of the keys from a static substrate to and from computer, and in relation to the management of the key material while stored and consumed on the digital medium. At the least, a strict protocol must be defined around the transport of the keys and the computer offering strong guarantees that the key material may not be compromised during the process; one may also require that the digital copy of the keys is destroyed after the calculation of signatures.

Third, this operational model assumes that the wallet creator can be fully trusted — not only in terms of his good faith, but also in terms of its competence and respect for the security protocol. Given the possible long-term use of the wallet, the model also assumes that the wallet creator is trusted not only at the time of wallet creation but will remain trusted for the whole lifetime of the wallet. This situation is commonly qualified of *single point of trust* in that a single entity has the critical power to break the whole security model. Although this may be tolerated for a personal wallet, it is considered a terrible practice in an institutional environment where the wallet may store large amounts of wealth and manage the assets of third-party investors trusting the operational framework of the service provider. Indeed, the wallet creator may unpurposely remember part or all of the key, or he may be dishonest and steal a copy of the key, or he may be incompetent and diverge from the strict key management protocol. Whether honest or dishonest, such scenarios may be exploited later on by a the wallet creator or by an attacker bribing or coercing him to reveal information.

9.6.3 The Binance hack

In order to emphasize the threat of keys being managed on connected computing devices, we relate the attack of cryptocurrency exchange Binance in May 2019. The company, founded in 2017, quickly became the largest exchange with a yearly trading volume crossing the US$1 trillion threshold in 2020. Despite the apparently high quality of its technical infrastructure, and its respected status on the market, it was subject to a massive theft of more than 7,000 BTC, or 2% of its BTC holdings at the time. Interestingly, only a limited number of wallets were reportedly affected: so called hot wallets, which are connected to the Internet and often programmed to automatically process withdrawal transfers based on clients' requests, were compromised and instantly emptied by the attacker. Other wallets subject to higher security standards, less automation, and more isolation from public networks, were seemingly unaffected.

9.6.4 The QuadrigaCX fraud

At the same period, the bankruptcy of another exchange, Quadriga CX, provides a painful demonstration of what a central point of trust can lead to. On February 5, 2019, Canadian exchange Quadriga CX goes down owing 76,000 users around CA$215 million in total, or about US$163 million at the time. The wife of the exchange founder Gerald Cotten reports that her husband died in a trip to India and that only he knows the password allowing the decryption of the Quadriga main wallets. The valuable coins are still under the control of the exchange, she claims, but with no knowledge of the password, they cannot be accessed and distributed back to their clients. The specific conditions of Mr. Cotten's death, or the reasons of his trip to India, remain uncommunicated. In itself this alleged situation is already a powerful example of the risks of centralized trust: the founder of a company, for noble reasons or not, is the only individual to have access to the reserves of a whole company managing millions of dollars. Despite his apparent good intentions, this man may be subject to memory loss, health issues, an accident, or he could unexpectedly die as he allegedly did. These possibilities in themselves are enough to consider alternative implementations of a wallet.

However on June 19, 2019, the exchange's court-appointed monitor Stikeman Elliott release an interim report on the investigations into the business and affairs of Quadrica CX and its previous director Mr. Cotten. Their preliminary conclusions identify the following principal concerns, which affected the situation facing Quadriga CX and its users. We summarize the executive summary as published by the authors of the report below:

(1) The company activities were largely directed by a single individual, Mr. Cotten and as a result, typical segregation of duties and basic internal controls did not appear to exist;

(2) No accounting records have been identified and there appears to have been no segregation of assets between Quadriga CX's funds and user funds. Funds received from and held by Quadriga CX appear to have been used by the company for a number of purposes other than to fund user withdrawals.

(3) Users' cryptocurrencies were not maintained exclusively in Quadriga CX's wallets. Significant volumes were transferred off the platform to competitor exchanges into personal accounts controlled by Mr. Cotten. It appears that users' cryptocurrencies were traded on these exchanges and in some circumstances used as security for a margin trading account established by Mr. Cotten. Trading losses incurred and incremental fees charged by exchanges appear to have adversely affected the company reserves.

What started as a loss of access to the wallet actually transformed into an apparent elaborate fraud with Ponzi characteristics. Mr. Cotten apparently used his clients' deposits to trade for his own account long before the announcement of his alleged death.

His ability to defraud his customers has in fact very little to do with the specificities of cryptocurrencies: similar fraudulent activities may have been — and have been — applied in traditional, regulated financial institutions. They are nonetheless facilitated by the pseudo-anonymity inherent to cryptocurrencies, the peer-to-peerness of transactions, and the relative immaturity of regulations, proper risk frameworks and controls in this industry.

9.7 Key isolation and transaction approval

9.7.1 Security ingredients

Overall, the security of a wallet heavily depends on two factors:

(1) the protection of its keys;
(2) the protection of its transaction signature process.

Even when the keys are well protected, the process under which the system may be authorized to use the secret key to sign a transaction can become the weak point of the system. An HSM, for instance, provides strong guarantees over the confidentiality of keys; however, HSMs often expose an interface to which an external service connects to ask for a signature. Even assuming that the key is well protected in the HSM, the interface becomes an attack vector: the attacker may attempt to access this interface and ask the HSM to sign an arbitrary transaction. Therefore, such an interface must be well guarded, and must enforce good enough controls and process to alleviate the risk of a single point of failure.

9.7.2 Hot and cold storage

We like to define two broad classes of wallets often called hot wallets and cold wallets.

(1) *Hot wallet*: A hot wallet is a wallet whose secret keys are manipulated on a device that is, or may be, connected to a network (generally the Internet). Hot wallets are typically implemented for applications where access to the funds must be fast, practical, and inexpensive. Because the keys are exposed to remote attackers, the attack surface is large and such wallets are generally used only when high availability or automation is required.
(2) *Cold wallet*: A cold wallet is a wallet whose secret keys are under no circumstance stored or manipulated on a device that may be connected to a network. Because the keys are physically isolated, they are immune against remote attacks and may only be attacked with physical access to the infrastructure. Cold wallets are typically implemented for applications where security is critical (e.g., when assets under management have significant value) and where the access time and cost to the funds is a secondary concern.

Naturally, specific implementations of hot wallets may embed robust protocols mitigating the large attack surface. Equivalently, inconsiderate implementations of cold wallets may reveal critical weaknesses trumping their apparent resilience. A solid wallet implementation should model specific thread scenarios to maximize security under specific availability requirements of the wallet use case.

In practice, most cryptocurrency exchanges store 90% or more of their assets into cold storage; the rest of the assets are distributed into hot wallets to facilitate the processing of client withdrawal requests. The Binance hack has only affected hot wallets at the exchange; assets stored in cold storage were unaffected.

9.7.3 Four-eyes principle

In order to mitigate the risk of centralized trust, one often defines a transaction approval workflow such that multiple independent parties approve the operation. At minimum, a four-eyes principle is enforced: a submitted operation must gather the validation of two users before its execution. By extension, more elaborate schemes may be defined, such as a so-called $m-by-n$ multi-approval scheme where m parties must approve the operation out of a total of n potential approvers. This scheme generalizes the four-eyes principle to an m-eyes principle, and also defines the total number of users authorized to participate in the quorum.

In addition, the approval workflow is often conditioned to the risk of the transaction to be approved: the transaction amount, the destination of the transaction, its fees, the time of the day, etc. may be factors impacting the workflow as defined by the custodian.

Approval workflows may be implemented *on-chain*, that is, directly on a distributed ledger, or *off-chain*, that is, in a traditional service lying on top of a distributed ledger. We discuss the options below.

9.7.4 Off-chain validation

An approval workflows can be enforced by the wallet infrastructure, off-chain. This approach is advantageous in that it gives the ability to create a single governance model

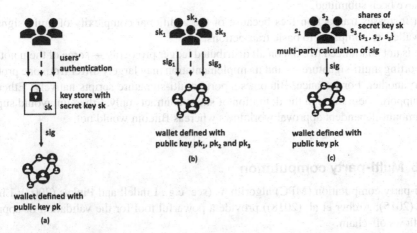

Figure 9.12 On-chain multi-signature versus hardware-based multi-approval versus multi-party computation. Own graphics.

guaranteed to work irrespective of the underlying ledger protocol and invisible to the other network participants. From the point of view of the distributed ledger, only a single key exists:

(1) This guarantees that transaction fees are as low as possible since no additional logic must be processed on chain through the use of a script or a smart contract.
(2) This obfuscates the governance processes entirely to the eyes of third-party blockchain observers because the ledger is unaware of the approval workflow details sitting off-chain.

However, the above security model concentrates risks on several critical off-chain components, which should be designed with a strict security architecture. If these components are compromised in any way that can leak the key material or bypass the security policies, the wallets may be emptied without additional controls. For these applications, hardware security modules and secure execution environments are often used to increase the trust in data and processes.

9.7.5 On-chain validation

Most distributed ledger protocols allow the definition of accounts controlled by more than one party, as illustrated in Figure 9.12. One can implement transaction approval workflows requiring on-chain validation of multiple signatures, such as an $m - by - n$ scheme. The approach has been common in the Bitcoin protocol, which natively supports multi-signature scripts; it can also be trivially implemented in Ethereum via smart contracts.

On the flip side, on-chain validation of non-trivial approval workflows suffers from a series of drawbacks:

(1) it makes the validation process visible to the blockchain observers, who can infer the security model of the wallet and track in real-time which of the respective approvals have been submitted;
(2) it increases transaction fees because of the additional complexity of multi-signature wallets as compared to basic transactions;
(3) it is not guaranteed to exist on all distributed ledger protocols — some of them not supporting multi-signature — and its implementation may largely differ from one protocol to another. For instance, Bitcoin supports multi-signature scripts natively; Ethereum supports them through the definition of smart contracts only. Ethereum would support amount-dependent approval workflows whereas Bitcoin would not.

9.7.6 Multi-party computation

Multi-party computation (MPC) algorithms (see, e.g., Lindell and Pinkas (2007); Lindell et al. (2015); Archer et al. (2018)) provide a powerful tool for the validation of approval workflows off-chain.

Multi-party computation is capable of maintaining distributed key shards stored in different locations that jointly can compute a signature, without ever reconstructing the wallet key. From the distributed ledger point of view, a single key is used to sign transactions; from the user point of view, a $m - by - n$ scheme may be implemented using MPC such that the key cannot be reconstructed if less than m fragments have been recovered. More elaborate workflows can even be introduced with MPC such that advanced workflows are validated without ever recentralizing the key.

Although elegant, this model imposes more communication between the signing nodes and can be computationally expensive depending on the complexity of the workflow rules. Promoters of MPC technology in this industry argue that the distribution of key shards removes the need for specialized hardware-secure environments, although the argument is debated.

9.8 Summary

In this chapter, we review the main risks associated with the management of cryptocurrencies as well as the best practices for key management. We lay out the typical architecture of a wallet, with its several components contributing to the overall service. We conclude that proper key management and governance is critical, and that best fitted implementations heavily depend on the specifics of the application. We introduce the notion of multi-signature, hardware security modules, and multi-party computation and their particularities.

9.9 Exercises

1. Consider a wallet leveraging the BIP32 specification. Explain how generating the seed on a traditional computing device may be dangerous. Assume that wallet keys are derived from the seed using nonhardened derivation. Explain the consequences of leaking the chain code. Explain the consequences of leaking one of the extended secret keys. Explain how you can leverage hardened derivation in this context.
2. Assume the seed has been generated according to BIP39 with 24 words. Assume one of the words has been lost and its position in the list is known. How much time would it take to brute force the seed, assuming 1 million attempts per second? What if the position is unknown? Assume eight words have been lost. How much time would it take to brute force under the same assumptions? Assume all of the words are known, but their ordering has been lost. How much time would it take to brute force under the same assumptions? Assume you have access to three safe deposits. Propose an archive model for the 24 words such that (i) breaking a single safe would not compromise the key, (ii) losing access to a single safe would not prevent rebuilding the key.
3. Study the possibility to shard the key into multiple independent fragments for better distribution of the risk and resiliency. Explain how Shamir's secret sharing can help. What is the weakness of this approach at the time of signature? Explain how multi-party computation can help. What are the advantages and disadvantages of this approach? Explain the application of hardware security modules and how they can help distribute risk.

CHAPTER
10

Cryptocurrencies and Quantitative Finance

10.1 Introduction

In addition to the fascinating technological aspects of distributed ledgers, the associated cryptoassets are interesting to study in the context of quantitative finance. Given the rapid development of DeFi and its mostly unregulated nature, one can observe some exciting and unusual phenomena. We can view the crypto universe as a financial supernova, similar to a star that suddenly increases brightness due to a catastrophic explosion ejecting most of its mass.

In this chapter, we briefly touch upon the very nature of cryptocurrencies and discuss what money-like features they possess, if any. We look at the elusive fair value of BTC and other cryptocurrencies. Given that the valuation of cryptocurrencies is a highly controversial topic, the estimation of their actual value is in the eye of the beholder. Further, we analyze historical returns of the big three — BTC, ETH, and XRP, as well as the total cryptoasset index, which we denote by MRKT, and show how to fit various popular distributions to these returns. In general, returns possess strongly non-Gaussian features and have extremely heavy tails. Exchanges, such as CME, wishing to support crypto futures trading, need to appreciate this fact in earnest. We use some interesting techniques borrowed from epidemiology to analyze BTC dominance as part of the cryptocurrency index. We also analyze automated market makers (AMMs) and study their profitability by using some tried and tested methods of mathematical finance. Many results presented in this Chapter are original and have not been published before.

The chapter proceeds as follows. Section 10.2 analyzes money-like aspects of cryptocurrencies. We argue that in their current form, one cannot view cryptos as money. Potentially, in the future, they can acquire some money-like features as part of their evolution. Section 10.3 presents our take on "fair" prices of cryptos. In Sections 10.4–10.6, which are rather technical, we study statistical characteristics of historical returns of the big-three — BTC, ETH, and XRP, as well as the market as a whole (MRKT). In all cases, the generalized hyperbolic distribution provides the best fit to the corresponding empirical returns. We also calculate the Value-at-Risk (VAR) for BTC returns. These quantities are of paramount importance for building a successful futures trading framework. In Section 10.7, we compare BTC, ETH, XRP, and MRKT among themselves, as well as against several currency pairs such as EUR/USD, GBP/USD, USD/JPY, among others. Section 10.8 studies the ecology of the cryptocurrency ecosystem as a whole. Section 10.9 describes the stock-to-flow valuation model for BTC. Section 10.10 uses epidemiological models, such as Susceptible-Infected-Removed (SIR) and Susceptible-Infected-Susceptible (SIS) to explain the behavior of the Bitcoin dominance index. Section 10.11 details an interesting concept of AMMs by using conventional mathematical finance tools. Section 10.12 summarizes our conclusions.

10.2 Can cryptocurrencies be viewed as money?

It is too early to say whether Bitcoin is a new form of money, but we think, at present, the answer is a qualified "no." Indeed, in the hard light of the day, Bitcoin does not satisfy most of the requirements listed in Chapter 2, which an instrument needs to fulfill to be money. Bitcoin is undoubtedly not a medium of exchange or a unit of account. However, one can probably view it as a form of treasure in the spirit of the asset classification developed by Greer, who divided assets into three "superclasses": capital, consumable/transformable (C/T), and store of value (SOV) assets; see Greer (1997). SOV assets, such as gold, art, and other collectibles, do have value, even though they cannot be consumed and do not generate income. Time will tell if Bitcoin is an honorary member of the SOV club; see Szabo (2017).

If Bitcoin were to acquire some monetary features at a later stage of its evolution, we could experimentally study how money obtains and retains value.[1] In particular, we shall be in a position to confirm or deny the following historical description from Mises of how commodities become money, see Mises (1998): "We will eventually arrive at a point in time when money was just an ordinary commodity where demand and supply set its price. The commodity had an exchange value in terms of other commodities... To put it simply, on the day a commodity becomes money, it already has an established purchasing power or price in terms of other goods. This purchasing power enables us to set up the demand for this commodity as money. This in turn, for a given supply, sets its purchasing power on the day the commodity starts to function as money. Once the price of money is fixed, it serves as input for the establishment of tomorrow's price of money."

At present, one can best interpret BTC's value if she thinks of the Bitcoin protocol as a game in a giant casino. As described in detail in Section 5.5, BTCs have a fixed supply, are divisible, and technically relatively easy to acquire.[2] Hence, they are ideal tokens, like casino chips. In addition to all other attributes, BTCs have entertainment value.

While one can argue about Bitcoin's money credentials, altcoins, even the biggest ones, such as ETH and XRP, are undoubtedly not money and, more likely than not, never will become one.

10.3 Do fair prices of cryptocurrencies exist?

Bitcoin price dynamic is unique. It exhibits all the familiar hallmarks of bubble-like behavior, but this bubble is on steroids because of its worldwide nature. Moreover, given the BTC notoriety, one can argue that a bubble can't last more than a decade. It is often claimed that

---

[1] Although researchers have extensively speculated about transforming an object into money, they were never able to come to a meaningful conclusion.

[2] For operational issues involving BTC trading and storage, see Chapter 5.

Bitcoin's design is such that anyone with money and access to the Internet everywhere in the world can buy BTCs in a decentralized fashion. Of course, in reality, nothing can be further from the truth — Bitcoin is being sold on centralized exchanges just undermining its *very premise of being decentralized*! Analysis of its price dynamics is very important for BTC trading on exchanges and BTC futures clearing performed by the Chicago Mercantile Exchange (CME) and other firms.

Part of the problem why it is so hard to assign a rational value to Bitcoin is not too difficult to discern, although only a few commentators manage to do so or to admit it in public. *Bitcoin has no value, and hence it can have any price!* Besides, to make things even harder to understand, Bitcoin ownership is characterized by extreme, truly feudal, inequality; see Table 10.1.

More often than not, Bitcoin whales, i.e., owners of large amounts of BTCs, are "hodlers," keeping their BTCs out of circulation, regardless of the price volatility. A smaller group of BTC owners are probably engaged in price manipulation.

Currently, there are three schools of thought regarding the long-term price of bitcoin — (a) BTC at US\$0; (b) BTC at US\$10,000,000 or more; (c) BTC at market price, whatever it might be. It would be an understatement to say that adherents to both the first and the second viewpoints have no idea of what they are talking about and, in many cases, have strong financial incentives to promote their opinions publicly. The third viewpoint is, of course, accurate but rather uninformative.[3]

Figure 10.1 shows cryptocurrency prices in USD. In Figure 10.2, we present them on a log scale, which is more useful to demonstrate that cryptocurrencies are subject to periodically occurring bubbles. For instance, BTC experienced at least three bubbles in its 12-year history. In all three cases, BTC lost about 80% of its value.

10.4 Distribution of daily BTC returns

The BTC log-returns and their probability density function (pdf) are shown in Figures 10.3 (a) and (b), respectively. The dataset consists of 2,562 daily returns between July 2013 and June 2020, which are denoted as $x_1, ..., x_n, ..., x_N$, $N = 2,562$.

For several decades, researchers have known that financial assets' daily returns, such as foreign exchange rates and bond and equity prices, are not Gaussian or normal. The corresponding empirical distributions are more peaked around the mean, exhibit heavier tails than the normal distribution, and often skewed toward the left tail, implying a higher probability of negative returns. As a result, large returns (both positive and negative) occur with a much higher frequency than under the normality assumption. The non-normality of

---

[3]One of the authors (AL) still vividly remembers his experience in January 2020 in Davos. Since Davos is a small town, ill-suited for hosting large international gatherings, space is at a premium. Before giving a scheduled interview, AL had to wait in a small room for the previous interview to finish. The interviewee, a prominent US professor, was very hostile to cryptocurrencies and used colorful language (perhaps best characterized as profane) to make his points. The longer the interview lasted, the clearer it became that this distinguished scholar had very little, if any, understanding of the topic he was opining on and compensated his lack of knowledge with the strength of his language. This situation, unfortunately, is not uncommon in the cryptocurrency space.

Table 10.1 Bitcoin wealth distribution as of October 18, 2020. Source: https://bitinfocharts.com/top-100-richest-bitcoin-addresses.html.

Balance	BTC Addresses	% Addresses	% Total	BTC	USD	% Coins	% Total
(0–0.001)	15748098	48.87	100.00	3,114	35,664,401	0.02	100.00
[0.001–0.01)	7888276	24.48	51.13	31,867	364,975,271	0.17	99.98
[0.01–0.1)	5430334	16.85	26.65	176,992	2,027,115,728	0.96	99.81
[0.1–1)	2339762	7.26	9.79	736,634	8,436,777,532	3.98	98.86
[1–10)	661728	2.05	2.53	1,719,911	19,698,396,718	9.29	94.88
[10–100)	137600	0.43	0.48	4,451,331	50,981,748,147	24.04	85.59
[100–1,000)	13925	0.04	0.05	3,494,690	40,025,193,505	18.87	61.55
[1,000–10,000)	2075	0.01	0.01	5,258,143	60,222,280,353	28.39	42.68
[10,000–100,000)	103	0	0	2,417,772	27,691,093,467	13.06	14.28
[100,000–1,000,000)	1	0	0	227,502	2,605,615,009	1.23	1.23

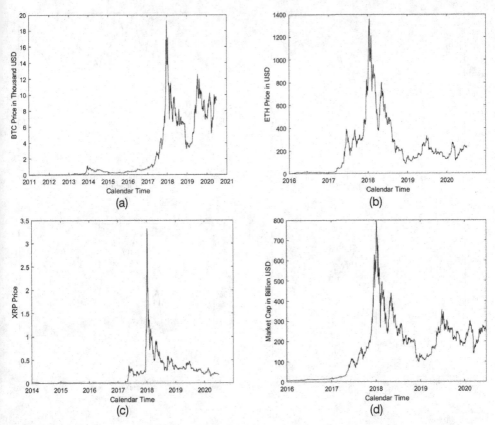

Figure 10.1 BTC, ETH, XRP, and MRKT price series. Own graphics. Source: coinmarket-cap.com.

returns on cryptocurrencies, such as Bitcoin, Ethereum, XRP, and others, is particularly pronounced.

This section discusses possible candidate distributions for BTC returns; ETH, XRP, and MRKT returns are analyzed in Section 10.6. We study several distributions, starting with the standard normal distribution and going all the way to the generalized hyperbolic distribution, initially introduced by Ole Eiler Barndorff-Nielsen to describe various physical phenomena; see Barndorff-Nielsen (1997). We show that the latter distribution provides an adequate, if not perfect, fit to empirical data. Several papers discuss fitting of BTC and other crypto returns as well as their qualitative behavior; see Chan et al. (2017); Chu et al. (2015); Scaillet et al. (2020), among others.

We use kernel distribution as a nonparametric representation of a random variable's pdf to avoid making any *a priori* assumptions. A kernel distribution depends on random samples from an unknown distribution, a smoothing function, and a bandwidth value, which controls the output curve smoothness.

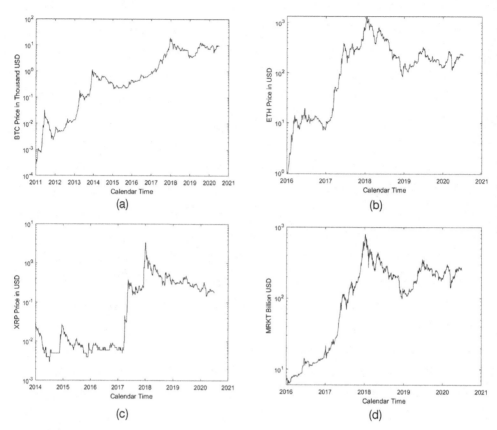

Figure 10.2 BTC, ETH, XRP, and MRKT price series on a log scale. Own graphics. Source: coinmarketcap.com.

The kernel density estimator generates a pdf for a random variable under consideration and is defined as follows:

$$\hat{f}_h(x) = \frac{1}{Nh} \sum_{n=1}^{N} K\left(\frac{x - x_n}{h}\right),\tag{10.1}$$

where $x_1, x_2, \ldots, x_N$ are random samples from an unknown distribution of BTC returns, N is the sample size, $K(\cdot)$ is the kernel smoothing function, and h is the bandwidth. The kernel distribution is similar to a histogram. On the one hand, it represents the probability distribution using the sample data. On the other hand, in contrast to a histogram, which places values into discrete bins, the kernel distribution sums the component smoothing functions and produces a continuous probability curve.

We choose the following Epanechnikov kernel:

$$K(x) = \frac{3}{4}\left(1 - x^2\right).\tag{10.2}$$

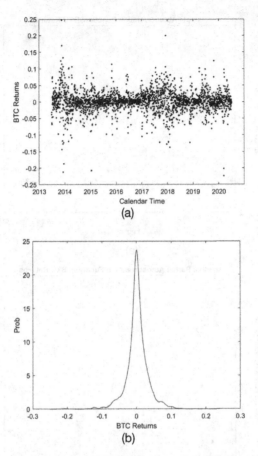

Figure 10.3 (a) BTC returns series; (b) BTC fitted kernel distribution. Own graphics.

It is well-known that, in the absence of further information, the Epanechnikov kernel produces superior fits compared to others. We choose the default bandwidth h, theoretically optimal for estimating densities for the normal distribution, as featured in Bowman and Azzalini (1997), and produce a reasonably smooth curve shown in Figure 10.3 (b).

Now, let's turn our attention to autocorrelations and partial autocorrelations, which play essential roles in time series analysis and forecasting. Autocorrelations measure a return's correlation with a delayed copy of itself as a function of delay lag k. Partial autocorrelations measure a signal's correlation with a delayed copy of itself as a function of delay lag k after the effects of correlations due to shorter lags $k' < k$ are removed; see Box et al. (2011). For BTC returns, autocorrelation and partial autocorrelation are shown in Figure 10.4.

This figure demonstrates that apart from short lags, BTC returns can be approximated as independent. Therefore, below we treat them as such for simplicity. We fit the corresponding empirical distribution with several well-known probability distributions via the log-likelihood maximization with penalties.

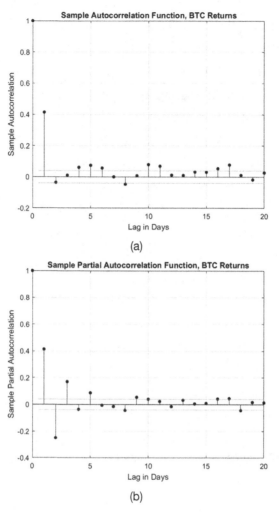

Figure 10.4 BTC autocorrelation and partial autocorrelation functions. Own graphics.

Recall that, subject to various technical conditions, any random variable X is characterized by the pdf $f(x)$, the cumulative density function (cdf) $F(x)$, and the characteristic function $\phi(\chi) = E\left\{e^{i\chi X}\right\}$. The relationships between $f(x)$ and $\phi(\chi)$ are as follows:

$$\phi(\chi) = \int_{-\infty}^{\infty} f(x) e^{i\chi x} dx,$$

$$(10.3)$$

$$f(x) = \frac{1}{2\pi} \int_{-\infty}^{\infty} \phi(\chi) e^{-i\chi x} d\chi.$$

We can think of several candidate distributions to fit the empirical distribution: the normal, Cauchy, stable, Student's t, normal-inverse Gaussian, and generalized hyperbolic distributions.

We start with the normal distribution, $x \sim \mathcal{N}\left(\mu, \sigma^2\right)$, invented by Carl Friedrich Gauss and Pierre-Simon Laplace. The corresponding $f_{\mathcal{N}}(x)$, $\phi_{\mathcal{N}}(\chi)$ have the form:

$$f_{\mathcal{N}}(x) = \frac{e^{-(x-\mu)^2/2\sigma^2}}{\sqrt{2\pi\sigma^2}},$$

(10.4)

$$\phi_{\mathcal{N}}(\chi) = \exp\left(i\mu\chi - \frac{\sigma^2\chi^2}{2}\right).$$

The meaning of the parameters μ, σ is as follows: μ is the mean or expectation of the distribution and its median and mode, σ is its standard deviation, and σ^2 — its variance. Normal distributions play an important role in applied mathematics, physics, and other scientific disciplines, including mathematical finance; see Abramowitz and Stegun (1948); Johnson et al. (1995). They are frequently used to represent random variables with unknown distributions by default. However, given the nature of BTC returns, it is unlikely that the normal distribution can be adequate.

For the Cauchy distribution $x \sim C(\delta, \mu)$, introduced by Siméon Denis Poisson, Augustin-Louis Cauchy, and Hendrik Lorentz, the corresponding $f_C(x)$, $\phi_C(\chi)$ have the form:

$$f_C(x) = \frac{\delta}{\pi\left(\delta^2 + (x-\mu)^2\right)},$$

(10.5)

$$\phi_C(\chi) = \exp(i\mu\chi - \delta|\chi|).$$

In Eq. (10.5) δ is the scale parameter, and μ is the location parameter. The Cauchy distribution is very unusual since it does not have finite moments of order greater than or equal to one, so its expected value and variance are undefined; see Johnson et al. (1995). Given the highly irregular nature of BTC returns, it is worth trying it as a fitting candidate.

For the stable distribution, $x \sim S(\alpha, \beta, \delta, \mu)$, discovered by Paul Lévy, an explicit expression for the pdf $f_S(x)$ is not known, while its characteristic function has the form:

$$\phi_S(\chi) = \exp\left(i\mu\chi - (\delta|\chi|)^\alpha(1 + i\beta\operatorname{sgn}(\chi)\psi(\alpha, \delta, \chi))\right),$$

(10.6)

where

$$\psi(\alpha, \delta, \chi) = \begin{cases} \tan\left(\dfrac{\pi\alpha}{2}\right)\left((\delta|\chi|)^{1-\alpha} - 1\right), & \alpha \neq 1, \\ \dfrac{2}{\pi}\ln(\delta|\chi|), & \alpha = 1. \end{cases}$$

(10.7)

The parameters $\alpha, \beta, \delta, \mu$, are defined as follows: α — stability, β — scewness, δ — scale, μ — location; see Lévy (1925); Zolotarev (1986). A linear combination of two independent stable random variables is a stable random variable. The Gaussian and Cauchy distributions are special cases of stable distributions. Many authors, such as Benoit Mandelbrot, think that stable distributions provide an excellent fit for financial instruments returns; see Mandelbrot (1997). Accordingly, we feel compelled to cover stable distributions in our analysis.

For the Student's t distribution, $x \sim \mathcal{ST}(\mu, v, \sigma)$, with $f_{ST}(x)$ and $\phi_{ST}(\chi)$ of the following form:

$$f_{ST}(x) = \frac{\Gamma\left(\dfrac{v+1}{2}\right)}{\sqrt{v\sigma^2\pi}\,\Gamma\left(\dfrac{v}{2}\right)}\left(1 + \frac{(x-\mu)^2}{v\sigma^2}\right)^{-(v+1)/2},$$

(10.8)

$$\phi_{ST}(\chi) = \frac{\sigma\left(\sqrt{v}\sigma\,|\chi|\right)^{v/2}\exp(i\mu\chi)K_{v/2}\left(\sqrt{v}\sigma\,|\chi|\right)}{2^{v/2-1}\Gamma(v/2)},$$

see Gosset (1908); Abramowitz and Stegun (1948); Johnson et al. (1995). The meaning of the parameters is as follows: μ — location, v — degrees of freedom, and σ — volatility. The t-distribution, being bell-shaped and symmetric regarding its mean, is similar to the normal distribution. An important distinction is that the t-distribution has heavy tails and produces more values from the mean than the normal one. Hence it is a suitable candidate for fitting BTC returns.

For the normal-inverse Gaussian distribution, $x \sim \mathcal{NIG}(\alpha, \beta, \delta, \mu), f_{\mathcal{NIG}}(x), \phi_{\mathcal{NIG}}(\chi)$ are as follows:

$$f_{\mathcal{NIG}}(x) = \frac{(\gamma/\delta)^{-1/2}\alpha e^{\beta(x-\mu)}}{\sqrt{2\pi}K_{-1/2}(\delta\gamma)}y^{-1}K_{-1}(\alpha y)$$

$$= \frac{\alpha\delta e^{\gamma\delta+\beta(x-\mu)}}{\pi}y^{-1}K_{-1}(\alpha y),$$

(10.9)

$$\phi_{\mathcal{NIG}}(\chi) = \exp(\gamma\delta + i\mu\chi - \delta v),$$

where

$$\gamma = \sqrt{\alpha^2 - \beta^2}, \quad y = \sqrt{\delta^2 + (x-\mu)^2}, \quad v = \sqrt{\alpha^2 - (\beta + i\chi)^2},$$

(10.10)

see Aas and Hobaek Haff (2005); Barndorff-Nielsen (1997). Here K is the modified Bessel function of the second kind. The meaning of the corresponding parameters is as follows: α — tail heaviness, β — asymmetry, δ — scale, and μ — location.

For the generalized hyperbolic distribution, $x \sim \mathcal{GH}(\alpha, \beta, \delta, \lambda, \mu)$, the corresponding $f_{\mathcal{GH}}(x)$ and $\phi_{\mathcal{GH}}(\chi)$ can be written as follows:

$$f_{\mathcal{GH}}(x) = \frac{(\gamma/\delta)^\lambda e^{\beta(x-\mu)}(y/\alpha)^{\lambda-1/2}K_{\lambda-1/2}(\alpha y)}{\sqrt{2\pi}K_\lambda(\delta\gamma)},$$

(10.11)

$$\phi_{\mathcal{GH}}(\chi) = \frac{\gamma^\lambda \exp(i\mu\chi)K_\lambda(\delta v)}{K_\lambda(\delta\gamma)v^\lambda},$$

see Barndorff-Nielsen (1977). In addition to $\alpha, \beta, \delta, \mu$, we add the parameter λ, which determines the shape of the distribution. The generalized hyperbolic distributions are used when large deviations from the mean are essential, for example, to model returns of financial instruments. It is the most general distribution we include in our analysis. The Student's t and the normal-inverse Gaussian distributions belong to the class of generalized hyperbolic (GH) distributions. Unfortunately, the class of generalized hyperbolic distributions is not closed under a linear combination of its members, making it very difficult to aggregate daily returns to monthly or annual, see next section.

The hyperbolic distribution, which is a special case of $\mathcal{GH}$ with $\lambda = 1$:

$$\mathcal{H}(\alpha, \beta, \delta, \mu) = \mathcal{GH}(\alpha, \beta, \delta, 1, \mu), \tag{10.12}$$

is often used as a candidate for fitting BTC returns; see Barndorff-Nielsen (1977). However, our experience suggests that the corresponding fit is generally not good, so, for brevity, we omit the corresponding calculations.

Straightforward comparison of the corresponding characteristic function confirms the well-known fact that the normal and Cauchy distributions are the special cases of the stable distribution:

$$\mathcal{N}(\mu, \sigma^2) = \mathcal{S}\left(2, 0, \frac{\sigma}{\sqrt{2}}, \mu\right),$$

$$\tag{10.13}$$

$$\mathcal{C}(\delta, \mu) = \mathcal{S}(1, 0, \delta, \mu).$$

The Student's t, and normal-inverse Gaussian are the special (or limit) cases of the generalized hyperbolic distribution:

$$\mathcal{ST}(\mu, v, \sigma) = \mathcal{GH}\left(0, 0, \sqrt{v}\sigma, -\frac{v}{2}, \mu\right),$$

$$\tag{10.14}$$

$$\mathcal{NIG}(\alpha, \beta, \delta, \mu) = \mathcal{GH}\left(\alpha, \beta, \delta, -\frac{1}{2}, \mu\right).$$

The normal and Cauchy distributions can be obtained in the limit as well, although neither of these distributions belong to the $\mathcal{GH}$ class.

Each distribution is fitted by the classical method of maximum likelihood; see, e.g., Hamilton (2020) among numerous others. We introduce the quantity $\ln L(\Theta)$ as follows:

$$\ln L(\Theta) = \sum_{n=1}^{N} \ln f(x_n; \Theta), \tag{10.15}$$

maximize it with respect to Θ, and choose the corresponding optimal value $\hat{\Theta} = (\hat{\theta}_1, ..., \hat{\theta}_k, ..., \hat{\theta}_K)$. The confidence intervals for $\hat{\Theta}$ are calculated by approximating the covariance matrix by the inverse of the empirical information matrix. Thus,

$$\mathrm{cov}\left(\hat{\Theta}\right) = \left(H\left(\hat{\Theta}\right)\right)^{-1}, \tag{10.16}$$

where $H\left(\hat{\Theta}\right)$ is the corresponding Hessian,

$$
H\left(\hat{\Theta}\right) = \left.\left(\begin{array}{ccccc}
\dfrac{\partial^2 \ln L\left(\Theta\right)}{\partial\theta_1^2} & \cdots & \dfrac{\partial^2 \ln L\left(\Theta\right)}{\partial\theta_1\partial\theta_k} & \cdots & \dfrac{\partial^2 \ln L\left(\Theta\right)}{\partial\theta_1\partial\theta_K} \\
\cdots & \cdots & \cdots & \cdots & \cdots \\
\dfrac{\partial^2 \ln L\left(\Theta\right)}{\partial\theta_k\partial\theta_1} & \cdots & \dfrac{\partial^2 \ln L\left(\Theta\right)}{\partial\theta_k^2} & \cdots & \dfrac{\partial^2 \ln L\left(\Theta\right)}{\partial\theta_k\partial\theta_K} \\
\cdots & \cdots & \cdots & \cdots & \cdots \\
\dfrac{\partial^2 \ln L\left(\Theta\right)}{\partial\theta_K\partial\theta_1} & \cdots & \dfrac{\partial^2 \ln L\left(\Theta\right)}{\partial\theta_K\partial\theta_k} & \cdots & \dfrac{\partial^2 \ln L\left(\Theta\right)}{\partial\theta_K^2}
\end{array}\right)\right|_{\Theta=\hat{\Theta}}. \tag{10.17}
$$

To account for the difference in the number of parameters in our choice of fitting candidates, we use several complementary criteria to pick the winner:

(1) The Akaike information criterion (AIC), see Akaike (1974):

$$
AIC = 2K - 2\ln L\left(\hat{\Theta}\right). \tag{10.18}
$$

The Akaike information criterion estimates the out-of-sample prediction error of different statistical models for a given set of data. One can increase the likelihood by adding additional parameters. For instance, the normal distribution has two parameters, μ, σ, while the generalized hyperbolic distribution has five, $\alpha, \beta, \delta, \lambda, \mu$. However, introducing extra parameters increases the risk of overfitting the data. By penalizing models with more parameters, AIC becomes a useful tool for model selection based on the relative quality of each model. The model with the lowest AIC wins.

(2) The Hannan–Quinn information criterion (HQIC), see Hannan and Quinn (1979):

$$
HQC = 2K\ln\left(\ln N\right) - 2\ln L\left(\hat{\Theta}\right). \tag{10.19}
$$

HQIC and AIC are broadly similar; however, the penalty term is larger in HQIC than in AIC.

(3) The Schwarz information criterion (SIC), also known as the Bayesian information criterion (BIC), see Schwarz (1978):

$$
SIC = K\ln N - 2\ln L\left(\hat{\Theta}\right). \tag{10.20}
$$

BIC and AIC are spiritually close, however, the penalty term is larger in BIC than in both AIC and HQIC.

For each of the above criteria, the objective is to choose the fitting distribution producing the lowest value.

In addition, we consider the following commonly used tests to evaluate the goodness of fit:

(1) The Anderson–Darling test, see Anderson and Darling (1954):

$$
ADT = -N - \frac{1}{N}\sum_{n=1}^{N}\left(2n-1\right)\left(\ln\left(\hat{F}\left(y_n\right)\right) + \ln\left(1 - \hat{F}\left(y_{N+1-n}\right)\right)\right), \tag{10.21}
$$

where $y_1 \leq \ldots \leq y_n \leq \ldots y_N$ are the observed returns arranged in increasing order. The Anderson–Darling test (AD test) estimates goodness of fit for a given probability distribution, with the parameters calculated via the maximum likelihood method.

(2) The Kolmogorov–Smirnov test, see Kolmogorov (1933); Smirnov (1948):

$$KST = \sup_x \left| F_N(x) - \hat{F}(x) \right|. \tag{10.22}$$

Here $F_N(x)$ is the empirical distribution, and $\hat{F}(x)$ is the maximum likelihood distribution,

$$F_N(x) = \frac{1}{N} \sum_{n=1}^{N} \mathbb{1}_{[-\infty, x]}(x_n), \tag{10.23}$$

where $\mathbb{1}_{[-\infty, x]}(.)$ is the indicator function, equal to 1 if $x_n \leq x$ and equal to 0 otherwise. The Kolmogorov–Smirnov test (KS test) is a nonparametric test for comparing a data sample with a given probability distribution. As for the AD test, the parameters of this distribution are calculated via the maximum likelihood method.

The objective of optimization is to choose the fitting distribution, which minimizes test metrics. Tables 10.2 and 10.3 summarize the calibration and maximization results for BTC returns.

The generalized hyperbolic distribution is a clear winner with regards to all the criteria and tests. We show the empirical distribution fitting with parametric distributions in Figure 10.5.

This figure clearly shows that the normal distribution provides the least adequate fit to the empirical distribution. In contrast, the generalized hyperbolic distribution fits the data well, particularly as far as heavy tails are concerned. Thus, large deviations from the mean are much more frequent for the actual returns than would have happened if they were normal. Figures 10.6 (a), (b) show the QQ plots for the fitted normal and generalized hyperbolic distribution vs. the kernel distribution.

An ideal fit should result in a straight line. Comparison of Figures 10.6 (a) and (b) shows that the generalized hyperbolic distribution produces an almost perfect QQ plot, while the normal distribution exhibits significant deviations. Assuming that the fitted generalized hyperbolic distribution adequately describes the price dynamics, we can calculate the Value-at-Risk (VAR) for both gains and losses, shown in Figure 10.7. These quantities are all-important for all risk management metrics, including initial margins for the Bitcoin futures clearing.

While there is a discernible difference between the empirical and $\mathcal{GH}$ VAR, it is relatively small. Moreover, VAR computations with $\mathcal{GH}$ provide more conservative estimates for both long and short positions.

Table 10.2 Parameters and confidence intervals for the fitted distributions of BTC returns.

Distr\Param	α	β	δ	λ	μ	ν	σ
Normal					0.0018		0.0318
					0.0005		0.0309
					0.0030		0.0326
Cauchy			0.0121		0.0014		
			0.0114		0.0008		
			0.0127		0.0021		
Stable	1.38014	0.04028	0.01396		0.00175		
	1.32871	−0.06098	0.01327		0.00083		
	1.44444	0.14099	0.01450		0.00267		
Student's t					0.00180	2.1782	0.01667
					0.00099	1.93373	0.01568
					0.00261	2.45356	0.01772
NIG	15.7736	0.1950	0.0168		0.0016		
	13.0000	−1.2967	0.0154	−0.5	0.0006		
	18.5472	1.6868	0.0182		0.0025		
Gen. Hyperbolic	31.8864	0.6838	0.0043	0.4186	0.0011		
	26.6530	−0.8018	0.0012	0.2009	0.0003		
	37.1197	2.1694	0.0073	0.6364	0.0019		

Table 10.3 Log-likelihoods and five other criterias to choose fitted distributions for BTC returns.

Distribution	−lnL	AIC	HQC	SIC	AD	KS
Normal	−5203	−10403	−10398	−10391	57.9241	0.1029
Cauchy	−5474	−10945	−10941	−10934	8.9495	0.0349
Stable	−5555	−11102	−11094	−11079	3.3234	0.0317
Student t	−5585	−11165	−11159	−11147	2.7447	0.0271
NIG	−5608	−11208	−11208	−11185	1.5633	0.0189
Gen. Hyperbolic	−5624	−11239	−11228	−11209	0.6441	0.0135

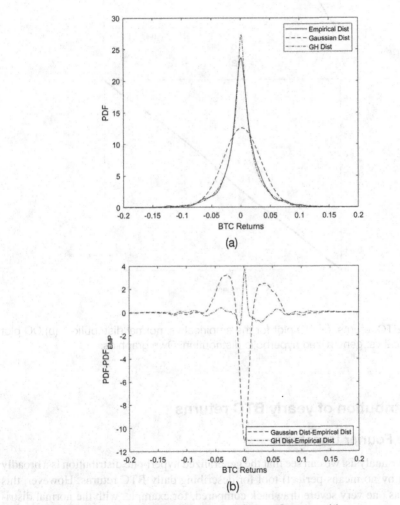

Figure 10.5 BTC empirical kernel distribution fitting. Own graphics.

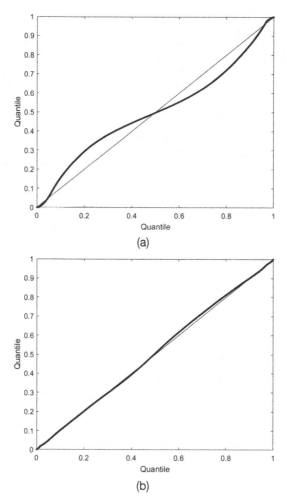

Figure 10.6 BTC returns: (a) QQ plot for the empirical vs. normal distribution; (b) QQ plot for the empirical vs. generalized hyperbolic distribution. Own graphics.

10.5 Distribution of yearly BTC returns

10.5.1 The Fourier transform

From the above analysis, we can see that the generalized hyperbolic distribution is a broadly adequate (but by no means perfect) tool for describing daily BTC returns. However, this distribution has one very severe drawback compared, for example, with the normal distribution and several others. In general, the generalized hyperbolic distribution is not closed with respect to a linear combination of independent random variables (RVs). To put it differently, a linear combination of two independent normal RVs is a normal RV. However, a linear combination of two independent generalized hyperbolic RV is not a generalized

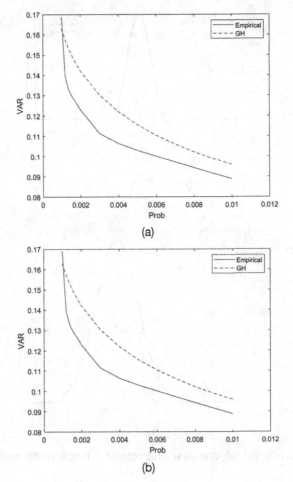

Figure 10.7 (a) BTC VAR for long positions; (b) BTC VAR for short positions. Own graphics.

hyperbolic RV. This fact makes it difficult to find the distribution of the cumulative returns over N days:

$$Y_N = X_1 + ...X_n + ... + X_N. \tag{10.24}$$

But this is what we need to do to aggregate daily to monthly or yearly returns. Here we disregard the fact that returns are not completely independent; see Figure 10.4.

Thankfully, not all is lost. Indeed, Eq. (10.11) gives an explicit expression for the characteristic function of the generalized hyperbolic distribution. One of the most fundamental theorems of probability theory states that the characteristic function of the sum of independent variables is the product of the characteristic functions of the summands:

$$\phi_{Y_N}(\chi) = \left(\phi_X(\chi)\right)^N. \tag{10.25}$$

For the set of calibration parameters $(\alpha, \beta, \delta, \lambda, \mu)$ shown in Table 10.2, and $N = 365$ (one year), we show the real and imaginary parts of $\phi_{Y_N}(\chi)$ in Figure 10.8 (a).

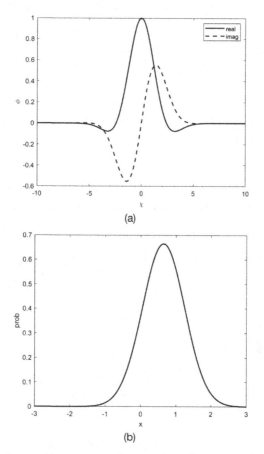

Figure 10.8 BTC returns: (a) one-year characteristic function; (b) one-year pdf. Own graphics.

Once an expression for $\phi_{Y_N}(\chi)$ is found, we can represent the density $f_{Y_N}(x)$ by virtue of the inverse Fourier transform:

$$f_{Y_N}(x) = \frac{1}{2\pi} \int_{-\infty}^{\infty} \phi_{Y_N}(\chi) e^{-ix\chi} d\chi$$

$$= \frac{1}{2\pi} \int_{-\infty}^{\infty} \left(\phi_X(\chi) \right)^N e^{-ix\chi} d\chi \tag{10.26}$$

$$= \frac{\gamma^{N\lambda}}{2\pi \left(K_\lambda(\delta\gamma) \right)^N} \int_{-\infty}^{\infty} \frac{\left(K_\lambda(\delta\upsilon) \right)^N e^{i(N\mu - x)\chi}}{\upsilon^{N\lambda}} d\chi.$$

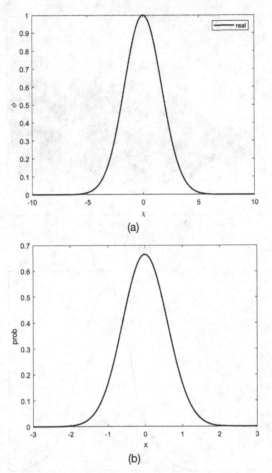

Figure 10.9 BTC returns: (a) one-year risk-neutralized characteristic function; (b) one-year risk-neutralized pdf. Own graphics.

By symmetry, the latter intergal is real-valued. It can be computed efficiently by using the Fast Fourier Transform (FFT). We show $f_{Y_N}(x)$ in Figure 10.8 (b).

It is clear that the expected value of daily BTC returns is positive, due to its *overall* upward sloping trajectory of the price. The expectation of the generalized hyperbolic distribution is given by

$$E = \mu + \frac{\beta \delta K_{\lambda+1}(\delta\gamma)}{\gamma K_\lambda(\delta\gamma)}. \tag{10.27}$$

For the set of calibrated parameters from Table 10.2, we get $E = 0.0018$.

However, in order to avoid a possible bias, as well as for pricing of options on BTC in the risk-neutral measure, we have to choose $\hat{\mu}$ in such a way that $E = 0$. The corresponding $\hat{\mu} = -0.0007$. For the risk neutralized parameter set $(\alpha, \beta, \delta, \lambda, \hat{\mu})$ we get $\hat{\phi}_{Y_N}(\chi), \hat{f}_{Y_N}(x)$ shown in Figure 10.9.

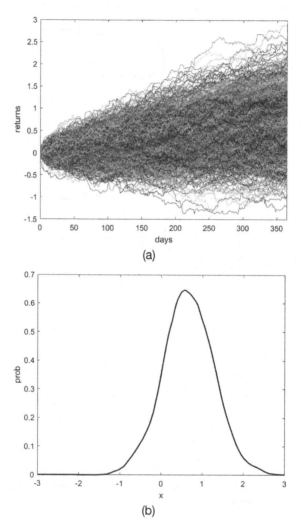

Figure 10.10 BTC returns: (a) one-year MC bunch; (b) one-year pdf. Own graphics.

10.5.2 Monte Carlo simulations

There is a viable alternative to the Fourier transform method. One can randomly draw RVs from the fitted generalized hyperbolic distribution and construct a Monte Carlo path set. The terminal values of these paths represent BTC returns for one year. We show the corresponding paths and their terminal distribution in Figure 10.10.

Comparison of Figures 10.8 (b) and 10.10 (b) shows that results are reassuringly close. In Figure 10.11, we present risk-neutralized results. Once again, graphs presented in 10.9 (b) and 10.11 (b) are in agreement.

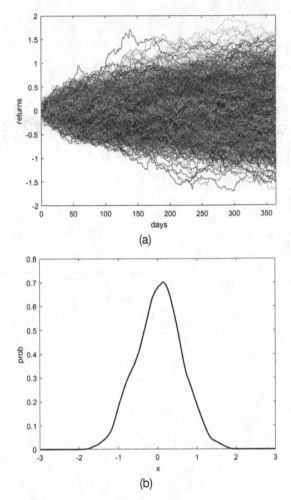

Figure 10.11 BTC returns: (a) one-year risk-neutralized MC bunch; (b) one-year risk-neutralized pdf. Own graphics.

10.6 Distribution of daily ETH, XRP, and MRKT returns

We study ETH, XRP, and MRKT returns along the same lines as before, so we can afford to be brief, skip small details, and present final results.

The corresponding returns and their Epanechnikov kernel distribution are shown in Figures 10.12–10.14. Figures 10.15–10.17 present the autocorrelation and partial autocorrelation. The calibration and maximization results for the returns are summarized in Tables 10.4–10.9. The fitting plots are shown in Figures 10.18–10.20. The QQ plots for the fitted normal and generalized hyperbolic distributions vs. the empirical distributions are demonstrated in Figures 10.21–10.23.

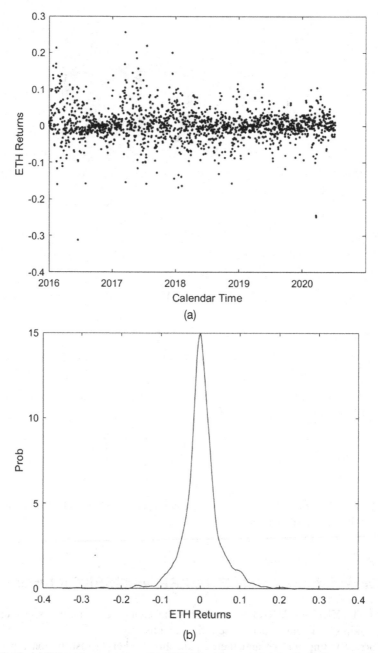

Figure 10.12 (a) ETH returns series; (b) ETH fitted kernel distribution. Own graphics.

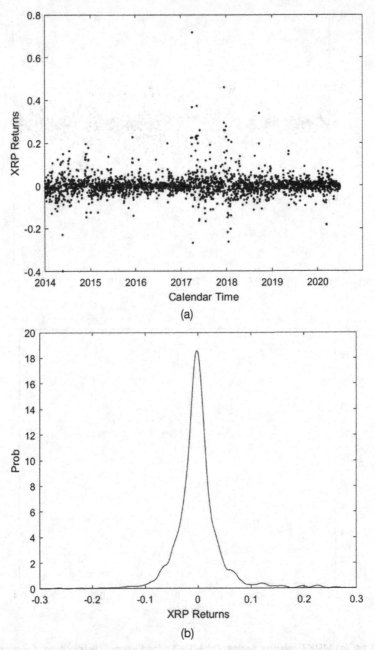

Figure 10.13 (a) XRP returns series; (b) XRP fitted kernel distribution. Own graphics.

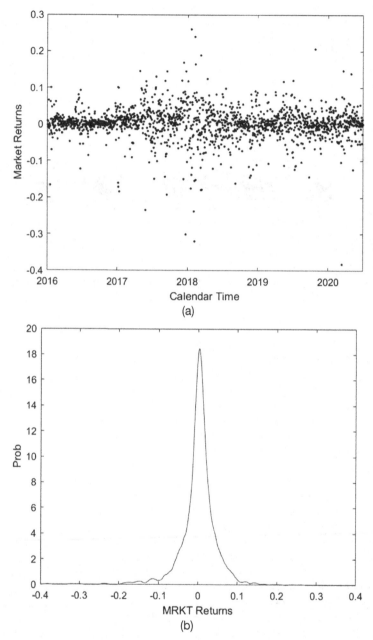

Figure 10.14 (a) MRKT returns series; (b) MRKT fitted kernel distribution. Own graphics.

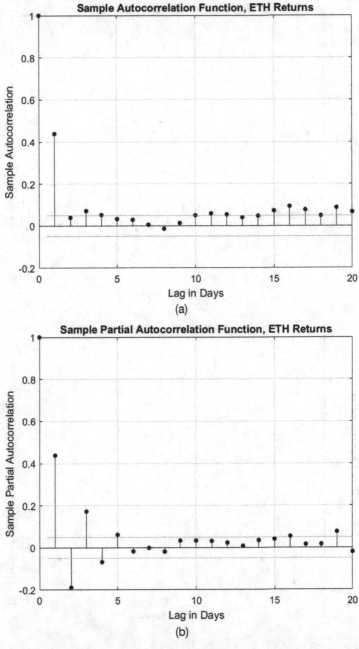

Figure 10.15 ETH autocorrelation and partial autocorrelation. Own graphics.

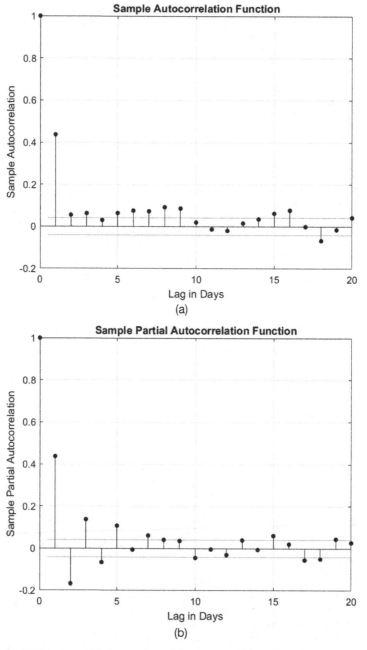

Figure 10.16 XRP autocorrelation and partial autocorrelation. Own graphics.

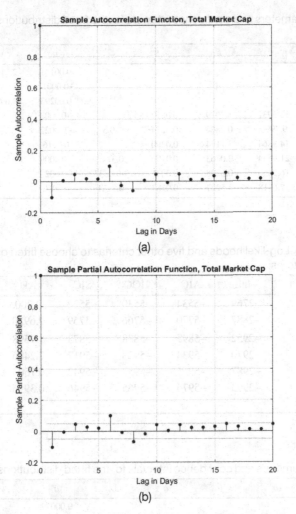

Figure 10.17 MRKT autocorrelation and partial autocorrelation. Own graphics.

Table 10.4 Parameters and confidence intervals for the fitted distributions of ETH returns.

Distr\Param	α	β	δ	λ	μ	ν	σ
					0.00333		0.04520
Normal					0.00114		0.04371
					0.00551		0.04679
			0.0182		0.0007		
Cauchy			0.0170		−0.0006		
			0.0194		0.0020		
	1.41225	0.13867	0.02073		0.00069		
Stable	1.34055	0.00973	0.01960		−0.00104		
	1.48395	0.26760	0.02185		0.00241		

(*Continued*)

Table 10.4 Parameters and confidence intervals for the fitted distributions of ETH returns. (*Continued*)

Distr\Param	α	β	δ	λ	μ	ν	σ
Student's t					0.00141	2.31658	0.02514
					−0.00011	1.98925	0.02334
					0.00293	2.69777	0.02708
NIG	12.3931	1.7029	0.0262		−0.0003		
	9.7998	0.2884	0.0234	-0.5	−0.0022		
	14.9864	3.1174	0.0290		0.0016		
Gen. Hyperbolic	21.9634	2.0763	0.0109	0.3142	−0.0009		
	16.2850	0.6298	0.0031	−0.0837	−0.0027		
	27.6418	3.5227	0.0187	0.7121	0.0010		

Table 10.5 Log-likelihoods and five other criterias to choose fitted distributions for ETH returns.

Distribution	−lnL	AIC	HQC	SIC	AD	KS
Normal	−2769	−5534	−5530	−5522	32.7003	0.1002
Cauchy	−2887	−5770	−5766	−5759	7.6930	0.0432
Stable	−2952	−5897	−5888	−5873	1.7598	0.0269
Student t	−2970	−5934	−5928	−5917	2.2486	0.0257
NIG	−2987	−5965	−5957	−5942	0.7503	0.0229
Gen. Hyperbolic	−2993	−5976	−5965	−5946	0.3956	0.0139

Table 10.6 Parameters and confidence intervals for the fitted distributions of XRP returns.

Distr\Param	α	β	δ	λ	μ	ν	σ
Normal					0.00079		0.04979
					−0.00122		0.04841
					0.00279		0.05125
Cauchy			0.0150		−0.0022		
			0.0142		−0.0031		
			0.0158		−0.0013		
Stable	1.31429	0.04990	0.01662		−0.00258		
	1.25715	−0.04637	0.01582		−0.00370		
	1.37142	0.14616	0.01742		−0.00146		
Student's t					−0.00231	1.78474	0.01916
					−0.00330	1.61088	0.01800
					−0.00131	1.97737	0.02039
NIG	8.3320	1.7494	0.0185		−0.0032		
	6.6253	0.7235	0.0172	-0.5	−0.0043		
	10.0387	2.7752	0.0199		−0.0021		

(*Continued*)

Table 10.6 Parameters and confidence intervals for the fitted distributions of XRP returns. (*Continued*)

Distr\Param	α	β	δ	λ	μ	ν	σ
	8.1767	1.7463	0.0187	−0.5114	−0.0032		
Gen.	4.6089	0.7192	0.0144	−0.7421	−0.0043		
Hyperbolic	11.7445	2.7735	0.0231	−0.2807	−0.0021		

Table 10.7 Log-likelihoods and five other criterias to choose fitted distributions for XRP returns.

Distribution	−lnL	AIC	HQC	SIC	AD	KS
Normal	−3752	−7500	−7496	−7489	120.3274	0.1616
Cauchy	−4499	−8994	−8990	−8983	6.2088	0.0392
Stable	−4556	−9103	−9095	−9080	1.9722	0.0234
Student t	−4572	−9137	−9131	−9120	1.8259	0.0211
NIG	−4586	−9164	−9156	−9141	1.0539	0.0189
Gen. Hyperbolic	−4586	−9162	−9152	−9133	1.0549	0.0188

Table 10.8 Parameters and confidence intervals for the fitted distributions of MRKT returns.

Distr\Param	α	β	δ	λ	μ	ν	σ
					0.002190		0.04465
Normal					0.00000		0.04317
					0.00435		0.04623
			0.0155		0.0037		
Cauchy			0.0144		0.0026		
			0.0165		0.0048		
	1.31429	−0.02968	0.01751		0.00418		
Stable	1.24545	−0.14570	0.01649		0.00276		
	1.38313	0.08634	0.01851		0.00560		
					0.00398	1.91052	0.02064
Student's t					0.00271	0.00271	0.01906
					0.00525	2.19241	0.02235
	9.7974	−1.1150	0.0203		0.0045		
NIG	7.5072	−2.4055	0.0183	−0.5	0.0031		
	12.0876	0.1756	0.0223		0.0060		

(*Continued*)

Table 10.8 Parameters and confidence intervals for the fitted distributions of MRKT returns. (*Continued*)

Distr\Param	α	β	δ	λ	μ	ν	σ
	19.1274	−0.8361	0.0077	0.1990	0.0038		
Gen.	14.7545	−2.1378	0.0034	−0.0473	0.0025		
Hyperbolic	23.5004	0.4657	0.0120	0.4452	0.0051		

Table 10.9 Log-likelihoods and five other criterias to choose fitted distributions for MRKT returns.

Distribution	−lnL	AIC	HQC	SIC	AD	KS
Normal	−2765	−5527	−5523	−5515	50.3679	0.1285
Cauchy	−3045	−6085	−6081	−6074	4.4793	0.0379
Stable	−3080	−6152	−6143	−6128	2.4748	0.0308
Student t	−3098	−6190	−6184	−6173	2.1966	0.0292
NIG	−3115	−5965	−6214	−6199	1.3853	0.0265
Gen. Hyperbolic	−3125	−6239	−6228	−6210	1.0311	0.0230

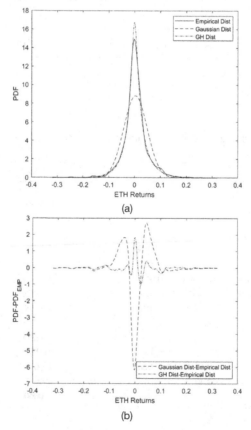

(a)

(b)

Figure 10.18 ETH empirical kernel distribution fitting. Own graphics.

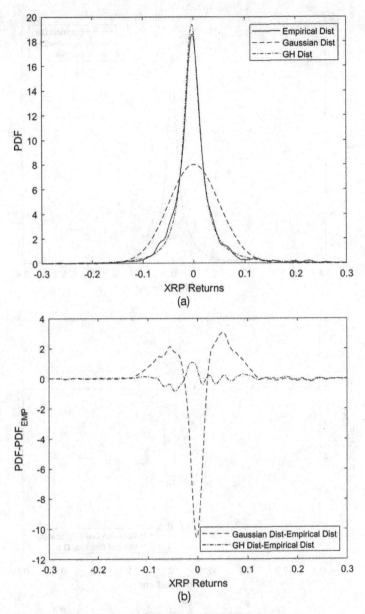

(a)

(b)

Figure 10.19 XRP empirical kernel distribution fitting. Own graphics.

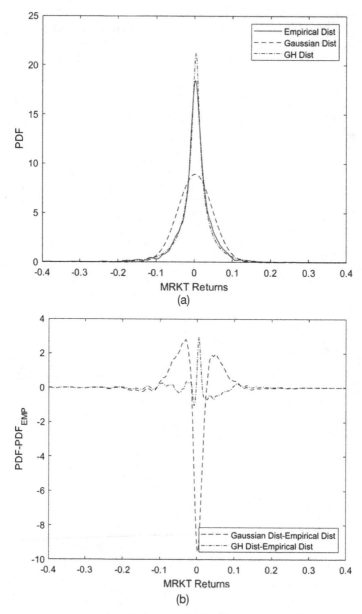

Figure 10.20 MRKT empirical distribution fitting. Own graphics.

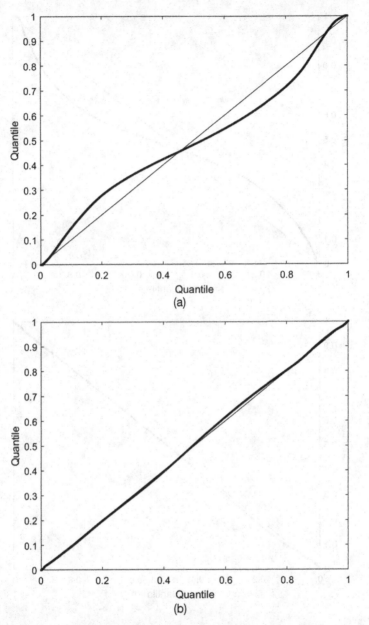

Figure 10.21 ETH returns: (a) QQ plot for the empirical vs. normal distribution; (b) QQ plot for the empirical vs. generalized hyperbolic distribution. Own graphics.

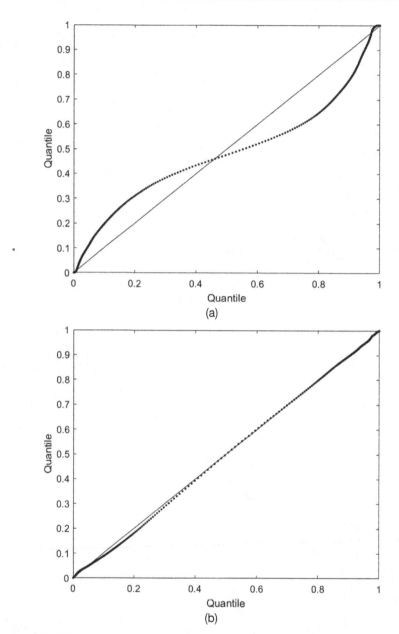

Figure 10.22 XRP returns: (a) QQ plot for the empirical vs. normal distribution; (b) QQ plot for the empirical vs. generalized hyperbolic distribution. Own graphics.

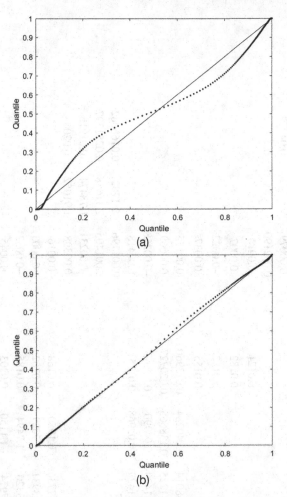

Figure 10.23 MRKT returns: (a) QQ plot for the empirical vs. normal distribution; (b) QQ plot for the empirical vs. generalized hyperbolic distribution. Own graphics.

We do not calculate VAR for these returns here. If necessary, one can perform the corresponding calculations along the same lines as for BTC.

10.7 Comparative statistics of cryptocurrency returns

Let us summarize our findings in a comprehensive Table 10.10, which presents fitting parameters for BTC, ETH, XRP, and MRKT returns. In all cases, the generalized hyperbolic distribution provides the best fit.

It is also instructive to compare BTC, ETH, XRP, and MRKT returns with returns on conventional currencies; see Table 10.11.

Table 10.10 Parameters and confidence intervals for the fitted distributions of BTC, ETH, XRP, and MRKT returns.

	Distr\Param	α	β	δ	λ	μ	ν	σ
BTC	Normal					0.0018		0.0318
ETH	Normal					0.00333		0.04520
XRP	Normal					0.00079		0.04979
MRKT	Normal					0.002190		0.04465
BTC	Cauchy			0.0121		0.0014		
ETH	Cauchy			0.0182		0.0007		
XRP	Cauchy			0.0150		-0.0022		
MRKT	Cauchy			0.0155		0.0037		
BTC	Stable	1.38014	0.040284	0.013960		0.00175		
ETH	Stable	1.41225	0.138665	0.020727		0.00069		
XRP	Stable	1.31429	0.04990	0.01662		-0.00258		
MRKT	Stable	1.31429	-0.02968	0.01751		0.00418		
BTC	Student's t					0.001796	2.1782	0.0166682
ETH	Student's t					0.001405	2.31658	0.0251397
XRP	Student's t					-0.00231	1.78474	0.01916
MRKT	Student's t					0.00398	1.91052	0.02064
BTC	NIG	15.7736	0.1950	0.0168		0.0016		
ETH	NIG	12.3931	1.7029	0.0262		-0.0003		
XRP	NIG	8.3320	1.7494	0.0185		-0.0032		
MRKT	NIG	9.7974	-1.1150	0.0203		0.0045		
BTC	Gen. Hyperbolic	31.8864	0.6838	0.0043	0.4186	0.0011		
ETH	Gen. Hyperbolic	21.9634	2.0763	0.0109	0.3142	-0.0009		
XRP	Gen. Hyperbolic	8.1767	1.7463	0.0187	-0.5114	-0.0032		
MRKT	Gen. Hyperbolic	19.1274	-0.8361	0.0077	0.1990	0.0038		

Table 10.11 Summary statistics log-returns of the exchange rate for BTC, ETH, XRP, and MRKT versus those of the Australian Dollar (AUD), Swiss Franc (CHF), Euro (EUR), British Pound (GBP), Japanese Yen (JPY), and Mexican Peso (MXN). Here, $CV = \frac{StDev}{Mean}$ is the so-called coefficient of variation, also known as relative standard deviation.

	BTC	ETH	XRP	MRKT	AUD	CHF	EUR	GBP	JPY	MXN
Min	-0.2120	-0.3120	-0.3983	-0.3478	-0.0785	-0.1719	-0.0306	-0.0840	-0.0522	-0.0616
$q_{1/4}$	-0.0107	-0.0160	-0.0177	-0.0172	-0.0039	-0.0035	-0.0034	-0.0047	-0.0033	-0.0035
Median	0.0015	0.0010	-0.0022	0.0033	0.0002	0.0000	0.0000	0.0000	0.0000	0.0000
$q_{3/4}$	0.0154	0.0220	0.0136	0.0251	0.0042	0.0035	0.0033	0.0046	0.0034	0.0034
Max	0.2470	0.2560	0.7163	0.3852	0.0818	0.0876	0.0373	0.0439	0.0334	0.0862
$q_{3/4} - q_{1/4}$	0.0261	0.0380	0.0313	0.0423	0.0081	0.0070	0.0067	0.0094	0.0066	0.0069
Max-Min	0.4590	0.5680	1.1146	0.7330	0.1603	0.2595	0.0678	0.1278	0.0856	0.1478
Mean	0.0018	0.0033	0.0007	0.0030	0.0000	-0.0001	0.0000	0.0000	0.0000	0.0001
StDev	0.0317	0.0452	0.0498	0.0548	0.0079	0.0069	0.0061	0.0062	0.0062	0.0070
Var	0.0010	0.0020	0.0025	0.0030	0.0001	0.0000	0.0000	0.0000	0.0000	0.0000
Skewness	-0.1207	0.0847	2.5675	-0.3986	-0.4211	-2.5393	-0.0037	-0.5058	-0.3245	1.0668
Kurtosis	6.54	8.23	33.73	10.88	11.18	77.54	1.91	9.01	4.39	14.64
CV	17.39	13.59	63.37	18.18	385.94	110.51	578.81	143.15	926.56	59.41

It is clear that cryptos are much more volatile than fiat currencies, so that conventional forex trading techniques tried and tested for years are hardly applicable to cryptos at this stage.

10.8 Cryptocurrency ecosystem and its ecology

BTC, ETH, and XRP are not the only cryptocurrencies around. In fact, they have thousands imitators. The largest coins and tokens are shown in Figure 10.24.

The altcoin universe is shown in Figures 10.25 (a) and (b), while the biggest tokens are shown in Figures 10.25 (c) and (d). It is interesting to note that in both cases, the capitalization has a power-law distribution. A typical price trajectories for a coin and a token are shown in Figure 10.26. Market cap of all cryptocurrencies and tokens with and without BitcoinMarket share of alt cryptocurrencies are shown in Figure 10.32.

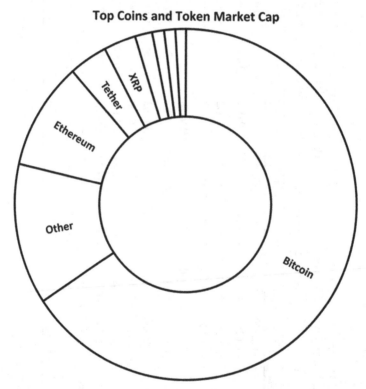

Figure 10.24 Distribution of the market capitalization of the biggest coins and tokens. Own graphics. Source: coinmarketcap.com.

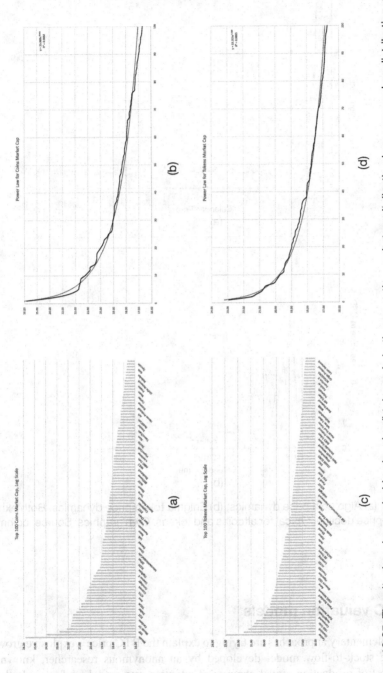

Figure 10.25 The altcoin universe: (a, b) coins, (c, d) tokens. In both cases, the market capitalization has a power-law distribution. Own graphics. Source: coinmarketcap.com.

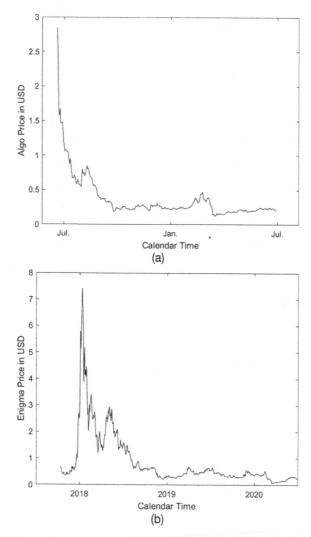

Figure 10.26 (a) Algo coin price dynamics; (b) Enigma token price dynamics. Both exhibit a substantial price decline, typical for altcoins and tokens. Own graphics. Source: coinmarketcap.com.

10.9 BTC valuation models

Several complementary approaches are trying to explain the BTC price's explosive growth, including the stock-to-flow model developed by an anonymous researcher, known as PlanB, the cost-of-production model, the rate-of-adoption rate model, and several others; see, for example, PlanB (2019).

Although none of these models are precise, we feel that a brief discussion of the stock-to-flow model is useful. As was mentioned earlier, BTC has no value. However, it does not mean that it should have zero price, as is occasionally claimed by its detractors. On the contrary, it can have any price (theoretically including zero and infinity).

It is worth articulating distinctions between BTC and conventional financial instruments, such as stocks and bonds. The valuation of bonds is straightforward in theory, if not in practice. In the absence of credit risk, their price is equal to the discounted value of the coupon stream and the principal. In the presence of credit risk, the valuation is more nuanced and considers the probability that the principal and some of the coupons would not be repaid. The valuation of stocks depends on their current and future dividends, quarterly earnings, and other financial metrics. It is not our intention to discuss the valuation of conventional financial instruments in detail. Instead, we want to mention that nothing is expected from BTC as far as financial metrics are concerned. And this is its greatest strength! It cannot fail by definition, either on its own or compared to other instruments. Hence its price can meander without restrictions. However, external events can impact the price both positively and negatively. For instance, adoption by a major financial institution might result in a price increase, while the imposition of adverse regulations might cause a price drop.

BTC scarcity, baked into the Bitcoin protocol from the start, makes it loosely similar to gold, diamonds, and art. The reader should not put too much credence into this similarity. Not all scarce objects acquire or retain their value; public tastes are fickle and can quickly move from one item of value to the next. Besides, the BTC cost of production depends on the number of miners involved in the process via the difficulty's dynamic adjustment. This distinction between BTC and gold or diamonds, whose production cost is mostly insensitive to the miners' activity as a group, is essential if seldom emphasized.

The actual growth of the number of BTCs in circulation is shown in Figure 5.19 (a), while the growth rate is shown in Figure 5.19 (b). We define the important stock-to-flow (S2F) ratio as follows:

$$S2F(t) = \frac{N_{BTC}(t)}{N_{BTC}(t) - N_{BTC}(t - \Delta t)}. \tag{10.28}$$

Here t is the calendar time in years, and $\Delta t = 1$. We define a backward-looking quantity since it reflects the actual production history, which was rather chaotic in the beginning. By now, BTC is produced at a relatively steady clip so that future values of $S2F(t)$ can be calculated with a high degree of certainty. Given that block rewards are halved every four years, $S2F$ is an increasing function of t. Figure 10.27 shows the behavior of $S2F$ between January 2011 and June 2020 as well as its hypothetical future behavior. Kinks reflect regular block rewards halving.

Figure 10.27 depicting a financial asset's future behavior is one of the very few which can be presented with certainty since it depends solely on the Bitcoin protocol specifications. However, the reader should keep in mind that if the protocol fails, so would the graph.

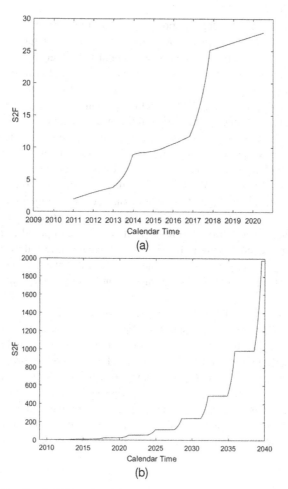

Figure 10.27 (a) Realized S2F ratio; (b) Theoretical S2F ratio. Own graphics. Source: blockchain.com.

We consider now three log-log regressions, which can help us to understand the price behavior for BTC. We start with a linear regression of log-price $\ln(\Pi)$ on log-time $\ln(T)$. Next, we consider a linear regression of log-price on log-S2F. Finally, we study a quadratic regression of log-price on log-S2F. The corresponding regressions can be written as follows:

$$\ln(\Pi) = a_1 \ln(T) + a_2,$$

$$a_1 = 5.524(5.428, 5.619), \quad a_2 = -4.104(-4.282, -3.926), \qquad (10.29)$$

$$R^2 = 0.9183, \quad RMSE = 0.7750,$$

$$\ln(\Pi) = b_1 \ln(S2F) + b_2,$$

$$b_1 = 3.229(3.190, 3.268), \quad b_2 = -1.569(-1.665, -1.473), \tag{10.30}$$

$$R_l^2 = 0.9578, \quad RMSE_l = 0.5566,$$

$$\ln(\Pi) = c_1 \ln(S2F)^2 + c_2 \ln(S2F) + c_3,$$

$$c_1 = -0.3156(-0.3659, -0.2654), \quad c_2 = 4.598(4.377, 4.819),$$

$$c_3 = -2.833(-3.053, -2.612) \tag{10.31}$$

$$R_q^2 = 0.9627, \quad RMSE_q = 0.5234.$$

Here coefficients are shown with their 95% confidence bounds in parentheses. All regressions produce decent results as far as their R-squared and Root Mean Square Error (RMSE) are concerned. We use Eqs (10.29)–(10.31) to represent $\ln(\Pi)$ as follows:

$$\ln(\Pi) \sim f(T) \equiv a_1 \ln(T) + a_2$$

$$= 5.524 \ln(T) - 4.104,$$

$$\ln(\Pi) \sim g(T) \equiv b_1 \ln(S2F(T)) + b_2$$

$$= 3.229 \ln(S2F(T)) - 1.569, \tag{10.32}$$

$$\ln(\Pi) \sim g(T) \equiv c_1 \ln(S2F(T))^2 + c_2 \ln(S2F(T)) + c_3$$

$$= -0.3156 \ln(S2F(T))^2 + 4.598 \ln(S2F(T)) - 2.833.$$

BTC prices and the corresponding fitting curves are plotted in Figures 10.28–10.30.

These figures demonstrate that the linear regression of log-price on log-time does not work too well. In contrast, the other two — the linear regression of log-price on log-S2F and the quadratic regression of log-price on log-S2F — offer qualitatively good fits. However, empirical values can significantly deviate from their theoretical estimates both up and down.

The first two of Eq. (10.32) can be written as power laws:

$$\Pi \sim 0.0165 \times T^{5.524},$$

$$\Pi \sim 0.2083 \times S2F(T)^{3.229}. \tag{10.33}$$

Although power laws frequently occur in nature, the reader should not become overly enthusiastic. In natural sciences, log-log regressions often produce decent results, even if the

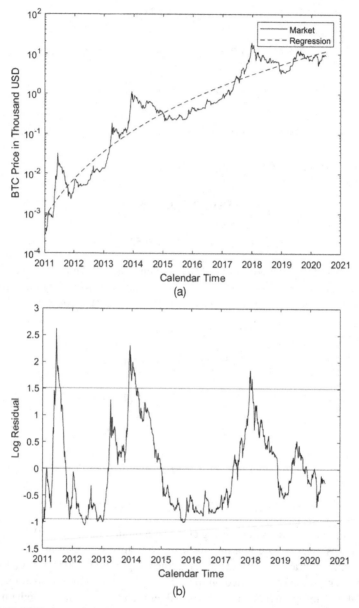

Figure 10.28 BTC log-price time series linearly regressed on log-time. (a) Price and regressed price as functions of calendar time; (b) The corresponding log-residual. Own graphics. Source: coinmarketcap.com.

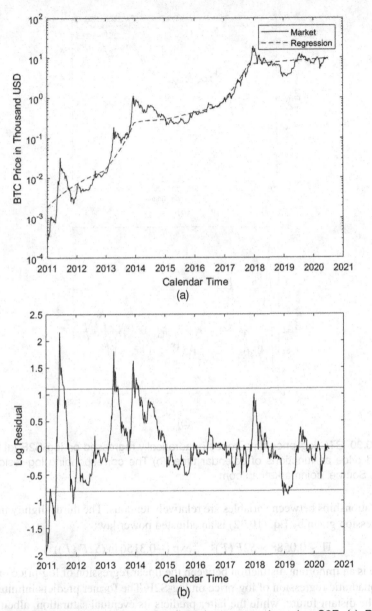

Figure 10.29 BTC log-price time series linearly regressed on log-S2F. (a) Price and regressed price as functions of calendar time; (b) The corresponding log-residual. Own graphics. Source: coinmarketcap.com.

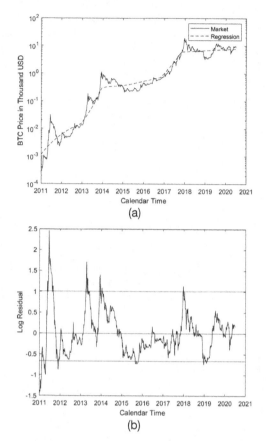

Figure 10.30 BTC log-price time series quadratically regressed on log-S2F. (a) Price and regressed price as functions of calendar time; (b) The corresponding log-residual. Own graphics. Source: coinmarketcap.com.

actual relationships between variables are relatively tenuous. The third, slightly more accurate, regression given by Eq. (10.32) is an adjusted power law:

$$\Pi \sim 0.0588 \times S2F(T)^{4.598} \exp\left(-0.3156\ln\left(S2F\left(T\right)\right)^2\right). \tag{10.34}$$

There is an important distinction between the linear regression of log-price on log-S2F and the quadratic regression of log-price on log-S2F. The former predicts infinite price for BTC in the distant future, while the latter predicts its eventual saturation, albeit at a very high level. Viewed as a function of $S2F$, the function g reaches its maximum at

$$\ln\left(S2F^*\right) = -\frac{c_2}{2c_1} = 7.2845, \quad S2F^* = \exp\left(7.2845\right) = 1457.5,$$
$$g^* = c_1 \ln\left(S2F^*\right)^2 + c_2 \ln\left(S2F\left(T\right)\right) + c_3 = 13.9142, \tag{10.35}$$
$$\Pi^* = \exp\left(13.9142\right) = \$1,103,700.$$

Figure 10.31 illustrates what can happen with the price in the long run according to these regressions. The reader should not get overly excited or worried about the total dominance

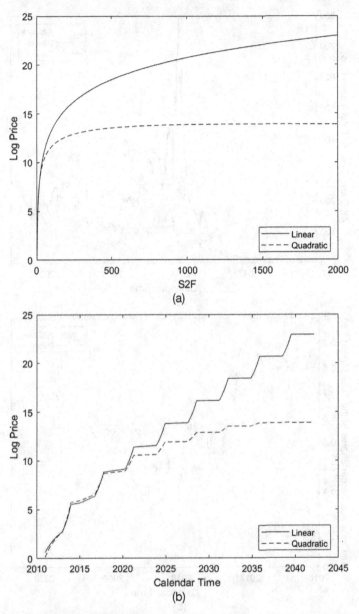

Figure 10.31 Hypothetical limiting values of BTC log price as functions of (a) S2F ratio, (b) calendar time. Own graphics.

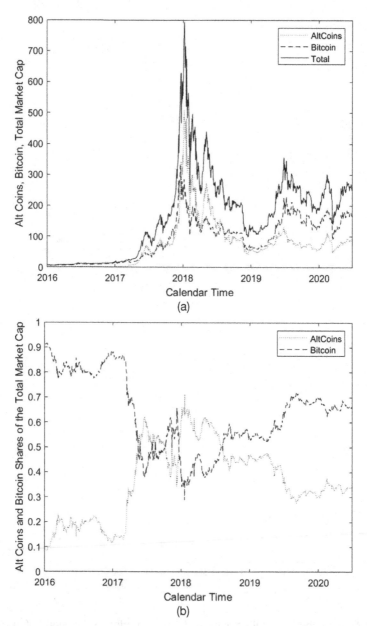

Figure 10.32 (a) The total market capitalization of all cryptoassets (coins and tokens) with and without Bitcoin; (b) Altcoins and Bitcoin shares of the total market capitalization. Own graphics. Source: coin.dance.

of BTC in the future. Regardless of what these figures show, there would be reasons, both political and practical, why these giddy levels will not be reached in reality. Moreover, as mentioned above, while the model offers a reasonable qualitative agreement with empirical data, quantitatively it remains inadequate.

For additional information concerning the BTC price dynamic and its modeling, see Akcora et al. (2018); Antonakakis et al. (2019); Bartolucci and Kirilenko (2020); Baur et al. (2018); Bovet et al. (2018); Gkillas and Katsiampa (2018); Kajtazi and Moro (2019); Sharma et al. (2019).

10.10 Bitcoin dominance index

10.10.1 The SIR model

The famous Susceptible-Infected-Removed (SIR) and Susceptible-Exposed-Infected-Removed (SEIR) models, originated by William Ogilvy Kemrack and Anderson Gray McKenrick in 1927, are the main workhorses of mathematical epidemiology; see Anderson et al. (1992); Brauer et al. (2008). The models are adequate when the entire pool of the susceptible population is well-mixed and reacts similarly to the infection. Unfortunately, these models proved to be grossly inadequate in the context of COVID-19 pandemic. One of the authors spent time at the beginning of COVID-19 pandemic on building the K-SEIR model, much better aligned with reality than its classical precursors; see Gershon et al. (2020); Lipton and Lopez de Prado (2020).

To explain the evolution of Bitcoin dominance, we consider the standard SIR model. Let S, I, R, be the number of susceptible, infected, and removed individuals, and N be the total number of individuals alive at time t. Assuming that N is approximately constant, $N \approx N(0)$, we can write the SIR equations in terms of $s = S/N$, $i = I/N$, $r = R/N$:

$$\frac{ds}{dt} = -\frac{\mathfrak{R}_0}{\tau_I} is,$$

$$\frac{di}{dt} = \frac{\mathfrak{R}_0}{\tau_I} is - \frac{1}{\tau_I} i, \tag{10.36}$$

$$\frac{dr}{dt} = \frac{1}{\tau_I} i.$$

Here $\mathfrak{R}_0$, called the basic reproduction number, is the expected number of secondary cases produced by a single infected individual in a completely susceptible population, absent of any intervention. Initially, an epidemic grows if $\mathfrak{R}_0 > 1$. Eventually, when the number of susceptibles goes down either due to infections or vaccinations, the so-called herd immunity is reached. It is clear that

$$s + i + 1 = 1. \tag{10.37}$$

For the case of BTC, we interpret s, i, r as follows: s is the fraction of economic agents already exposed to the "Bitcoin bug," i is the fraction of those, who move some of their holdings to altcoins, and r is the fraction of agents who return back to bitcoin as an asset

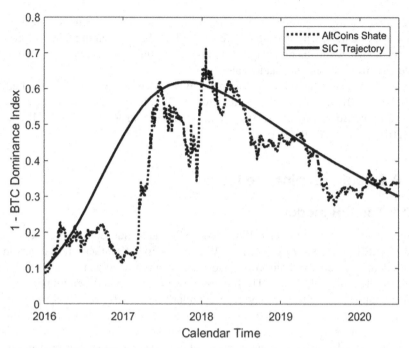

Figure 10.33 The fitted SIR model qualitatively explains the altcoins' fraction in the total capitalization of cryptoassets. Own graphics.

class. After scanning the parameter space, we have choosen $\mathfrak{R}_0 = 7.71$, $\tau_I = 2.86$, $s(0) = 0.9$, $i(0) = 0.1$, $r(0) = 0.0$. Details are shown in Figure 10.33, which qualitatively explains the Bitcoin dominance.

10.10.2 The SIS model

Somewhat less known, but possibly more efficient is the Susceptible-Infected-Susceptible (SIS) model. This class of models describe recurrent diseases, for instance, sexually transmitted ones; see Hethcote and Yorke (2014). The classical SIS model is deterministic. To make it realistic, one has to use the parameter perturbation technique to examine the effect of environmental stochasticity; see, e.g., Gray et al. (2011).

It is clear that the behavior of the Bitcoin dominance index is highly stochastic. Therefore, we apply the stochastic version of the SIS model to our needs. The main dynamic equation is as follows:

$$\frac{dx}{x} = \beta(m - x)\,dt + \sigma(1 - x)\,dW(t), \tag{10.38}$$

where x is the altcoin fraction in the total capitalization, and m, $0 < m < 1$ is its equilibrium value. Parameters of the model are dimensional:

$$[x] = [m] = 1, \quad [\beta] = \frac{1}{T}, \quad [\sigma] = \frac{1}{\sqrt{T}}. \tag{10.39}$$

For the model to make sense, it has to have a unique global solution $x(t)$, which will remain within $(0, 1)$ whenever it starts from this interval. A rather complicated indirect proof is given in Gray et al. (2011). However, the fact that $0 < x(t) < 1$ can be verified directly by using the following change of variables:

To model things numerically, it is convenient to map $(0, 1)$ to $(-\infty, \infty)$. To this end, we introduce y, such that

$$x = \frac{1}{1 + e^y}, \quad y = \left(\frac{1}{x} - 1\right). \tag{10.40}$$

Further,

$$dy = -\frac{dx}{x(1-x)} + \frac{(1-2x)\,dx^2}{2x^2(1-x)^2}$$

$$= -\frac{\beta\,(m-x)\,dt + \sigma\,(1-x)\,dW(t)}{(1-x)} + \frac{\sigma^2\,(1-2x)\,x^2(1-x)^2 dt}{2x^2(1-x)^2}$$

$$= -\frac{\beta\,(m-x)\,dt + \sigma\,(1-x)\,dW(t)}{(1-x)} + \frac{\sigma^2\,(1-2x)\,dt}{2}$$

$$= \left(-\frac{\beta\,(m-x)}{(1-x)} + \frac{\sigma^2\,(1-2x)}{2}\right)dt - \sigma dW(t)$$

$$= \left(-\frac{\beta\left(m - \dfrac{1}{1+e^y}\right)}{\left(1 - \dfrac{1}{1+e^y}\right)} + \frac{\sigma^2\left(1 - \dfrac{2}{1+e^y}\right)}{2}\right)dt - \sigma dW(t) \tag{10.41}$$

$$= \left(-\beta\,(m + (m-1)\,e^{-y}) + \frac{\sigma^2\,(1 - e^{-y})}{2\,(1 + e^{-y})}\right)dt - \sigma dW(t).$$

Finally,

$$dy = \beta\left(-m + (1-m)\,e^y + \frac{\sigma^2}{2\beta}\frac{(1 - e^{-y})}{(1 + e^y)}\right)dt - \sigma dW(t). \tag{10.42}$$

Since this equation is non-singular, the solution $x(t)$ of the original equation is always sandwiched between zero and one.

To understand what is going to happen with Bitcoin dominance in the long run, we study the behavior of $x(t)$ for $t \to \infty$. Following Gray et al. (2011), it can be shown that $x(t)$ goes to zero for $t \to \infty$ (extinction), provided that

$$\frac{\sigma^2}{\beta} > \begin{cases} 2m, & 0 < m < \frac{1}{2}, \\ \dfrac{1}{2(1-m)}, & \dfrac{1}{2} \le m < 1, \end{cases} \tag{10.43}$$

while $x(t)$ stays away from zero for $t \to \infty$ (persistence) when

$$\frac{\sigma^2}{\beta} < 2m. \tag{10.44}$$

What happens for

$$2m < \frac{\sigma^2}{\beta} < \frac{1}{2(1-m)}, \quad \frac{1}{2} < m < 1, \tag{10.45}$$

is unclear, but extinciton is the probable outcome.

It is well-known that the static asymptotic distribution is described by the stationary Fokker–Planck equation for the pdf $P(x)$ as follows:

$$((m-x)xP)_x - \frac{\sigma^2}{2\beta}\left((1-x)^2 x^2 P\right)_{xx} = 0, \tag{10.46}$$

see, e.g., Lipton (2001). The change of dependent variable:

$$Q = (1-x)^2 x^2 P, \tag{10.47}$$

yields the following modified equation for the stationary distribution:

$$\frac{(m-x)}{x(1-x)^2}Q - \frac{\sigma^2}{2\beta}Q_x = 0. \tag{10.48}$$

Thus,

$$Q = e^{\Phi}, \tag{10.49}$$

where

$$\Phi = \frac{2\beta}{\sigma^2}\int \frac{(m-x)}{x(1-x)^2}dx = \frac{2\beta}{\sigma^2}\left(m\ln\left(\frac{x}{1-x}\right) + \frac{(m-1)}{(1-x)}\right) + const. \tag{10.50}$$

Accordingly,

$$Q = \frac{x^{2\beta m/\sigma^2}}{(1-x)^{2\beta m/\sigma^2}}\exp\left(\frac{2\beta(m-1)}{\sigma^2(1-x)} + const\right). \tag{10.51}$$

Finally,

$$P = \varpi\frac{x^{2\beta m/\sigma^2 - 2}}{(1-x)^{2\beta m/\sigma^2 + 2}}\exp\left(\frac{2\beta(m-1)}{\sigma^2(1-x)}\right), \tag{10.52}$$

where ϖ is a normalizing factor. The distribution P is integrable at $x = 0$ provided that inequality (10.44) is valid. By its very nature, it is independent on the choice of $x(0)$.

The fitted SIS model with $m = 0.6, \beta = 10.0, \sigma = 1.0, x(0) = 0.1$, provides a reasonable explanation of the stochastic behavior of the proportion of altcoins in the total capitalization of all cryptos; see Figure 10.34.

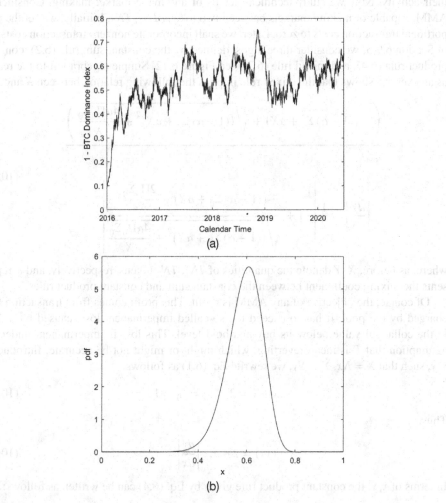

Figure 10.34 (a) The altcoins' fraction of the total capitalization of cryptoassets; (b) The stationary probability density function for the altcoins fraction of the total. Here $\beta = 10.0$, $\sigma = 1.0$, $m = 0.6$, $x_0 = 0.1$. Own graphics.

This figure shows that for sufficiently large volatility σ, the Bitcoin dominance index will hover around its initially chosen value $m = 0.6$. However, one can show that it will go to zero for smaller volatility values, similarly to the SIR model. Thus, the future of the entire cryptocurrency ecosystem depends on how volatile it is going to be.

10.11 Automated market makers

We already introduced AMMs in Section 6.9.5 and discussed some qualitative aspects of their activity. Now we study technical details of automatic market making. Consider an AMM, capable of making markets between two tokens TN_1, TN_2. Initially, we set the proportional transaction costs to zero. Later, we shall incorporate nonzero transaction costs. As in Section 6.9.5, we consider three algos, defined by the constant sum rule (6.2), constant product rule (6.4), and mixed rule (6.7); see Figure 6.12. Simple algebra left to the reader as an exercise shows that the mixed rule implies the following relation between X and Y:

$$Y = \frac{\left(-\left((1-\alpha)\Sigma_0 + \alpha X\right) + \sqrt{\left((1-\alpha)\Sigma_0 + \alpha X\right)^2 + \frac{4\alpha\Pi_0\Sigma_0}{X}} \right)}{2\alpha},$$

(10.53)

$$\left| \frac{dY}{dX} \right| = \frac{1}{2} \left| -1 + \frac{\left((1-\alpha)\Sigma_0 + \alpha X\right) - \frac{2\Pi_0\Sigma_0}{X^2}}{\sqrt{\left((1-\alpha)\Sigma_0 + \alpha X\right)^2 + \frac{4\alpha\Pi_0\Sigma_0}{X}}} \right|,$$

where, as before, X, Y denote the quantities of TN_1, TN_2 tokens, respectively, and α represents the mixing coefficient between the constant sum and constant product rules.

Of course, the objective of any AMM is profit. This profit comes from transaction fees charged by the pool. It has to exceed the so-called impermanent loss, caused by a drop in the collateral value below its buy-and-hold level. This loss is impermanent under the assumption that P is mean-reverting, which might or might not be accurate. Introducing x, y, such that $X = Nx$, $Y = Ny$, we rewrite Eq. (6.1) as follows:

$$x + y = 2, \quad x_0 = y_0 = 1.$$

(10.54)

Thus,

$$y(x) = 2 - x, \quad \left| \frac{dy}{dx} \right| = 1.$$

(10.55)

In terms of x, y, the constant product rule given by Eq. (6.4) can be written as follows:

$$xy = 1, \quad x_0 = y_0 = 1,$$

(10.56)

so that

$$y(x) = \frac{1}{x}, \quad \left| \frac{dy}{dx} \right| = \frac{1}{x^2}.$$

(10.57)

Finally, Eq. (6.7) written in terms of x, y becomes:

$$\left(\frac{1}{xy} - 1 \right) - \alpha \left(\frac{x+y}{2} - 1 \right) = 0,$$

(10.58)

The mixed rule implies the following relation between x and y:

$$y_\alpha = \frac{1}{2\alpha}\left(-(2(1-\alpha)+\alpha x)+\sqrt{(2(1-\alpha)+\alpha x)^2+\frac{8\alpha}{x}}\right),$$

(10.59)

$$\frac{dy_\alpha}{dx} = \frac{1}{2}\left(-1+\frac{2(1-\alpha)+\alpha x-4/x^2}{\sqrt{(2(1-\alpha)+\alpha x)^2+8\alpha/x}}\right).$$

We now analyze the situation, when P moves away from its equilibrium value $P_0 = 1$. Let $P > 1$. For the constant sum contract, an arbitrageur can choose a number x, $1 < x \le 2$, and deliver $(x-1)N$ TN$_1$ tokens to the pool in exchange for getting $(x-1)N$ TN$_2$ tokens. The arbitrageur's profit or loss is given by

$$\Omega(x) = (P-1)(x-1)N.$$

(10.60)

Since Ω is a linear function of x, the profit is maximized by exhausting the entire pool, i.e., by choosing the following optimal values (x^*, y^*, Ω^*):

$$x^* = 2, \quad y^* = 0, \quad \Omega^* = (P-1)N.$$

(10.61)

Similarly, when $P < 1$, we have

$$x^* = 0, \quad y^* = 2, \quad \Omega^* = (1-P)N.$$

(10.62)

The value of the arbitraged portfolio is $\pi^*(P)N$, where

$$\pi^*(P) = \begin{cases} 2, & P \ge 1, \\ 2P, & P < 1. \end{cases}$$

(10.63)

while the buy-and-hold portfolio's value is $(P+1)N$. The difference is given by

$$\omega = (P+1)N - \pi^*(P)N.$$

(10.64)

In the DeFi parlance, ω is called the impermanent loss. Of course, this terminology is misleading because the loss can quickly become permanent if P drifts away from its "equilibrium" value of one. The percentage loss of the realized portfolio compared to the buy-and-hold portfolio is

$$\lambda = 1 - \frac{|P-1|}{P+1}.$$

(10.65)

For the constant product contract, an arbitrageur can choose a number $x > 1$ and deliver $(x-1)N$ TN$_1$ tokens to the pool in exchange for getting $(1-y)N$ TN$_2$ tokens, where $y = 1/x$. In this case, the arbitrageur's profit or loss is given by

$$\Omega(x) = \left(P\left(1-\frac{1}{x}\right)-(x-1)\right)N.$$

(10.66)

The profit is maximized when

$$\Omega'(x) = \left(\frac{P}{x^2}-1\right)N = 0.$$

(10.67)

The corresponding optimal values (x^*, y^*, Ω^*) are

$$x^* = \sqrt{P}, \quad y^* = \frac{1}{\sqrt{P}}, \quad \Omega^* = \left(\sqrt{P} - 1\right)^2 N. \tag{10.68}$$

This equation shows that a constant product collateral pool can never be exhausted. At every stage, the values of TN_1 and TN_2 held in the portfolio are equal to $\sqrt{PN}$, each. Since the value of both tokens in the portfolio has to be equal, the implied optimal value of TN_2 expressed in terms of TN_1 is $P^* = x^*/y^* = P$, as a consequence of arbitrage. The value of the arbitraged portfolio is $\pi^* = 2\sqrt{PN}$, while the buy-and-hold portfolio value is $(P + 1)N$. The difference is given by

$$\omega = (P + 1)N - 2\sqrt{PN}. \tag{10.69}$$

The percentage loss of the realized portfolio compared to the buy-and-hold portfolio is

$$\lambda = 1 - \frac{2\sqrt{P}}{(P + 1)} = \frac{\left(\sqrt{P} - 1\right)^2}{(P + 1)}. \tag{10.70}$$

For the mixed rule AMM, the arbitrageur's profit for $P > 1$ has the form:

$$\Omega(x) = \left(P\left(1 - y_\alpha(x)\right) - (x - 1)\right)N, \tag{10.71}$$

with the optimum achieved at $x_\alpha^*, y_\alpha^*, \Omega_\alpha^*$, such that

$$y_\alpha'\left(x_\alpha^*\right) = -\frac{1}{P}, \quad y_\alpha^* = y_\alpha\left(x_\alpha^*\right), \quad \Omega_\alpha^* = \left(P\left(1 - y_\alpha^*\right) - \left(x_\alpha^* - 1\right)\right)N. \tag{10.72}$$

We find x_α^* via the Newton–Raphson method starting with a suitable $x_\alpha^{(0)}$:

$$x_\alpha^{(n+1)} = x_\alpha^{(n)} - \frac{y_\alpha'\left(x_\alpha^{(n)}\right) + \frac{1}{P}}{y_\alpha''\left(x_\alpha^{(n)}\right)}. \tag{10.73}$$

Since the Newton–Raphson method has quadratic convergence, we can choose 10 iterations and set $x_\alpha^* = x_\alpha^{(10)}$. The value of the arbitraged portfolio is

$$\pi^* = x_\alpha^* + Py_\alpha\left(x_\alpha^*\right). \tag{10.74}$$

The constant sum, constant product, and mixed rule composition curves are shown in Figure 10.35, which is a scaled version of Figure 6.12.

The constant sum, constant product, and mixed rule composition and loss curves are shown in Figures 10.36 (a), (b). This figure indicates that the impermanent loss is relatively mild for the constant product rules, moderate for the mixed rule, and very high for the constant sum rule. Even after a fivefold change in price P relative to its equilibrium value, the impermanent loss for the constant product rule is tolerable, especially compared to the mixed rule.

Of course, the objective of any market maker, including the funder of an AMM, is profit. This profit has to cover the impermanent loss and other costs. One possibilitiy is to charge the proportional transaction fees every time an arbitrageuer or other market participant removes one token and adds the other one. Typically, it is necessary to add more tokens to the pool that is required by its constituent rule. To simplify the exposition, we abstract details by assuming that fees are paid directly to the pool's operator and distributed appropriately. Consider an AMM defined by the constant product rule. Let ε be a percentage fee. Let T_0 and T_1 be two time slices. At time T_0, the price is P_0, and the pool's composition is (x_0, y_0). At time T_1, the price is P_1 and we need to find the arbitraged pool's composition (x_1, y_1). First, we assume that $P_1 > P_0$, so that, with zero transaction costs, it is beneficial to arbitrageuers to withdraw TN_2 and add TN_1. With non-zero transaction costs the situation is more complicated. The arbitrageur can choose a number $x_1 > x_0$ and deliver $(1 + \varepsilon)(x - x_0) N \, TN_1$ tokens to the pool in exchange for getting $(y_0 - y_1) N \, TN_2$ tokens from the pool, where $y_1 = 1/x_1$. In this case, the arbitrageur's profit or loss is as follows:

$$\Omega(x) = \left(P_1 \left(y_0 - \frac{1}{x} \right) - (1 + \varepsilon)(x - x_0) \right) N. \tag{10.75}$$

The profit is maximized when

$$\Omega'(x) = \left(\frac{P_1}{x^2} - (1 + \varepsilon) \right) N = 0, \tag{10.76}$$

so that

$$x_1^* = \max \left(\sqrt{\frac{P_1}{(1 + \varepsilon)}}, x_0 \right), \quad y_1^* = \frac{1}{x_1^*}. \tag{10.77}$$

Here, we take the maximum in order to ensure that $x_1^* \geq x_0$, so that TN_1 tokens are added to the pool, rather than withdrawn from it. Thus, for $P_1 > P_0$, adjustment occurs only when

$$P_1 > (1 + \varepsilon) x_0^2. \tag{10.78}$$

Similarly, for $P_1 < P_0$,

$$x_1^* = \min \left(\sqrt{(1 + \varepsilon) P_1}, x_0 \right), \quad y_1^* = \frac{1}{x_1^*}, \tag{10.79}$$

so that adjustment happens when

$$P_1 < \frac{x_0^2}{(1 + \varepsilon)}. \tag{10.80}$$

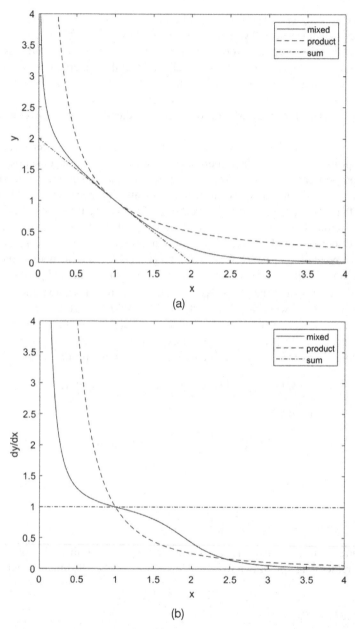

Figure 10.35 (a) y as a function of x for three different types of AMM: constant sum, constant product, mixed rule; (b) Relative price of TN_2 expressed in terms of TN_1 is equal to $|dy/dx|$: for constant sum, constant product, mixed rule. This figure is a scaled version of Figure 6.12. Own graphics.

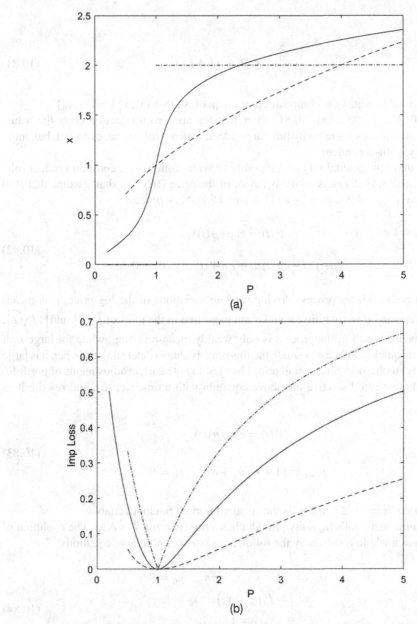

Figure 10.36 (a) x as a function of P for three different types of AMM: constant sum, constant product, mixed rule; (b) Impermanent loss as a function of P: constant sum, constant product, mixed rule. Own graphics.

Finally, when

$$\frac{x_0^2}{(1+\varepsilon)} \leq P_1 \leq (1+\varepsilon)x_0^2, \tag{10.81}$$

it is suboptimal to adjust the composition of the pool, so that $(x_1^*, y_1^*) = (x_0, y_0)$.

Eqs (10.78), (10.80), and (10.81) show that for nonzero transaction costs the actual composition of the pool is not only time-dependent, which is, of course, expected, but, more surprisingly, path-dependent.

Let us study the profitability of a liquidity provider following a constant product rule. Clearly, profitability depends on the behavior of the price $P(t)$. We shall assume that $P(t)$ is mean-reverting and is driven by an Ornstein–Uhlenbeck process:

$$P(t) = \exp(p(t)),$$

$$dp(t) = -\kappa p(t)\,dt + \sigma dW(t), \quad p(0) = 0. \tag{10.82}$$

Here $W(t)$ is the Wiener process driving random variations of the log-price, κ is mean-reversion rate, and σ is volatility; κ and σ are measured in the units of $\left[1/T\right]$ and $\left[1/\sqrt{T}\right]$, respectively. For small κ, the process is only weakly mean-reverting, while for large κ, it mean-reverts quickly. When σ is small, the log-price is almost deterministic, when it is large it is strongly stochastic.[4] Since small price changes do not result in adjustments of portfolio composition, we can discretize the above equation with a time step Δt, and rewrite it as follows:

$$P(t) = \exp(p(t)),$$

$$p_{n+1} = (1-\overline{\kappa})p_n + \overline{\sigma}\eta_n, \quad p_0 = 0. \tag{10.83}$$

Here $\overline{\kappa} = \kappa\Delta t$, $\overline{\sigma} = \sigma\sqrt{\Delta t}$, and η_n is the standard normal random variable.

We assume that pool's liquidity provider has a time horizon $T = N\Delta t$. The evolution of the system as a whole is driven by the following system of nonlinear equations:

$$\begin{aligned} p_{n+1} &= (1-\overline{\kappa})p_n + \overline{\sigma}\eta_n, \quad p_0 = 0, \\ x_{n+1} &= f(p_{n+1}, x_n), \quad x_0 = 1, \\ y_{n+1} &= 1/x_{n+1}, \\ P_n &= \exp(p_n). \end{aligned} \tag{10.84}$$

---

[4]Of course, since κ and σ are dimensional quantities, their size should be measured relative to the time period of interest, T, and their relative magnitude, κ/σ^2.

Here

$$
f\left(p_{n+1}, x_n\right) =
\begin{cases}
\sqrt{\dfrac{P_{n+1}}{(1+\varepsilon)}}, & (1+\varepsilon)\,x_n^2 < P_{n+1}, \\[2.5ex]
x_n, & \dfrac{x_n^2}{(1+\varepsilon)} \le P_{n+1} \le (1+\varepsilon)\,x_n^2, \\[2.5ex]
\sqrt{(1+\varepsilon)\,P_{n+1}}, & P_{n+1} < \dfrac{x_n^2}{(1+\varepsilon)}.
\end{cases}
\tag{10.85}
$$

The corresponding *P&L* versus the buy-and-hold strategy is a path-dependent quantity given by the following formula:

$$
P\&L = \varepsilon \sum_{n=0}^{N-1} \left(\mathbb{1}_{(1+\varepsilon)x_n^2 < P_{n+1}} \left(x_{n+1} - x_n\right) + \mathbb{1}_{P_{n+1} < \frac{x_n^2}{(1+\varepsilon)}} P_{n+1}\left(\frac{1}{x_{n+1}} - \frac{1}{x_n}\right) \right)
\tag{10.86}
$$

$$
+ x_N + \frac{P_N}{x_N} - \left(1 + P_N\right).
$$

Here $\mathbb{1}$, is the indicator function equal to one if the condition is satisfied, and zero otherwise. The market making activity makes sense only when *P&L* > 0. Perhaps, a more informative quantity is the relative $\overline{P\&L}$, which shows the percentage return on market making, compared with the buy-and-hold strategy:

$$
\overline{P\&L} = \frac{P\&L}{\left(1 + P_N\right)}.
\tag{10.87}
$$

Of course, due to the fact that the log-price is stochastic, we can only analyze $\overline{P\&L}$ in probabilistic sense by running Monte Carlo simulations. To this end, we consider M MC paths, which are characterized by a random matrix $\left(\eta_{mn}\right)$, calculate a set of M P&L values, $\left\{\overline{P\&L_m}\right\}$, and study their statistical properties. We present the corresponding results in Figure 10.37.

These figures illustrate the complicated dependence of $\overline{P\&L}$ on three parameters: $\kappa, \sigma, \varepsilon$. The reader, wishing to become an AMM, needs to explore this dependence in detail.

One can analyze a mixed rule AMM by using a similar technique. We leave this analysis to an experienced reader as an exercise.

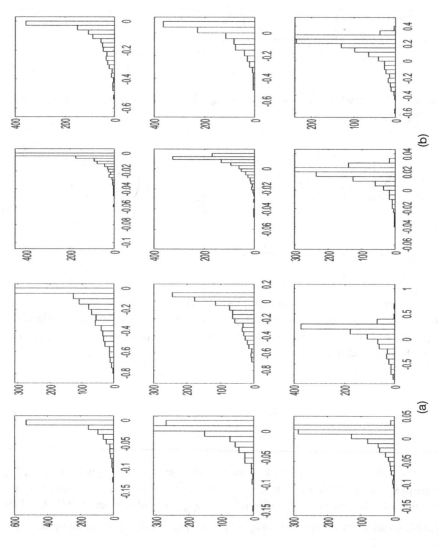

Figure 10.37 Histograms for the Profit-and-Loss distribution: (a) low mean-reversion; (b) high mean-reversion. Own graphics.

Most of the material covered in this section is original. Additional insights can be found in several recent papers; see, e.g., Angeris et al. (2019); Egorov (2019); Schär (2020); Zhang et al. (2018).

10.12 Summary

This chapter covers some important topics related to the behavior of cryptocurrencies as assets. We have shown that their returns are significantly more volatile than conventional currencies, such as EUR/USD and USD/JPY. Even more importantly, these returns have profound non-Gaussian features, including heavy tails. This observation is of paramount importance for risk managers in general and clearing houses wishing to engage in cryptocurrency futures trading, in particular.

Next, we covered some possible approaches to BTC pricing. Given the fact that BTC is an SoV asset, conventional financial tools are not useful for its valuation. While there are several approaches tried in the literature, none is particularly convincing. Still, we described the well-known S2F model and proposed its improvements. The latter enabled us to demonstrate that, in contrast with predictions of the standard model, BTC price does reach saturation in the long run, albeit at a very high level. We presented an analysis of the Bitcoin dominance index's behavior, based on two complementary approaches borrowed from epidemiology, and concluded that BTC would dominate all other cryptocurrencies in the long run. Finally, we discussed AMMs and studied their profitability (or lack thereof) using the standard tools of quantitative finance, such as Monte Carlo simulation. We have shown that AMM could be profitable, but not as much as is occasionally claimed.

Studying cryptocurrencies via quantitative finance methods is becoming more popular by the day. We expect a lot of exciting results in this research area going forward.

10.13 Exercises

1. Parameters shown in Tables 10.2 and 10.3 are calculated using market data for six years, from July 1, 2013 to July 1, 2020. Extend the data series until the present by using a suitable source of information, such as coinmarketcap.com, and redo the calculations. Which distribution provides the best fit for your data?
2. Same question for parameters shown in Tables 10.4–10.10.
3. At present, the Bitcoin protocol generates approximately 900 BTC per day. Given that, do you think that the S2F model is suitable for studying BTC price dynamics?
4. Why is it not possible to use an AMM based on the constant sum rule? What are the pros and cons of the constant product rule? What motivates designers of the DeFi protocols to consider the mixed rule for AMMs?

Current Research Topics

11.1 Introduction

DLT is a field of rapid innovation and constant development, but despite frantic research efforts, it remains plagued by many fundamental issues that must be resolved in order for the underlying technology to achieve its full potential. Comprehensive solutions for these problems have yet to be identified, which is not particularly surprising given their intrinsic complexity. The current chapter examines some of these problems, including intra- and interoperability, privacy, and the so-called scalability trilemma. Since our goal is to give a bird's eye overview of the problems and sketch the framework for their solutions, we will omit most technical details and refer the reader to the relevant references.

The chapter is organized as follows. In Section 11.2, we discuss intra- and interoperability of distributed ledgers. Intraoperability allows automatically exchanging different tokens defined on the same blockchain, such as USDC and TUSD. Interoperability is broadly defined as the ability to move any information between different blockchains. Narrowly, it means the capacity to exchange coins and tokens hosted on diverse blockchains.

Section 11.3 covers a vital topic of blockchain privacy, which is occasionally misunderstood. In the most common Bitcoin transactions, publicly known amounts of BTCs are moved between hashed public addresses. For the untrained eye, little information about the actual owners of the accounts is exposed. However, with some efforts, Bitcoin addresses can be associated with real-world owners, especially if the original BTCs are purchased on an exchange. Thus, to achieve better, if not perfect, privacy, additional cryptographic tools must be deployed. We discuss several approaches, including payment channels, mixing-based approaches, and zero-knowledge-based privacy.

Section 11.4 discusses the so-called scalability trilemma, which states that only two out of three properties — decentralization, scalability, and security — can be achieved simultaneously. For instance, the existing banking system is scalable and secure (within reason), but it is not decentralized since it relies on a small number of commercial banks. The crypto protocols, such as Bitcoin and Ethereum, are secure and decentralized (within reason), but not scalable since they are processing very few transactions per second. We cover different approaches to solving the scalability trilemma, such as introducing larger blocks, using sharding, applying different consensus mechanisms, utilizing lightning networks, and several other possibilities. Conclusions are drawn in Section 11.5.

11.2 Intra- and interoperability

11.2.1 Background

A well-designed blockchain, viewed in isolation, provides intrinsic decentralization. However, for the DLT to be truly successful, it is essential to ensure that decentralization occurs across different blockchains. Such decentralization offers an application developer a single view regardless of the number of distinct blockchain systems involved. Thus, true decentralization should allow functions to be spread across separate blockchain systems using different consensus mechanisms and other ledger data structures, with nodes implementing different software stacks; see Hardjono et al. (2019a); Lipton and Hardjono (2021). The TCP/IP Internet is architected along these lines. This design explains why the Internet has grown in size and traffic capacity to serve end-users globally: rather than being a single contiguous IP network, the Internet is a collection of interconnected Autonomous Systems (ASs); see Clark (1988). Having a well-defined physical boundary, each AS is operated by an Internet Service Provider (ISP).

In this section, we discuss approaches to blockchain intra- and interoperability. We define blockchain intraoperability as an ability for automatically exchanging different assets defined on the same blockchain, supporting smart contracts, for instance, Ethereum. We define interoperability as a capability to move assets between different blockchains.

Currently, it is not known how to fully achieve either intraoperability or interoperability. Below we argue that we can reach intraoperability by using automated market makers (AMMs) discussed in Section 10.11, while for interoperability the gateway-to-gateway transfer protocols can be utilized. The reader can find additional information in Abadi and Brunnermeier (2018); Buterin (2016); Hardjono et al. (2019a); Hope-Bailie and Thomas (2016); Koens and Poll (2019); Lima (2018); Lipton and Hardjono (2021); Pillai et al. (2019); Thomas and Schwartz (2015), and references therein.

11.2.2 Intraoperability

Lipton and Hardjono (2021) show that intraoperability is achievable by using AMMs, which currently operate as dark pools. Recall that dark pools of liquidity (also dark liquidity) operate as crossing networks for large buy and sell orders providing liquidity not displayed on exchanges. By using dark pools, traders can buy or sell large numbers of shares without revealing their positions or impacting the market prices because neither the price nor the trading company's identity is displayed. Dark pools transact at prices established at public exchanges; hence they can be viewed as price takers. AMMs operate along similar lines. Arbitrageurs and other agents add/remove coins to the pools based on the underlying coins' prices established on centralized exchanges. Thus, despite their name, AMMs cannot operate autonomously. To achieve intraoperability, one needs to reach a state when AMMs become price makers, i.e., their composition defines the relative price of constituent tokens and not the other way around. Intraoperability requires AMMs to become market players, dominating centralized exchanges. While appealing in theory, time will tell if such dominance is achievable in practice.

11.2.3 Interoperability

According to Hardjono et al. (2019a); Lipton and Hardjono (2021), in order to use DLT to represent economic value through tokens or ledger entries, one needs to achieve interoperability at two levels:

(1) Interoperability at the value level requires the notion of value and the standardization of mechanisms to represent it via tokens within blockchain systems. At this level, a core part of interoperability is the standardization of assets' definitions, which permits transacting parties to refer to the same definition of the virtual asset they wish to exchange.

(2) Interoperability at the mechanical protocol level requires developing standard protocols performing value conversions and token-transfers across blockchain networks.

Figure 11.1 shows an interoperable blockchain architecture relying on value-representation conversion points in the blockchain network and representation translation points across blockchain networks.

The essential ingredients are:

(1) Ingress and egress points: Value-representation conversion points are the ingress and egress points where value enters and departs the blockchain system, respectively. When an economic value from an external source enters the blockchain via an ingress point, it is represented in the form of the token data-structure defined in that blockchain system; see Step 1 in Figure 11.1. When a DLT value-representation is removed from a blockchain, it leaves the system via an egress point; see Step 5 in Figure 11.1. The ledger is marked accordingly to indicate that the token no longer exists and has been invalidated. We emphasize that in some blockchain protocols, such as Bitcoin, the tokens are an inherent part of the system, are not a derivative of any other asset, and never enter or leave it.

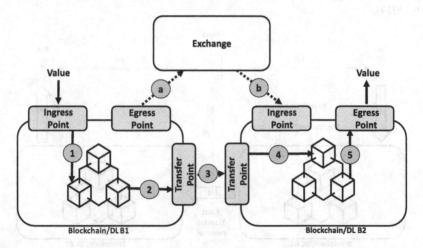

Figure 11.1 Generic interoperability architecture. Own graphics. Source: Lipton and Hardjono (2021).

(2) Token transfer points: These are the points allowing direct transfer of value from one blockchain to another without any change to the economic value represented by the token; see Step 3 in Figure 11.1. Such a transfer requires a data structure (format) translation if the two blockchain infrastructures employ differing interior ledger data-structures. The token data-structure in the origin blockchain (Step 2) is destroyed (or marked as invalid), while a new token data-structure is created (added) in the destination blockchain (Step 4). The transfer does not change the economic value presented by either of the tokens. Suppose the traversal by a token across blockchain networks is impossible. In that case, both value-representation conversion and token-format translation occur via a mediating third party, such as a crypto-exchange or a similar Virtual Asset Service Provider (VASP). Steps (a), (b) in Figure 11.1 illustrate the idea. The third-party exchange entity must participate in both blockchain systems and have the means to perform the exchange.

Figure 11.2 illustrates the two levels of interoperability.

Alice, who uses a blockchain BC1, wishes to transfer a virtual asset to Bob, who uses a different blockchain BC2. Steps (a) and (b) allow them to agree upon the definition of the corresponding asset. Alice initiates the transfer protocol (e.g., smart contract, application, etc.) in Step (c), which results in the asset transfer protocol executing between the two blockchain systems B1 and B2 in Step (d). Eventually, Bob becomes the new owner of the asset, Step (e). The corresponding protocol understands the definition of the asset being transferred, Step (f).

Figure 11.3 provides a schematic illustration of gateway nodes G within two blockchain domains.

Typically, a small number of nodes are designated as inter-domain gateway nodes and possess proper software, hardware, and trusted computing base. Ideally, all nodes in a given blockchain system should be capable of becoming gateway nodes and forming gateway groups collaborating on behalf of the blockchain system as a whole; see Hardjono and Smith (2019).

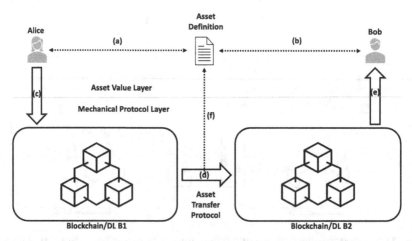

Figure 11.2 Asset transfer using a standard asset definition profile. Own graphics. Source: Lipton and Hardjono (2021).

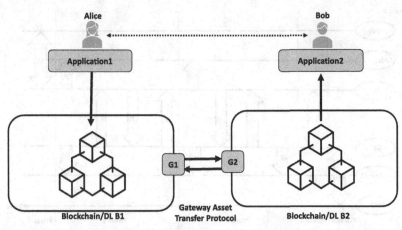

Figure 11.3 Gateway-to-gateway transfer protocol. Own graphics. Source: Lipton and Hardjono (2021).

Hardjono et al. (2019a) proposed the interoperability framework based on the following assumptions and principles that they correspond to the design principles of the Internet architecture:

(1) Opaque blockchain-resources principle: Each blockchain system's internal resources should be opaque to (hidden from) external entities. Any internal resources can be accessed by an external entity only via a gateway node with proper authorization. The opaque resources principle permits interoperability even if one (or both) blockchain systems are permissioned (private). It is similar to the autonomous systems principle in IP networking, assuming that interior routes in local subnets are not visible to external autonomous subnets; see Clark (1988).

(2) Externalization of value principle: The gateway-to-gateway protocol is indifferent (oblivious) to the economic or monetary value of the virtual asset it transfers. This principle allows designing asset transfer protocols for efficiency, speed, and reliability, ignoring the changes in the virtual asset's perceived economic value. It is similar to the end-to-end principle in the Internet architecture, placing contextual information (economic value) at the endpoints of the transaction; see Saltzer et al. (1984). The originator and beneficiary at the respective blockchain systems are assumed to agree on the asset's economic value. Using gateways is necessary because the native technical constructs in interacting blockchains are incompatible with one another. The blockchain interoperability framework based on interoperable gateways permits two gateway nodes belonging to different blockchain systems to conduct a virtual asset transfer between them in a secure and non-repudiable manner. Thus, gateways ensure that the asset does not exist simultaneously on both blockchains, thus solving the double-spend problem. Gateways "hide" the complexity of their underlying blockchains and can cooperate among themselves via standard APIs.

Different phases of gateway-to-gateway asset transfer are shown in Figure 11.4.

A gateway-to-gateway asset transfer protocol between two blockchain systems is executed by gateway nodes G1 and G2, representing the two respective blockchain systems,

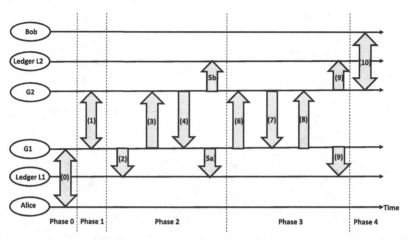

Figure 11.4 Five phases of the gateway asset-transfer protocol. Own graphics. Source: Lipton and Hardjono (2021).

B1 and B2. After a successful transfer from B1, the asset is extinguished on B1 by the origin-gateway G1 and introduced on B2 by the destination-gateway G2. The extinguish/introduce mechanism is ledger specific. Gateways negotiate the choice of the commitment protocol (type/version) and the corresponding commitment evidence during Phase 1; they implement a transactional commitment protocol coordinating their actions in Phase 2 and the final commitment protocol transferring the asset in Phase 3. The gateways G1 and G2 may use the classic 2-Phase Commit (2PC) protocol (Traiger et al., 1979), to ensure efficient and non-disputable commitments to the asset transfer. Different phases of gateway-to-gateway asset transfer are shown in Figure 11.4: (0) originator asset-transfer request; (1) identity and asset validation; (2) asset lock/escrow; (3) lock/escrow evidence; (4) evidence receipt; (5a) receipt logged; (5b) asset-lock logged; (6) commit-prepare; (7) prepare-acknowledged; (8) commit; (9) commit final; (10) destination asset-transfer commit.

11.3 Privacy

11.3.1 Background

Any financial infrastructure worth its salt has to preserve the user's privacy. In particular, this is true for the crypto ecosystem. There are at least three blockchain-related privacy aspects, associated with a given transaction:

(1) the identity of transacting entities;
(2) the transaction data;
(3) the change in the global state of the blockchain caused by the transaction.

Recall that Bitcoin users' identities are pseudonymous and represented by the hashes of their public keys. Thus, if Alice mines her BTCs, she can keep her identity private. However, when Alice acquires her BTCs on an exchange, her real identity can be uncovered, at least by the exchange operator. Moreover, chain analysis enables to inspect complicated sequences of transactions and extract real-world identities. For instance, when Alice sends BTCs to Bob, hashes of their public keys, the link between these hashes, and the amount of BTCs become public.

To make chain analysis more difficult (but not impossible), one can contemplate reducing the amount of information revealed by a particular transaction by keeping the sender's and recipient's public keys private. In addition, the transaction size can be kept private, too. However, the idea of making the global state of the blockchain completely private jeopardizes integrity and is self-defeating: If the total number of coins in circulation were to be shielded, users could not be sure that there is no surreptitious issuance of additional coins.

In order to enforce user privacy, several methodologies have been developed. Below, we describe three of them, payment channels, coin mixing, and cryptographic obfuscation. The reader can find discussion and additional useful information in a recent survey of the blockchain privacy; see Peng et al. (2020).

11.3.2 Payment channels

A straightforward way for a group of two or more agents, who wish to transact anonymously with each other, is to establish a payment channel among themselves. Figure 11.5 shows the simplest possibility — a payment channel between Alice and Bob.

To open the channel, both deposit 10 BTC to a particular address. Then, they trade with each other off-blockchain until such time that one of them (or both) wishes to close

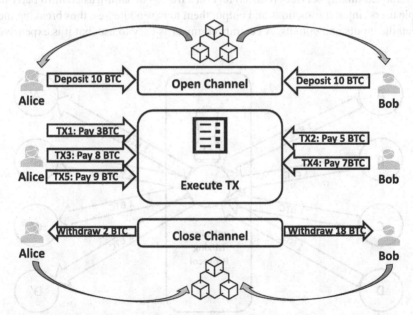

Figure 11.5 A lightning channel between Alice and Bob is used to ensure privacy of their transactions. Own graphics.

the channel, or the pre-funded buffer gets exhausted. The final settlement occurs on the blockchain. In the example shown in Figure 11.5, after a series of transactions, which are kept private, Alice ends up with 2 BTC, and Bob with 18 BTC. In Section 11.4.4, we show that, in addition to enhancing privacy, payment channels improve the underlying protocol's scalability. However, strictly speaking, such transactions cannot be qualified as blockchain-based and are prone to various risks, as all traditional transactions.

11.3.3 Coin mixing

Coin mixing is an alternative to establishing payment channels. The simplest centralized mixer is shown in Figure 11.6.

This figure shows four agents with Bitcoin addresses A, B, C, D, each sending 1 BTC to the mixing service address. In turn, the mixer's operator sends their BTCs to new addresses A', B', C', D' (minus a small fee). Assuming that the operator is honest and does not steal from its clients, agents end up with BTCs that are not directly linked to their addresses; instead, they are connected to the operator's address. Given that all participants send the same amounts, their anonymity set is four. An external observer cannot say, who is the owner of unspent transaction outputs (UTXOs) at A', B', C', D', although they can guess that the original owner's address belongs to the set {A, B, C, D}. While not without merits, this approach is centralized and relies on honesty and discreetness of the operator. Her honesty means she does not steal clients' BTCs, and discreetness assumes that she keeps the actual mapping between the sets {A, B, C, D} and {A', B', C', D'} to herself.

Centralized mixing services (CMSs) rely on a trusted or semi-trusted third party to mix multiple users' input transactions and output them to new addresses, thus breaking the link between the inputs and outputs. A centralized mixer is easy to use, but it is expensive, not

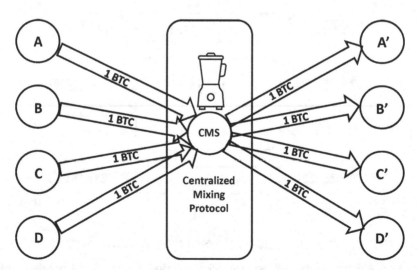

Figure 11.6 Centralized mixing service. Own graphics. Source: Peng et al. (2020).

to mention that its operator is a single point of failure. Hence, researchers proposed to use decentralized mixers instead. Decentralized mixers are cheap or even free; besides, they don't rely on a third party and don't have a single point of failure. A decentralized mixer is shown in Figure 11.7.

One of the first decentralized mixing services, called CoinJoin, was proposed by Greg Maxwell in 2013; see Maxwell (2013). Instead of relying on a third party, CoinJoin users mix their coin themselves. Mixing starts with negotiations allowing a group of payers to agree on the set of input and output addresses they wish to use. Then, a transaction containing all input/output pairs is generated and checked by users for accuracy. Finally, users jointly sign the transaction and publish it on the Bitcoin blockchain. Compared to centralized mixers, CoinJoin significantly reduces the risk of information leakage and eliminates the possibility of coin theft. However, it is far from perfect since the negotiation process can be abused to discover its participants' information. Besides, the CoinJoin process is vulnerable to the Denial of Service (DoS) attack since a single user's inability to sign the final transaction results in the failure of the whole process. Subsequently, several other protocols addressing these shortcomings, such as CoinShuffle and CoinParty, were designed; see Peng et al. (2020) and references therein.

The Monero protocol automatically mixes a sender's coin with randomly selected decoy coins belonging to other users; see Noether (2015). It uses the so-called ring signature produced by the actual sender via a one-time spend key, corresponding to the output from the sender's wallet. Properties of ring signatures are such that it is possible to prove that someone within a group of addresses has signed a transaction without revealing a specific signatory.

Mixing is particularly relevant for stablecoins since they are used as means of payment and hence have to preserve the transacting parties' privacy, just like cash. Since centralization is a feature of all practically-viable stablecoins from the start, it is natural to benefit

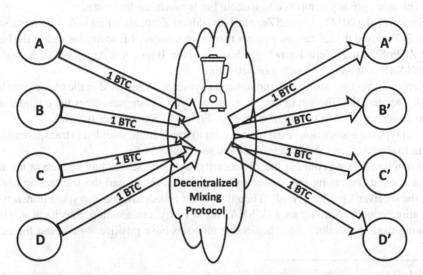

Figure 11.7 Decentralized mixing service. Own graphics. Source: Peng et al. (2020).

from this fact and use their originators as mixer operators. Upon request, they will allow routing of payments via blockchain addresses under their control, making the source of funds impossible to deduce for other economic agents. Ideally, legitimate authorities, armed with a legally binding request, should have access to the flow of funds trace with ease as if it was not mixed.

11.3.4 Built-in cryptographic obfuscation

It has been known since the 1980s that technology based on zero-knowledge proofs (ZKPs) can significantly enhance privacy; see, e.g., Goldwasser et al. (1989). A ZKP has to be complete, sound, and zero-knowledge. Completeness means that the prover can convince the verifier if she knows a certain fact, X. Soundness means that the prover cannot convince the verifier if she does not know X. Finally, zero-knowledge implies that the actual information about X is not leaked in the process of interaction between the prover and the verifier.

In Chapter 4, we already dealt with an example of a ZKP, namely, digital signature. Indeed, by signing a message with her secret key, sk, the prover can convince the verifier that she knows the corresponding key without revealing any information about the key itself.

The original ZKP-inspired protocol was Zerocoin, which extended the Bitcoin protocol by allowing users to convert their BTCs into zerocoins and redeem zerocoins into BTCs while breaking the input-output linkage; see Miers et al. (2013). Zerocoin is based on ZKP of Knowledge (ZKPoK) protocol. A user can enhance her anonymity by "depositing" BTCs with the system at large and receiving zerocoins; before a zerocoin is spent, it must be converted back to BTC. An astute reader can recognize an apparent influence of David Chaum's Digicash described in Section 4.5.7.

By construction, Zerocoin cannot hide the transaction amount and address balance. Moreover, it supports only fixed-denomination payments. Besides, the ZKPoK scheme is rather involved, so that Zerocoin is operationally much heavier than Bitcoin. Thus, Zerocoin enhances privacy compared to Bitcoin but is much harder to use.

Sasson et al. (2014) designed Zerocash to address Zerocoin's flaws. Zerocash is a standalone PoW-based digital currency rather than an extension of Bitcoin. Instead of the heavy-duty ZKPoK, it uses Zero-Knowledge Succinct Non-Interactive Argument of Knowledge (zk-SNARKs), which are much more efficient.[1]

Zerocash supports arbitrary denominated payments. When used in the transparent layer, Zcash operates like Bitcoin; in the ZK security layer, Zcash guarantees the confidentiality of the payer, payee, and transaction amount. Figure 11.8 illustrates the idea.

Four types of transactions exist: transparent to transparent, shielded to transparent, transparent to shielded, and shielded to shielded, see Figure 11.8.

When transferring coins in the ZK security layer, the sender has to encrypt the transaction content, including the sender and receiver addresses and the transaction amount, with the receiver's public key, pk. Then the sender broadcasts the encrypted transaction to the entire network. A miner uses zk-SNARKs to verify a transaction. Specifically, without knowing the actual values, she checks that all coins have positive values, and the sum of

---

[1]A word of caution is in order. The security of Zerocash relies on a trusted setup process to determine the parameters of zk-SNARKs. By compromising this process, an adversary can break the privacy of Zerocash.

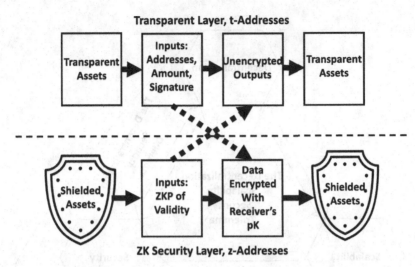

Figure 11.8 Zcash has a built-in privacy layer, called the ZK security layer, allowing participants to transact anonymously. Own graphics.

the inputs is equal to the sum of the outputs. This framework guarantees that transactions stay anonymous. Upon receiving the encrypted transaction, the receiver decrypts its content with her secret key and generates a unique one-time serial number for the newly acquired coins, which is used for preventing double-spending attacks. This number becomes public once the coin is spent. This technique is very similar to the one proposed by Chaum, see Section 4.5.7.

11.4 Scalability trilemma

11.4.1 Background

The Ethereum Wiki defines the scalability trilemma as follows: "This sounds like there's some kind of scalability trilemma at play. What is this trilemma and can we break through it? The trilemma claims that blockchain systems can only at most have two of the following three properties: decentralization (defined as the system being able to run in a scenario where each participant only has access to $O(c)$ resources, i.e., a regular laptop or small VPS), scalability (defined as being able to process $O(n) > O(c)$ transactions), security (defined as being secure against attackers with up to $O(n)$ resources)."[2] The trilemma is presented graphically in Figure 11.9.[3]

First, we translate the above definitions into layman terms. Being decentralized is the *raison d'être* for the entire crypto concept. Correctly implemented, decentralized protocol

---

[2]https://github.com/ethereum/wiki/wiki/Sharding-FAQ.

[3]The decentralized + secure dilemma claims that the foundational protocols, such as Bitcoin and Ethereum, can achieve decentralization and security. However, this claim is difficult to accept at face value. Indeed, as was

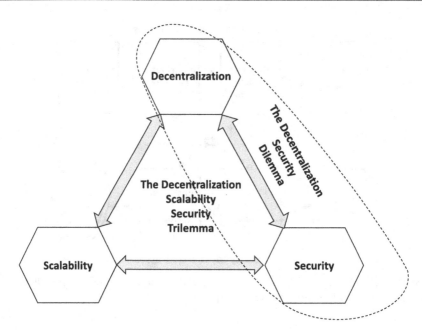

Figure 11.9 The scalability trilemma. Own graphics.

empowers censorship-resistance and allows anyone to participate in the ecosystem without discrimination. The scalability of a given blockchain protocol means its ability to process transactions generated by millions of participants. Security of a protocol signifies the ledger's immutability, consistency, and overall resistance to potential attacks, including 51%, DoS, Sybil, and other attacks; see Chapters 5 and 6.

Thus, disproving the trilemma, which, of course, is not a rigorous mathematical statement, and building blockchain protocols, which are simultaneously decentralized, scalable, and secure, has been a significant objective of both academics and practitioners. In this section, we discuss several approaches to solving the scalability problem. Further details can be found in Chu and Wang (2018); Kwon et al. (2019); Pappalardo et al. (2018); Zhou et al. (2020) among many others.

11.4.2 Current transactional banking model: scalable + secure

In Chapter 2, we mentioned that the current transactional banking is based on the secure + scalable paradigm. However, it is highly centralized, in agreement with the trilemma. As periodic financial crises, including the Global Financial Crisis (GFC) of 2007–2008 show, the rest of the financial system is far from being genuinely robust.

---

explained in Chapters 5 and 6, mining for these protocols is highly centralized, since miners coalesce into gigantic mining pools; see Figures 5.29 and 6.6. Besides, it consumes electricity on a prodigious scale. For example, in 2020, the Bitcoin protocol processed about 112 million transactions and consumed more than 70 TWh of electricity, about the same as Austria or Venezuela. In comparison, Visa processes 150 million transactions per day.

At the level of a single country, it operates as a centralized system shown in Figure 1.1 (a), while at a multi-county level, it works as a decentralized system shown in Figure 1.1 (b). In both cases, hubs correspond to central banks, while spokes to commercial banks. Provided that the trilemma is disproven, the centralized financial system may be replaced by a distributed one in the not too distant future.

11.4.3 Current crypto model: decentralized + secure

In Chapters 3, 5–7, we discussed some of the most popular blockchain protocols currently in operation — Bitcoin, Ethereum, and Ripple. These protocols aim at being decentralized and secure. They are unquestionably secure, having stayed so since inception, and being impervious to all possible direct attacks, including the dreaded 51%. At the same time, reasonable people can disagree about the precise degree of their decentralization.[4] Like any other capitalist effort of comparable size, mining operations result in the emergence of giant mining pools, which thoroughly dominate it. Besides, buying cryptocurrencies on the secondary market is possible only via gatekeepers, such as exchanges or investment pools.

Hence, one can argue that for cryptocurrencies even the decentralized + secure dilemma is not completely solved. In the next subsection, we discuss possible approaches to solving the trilemma, and hence the dilemma.

11.4.4 Future fintech model: decentralized + scalable + secure

11.4.4.1 Motivation

First, we categorize different approaches to solving the trilemma, based on the so-called layer protocols, which are summarized in Table 11.1.

Layer0 protocols aim at accelerating network communications and reducing their bandwidth footprint. Layer1 protocols investigate several options, namely: utilizing larger, more compactly designed blocks; applying transaction sharding; changing the underlying consensus algorithms; designing non-blockchain-based distributed ledgers, and similar approaches. Layer2 protocols build payment channels and side-chains above the existing blockchains. We briefly discuss some of them below.

11.4.4.2 Layer0 protocols

Layer0 protocols are designed to increase the usable bandwidth of the network and improve the overall connectivity. The Erlay protocol is a representative example; see Naumenko et al. (2019). When Alice broadcasts a transaction, it has to reach all full nodes in the network; see Figure 5.12. Broadcast results in multiple exchanges of the same transaction IDs among nodes and is characterized by high redundancy. This redundancy is highly significant, since about half of all the bandwidth consumed by a full node is used to share transactions.

---

[4] There is no question in our minds that, in relative terms, Bitcoin and Ethereum are much more decentralized than Ripple.

Table 11.1 Approaches to blockchain scalability. Source: Zhou et al. (2020).

Layer	Category	Protocol
Layer0	Rationalizing network utilization	bloXroute, Erlay, Kadcast, Velocity
Layer1	Block storage and TXs handling	Bitcoin Cash, Compact block relay, CUB, Jidar, SegWit, Txilm
	Transactions sharding	Elastico, Monoxide, OmniLedger, RapidChain, Zilliqa
	Changing consensus algorithm	Algorand, Bitcoin-NG, Ouroboros, Snow white
	Changing topology to DAG	Byteball, Conflux, Dagcoin, Hedera Hashgraph, Inclusive, IOTA, PHANTOM, SPECTRE
	Permissioning	Corda, Hyperledger Fabric
Layer2	Cross-chain	Cosmos, Polkadot
	Off-chain	Arbitrum, Truebit
	Payment channel	DMC, Lightning Network, Raiden Network, Sprites
	Side-chain	liquidity.network, Pegged Sidechain, Plasma

The Erlay protocol proposes reducing the redundancy by sharing transaction IDs and their sketches; see Naumenko et al. (2019). A sketch is a compact representation of all the transactions accepted by a particular node since the last reconciliation. First, the nodes share IDs, but only with eight peers or less, regardless of their overall connectivity. Second, they request sketches from their peers. A node processes the sketch it receives, identifies new transactions, and requests them from the peers. Such a design enables to materially reduce the amount of inter-node traffic, while consuming a fraction of the conventional method's bandwidth.

11.4.4.3 Layer1 protocols

Bitcoin Cash, Bitcoin SV, and Lightcoin, have implemented changes in the original Bitcoin protocol, allowing a substantial increase of the block size. This increase is very consequential since it results in a proportional increase in the number of transactions recorded and confirmed in a single block. Naturally, larger blocks reduce decentralization of the protocol since they impose heavy demands on storage capabilities of the full nodes.

The idea of sharding is straightforward, but its implementation is anything but. Sharding is a well-known technology initially designed to optimize large commercial databases. The idea is to divide the data fragments stored at separate servers, thus reducing the pressure of a centralized server, improving the search performance, and increasing the database's storage capacity. When applied to a distributed ledger, sharding produces several smaller sub-ledgers, or shards, each containing a fraction of the nodes. Transactions are assigned to different shards, which process them in parallel, thus dramatically increasing the overall throughput.

Figure 11.10 (a) demonstrates the process of splitting nodes into shards and allocating transactions to each shard. Of course, sharding requires solving two difficult problems: achieving consensus within a given shard and protecting it against possible attacks and handling cross-shard transactions efficiently and consistently.

Successfully implemented sharding results in a structure consisting of the main chain and several shard chains, which maintain their own subchains and commit their state to the main chain periodically. The main chain has to be validated to ensure that cross-shard transactions are treated consistently, see Figure 11.10 (b). Zilliqa and several other protocols

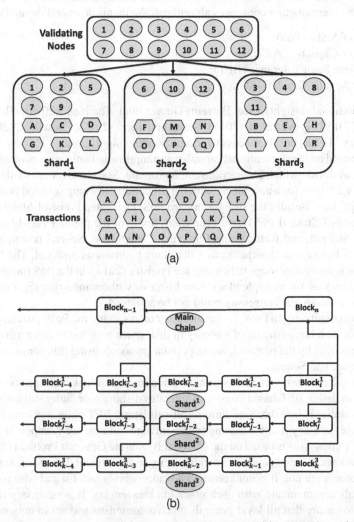

Figure 11.10 One can use sharding to scale transactions: (a) separation of nodes and transactions into shards; (b) validation of transactions by different sharded subchains. Here superscripts denote shards, and subscripts denote block numbers. Own graphics. Source: Zhou et al. (2020).

use sharding to achieve scalability and very high transactions per second (TpS) level; see Luu et al. (2016); Zilliqa (2018). In experiments, Zilliqa achieved about 2800 TpS, and reached nearly 300 TpS on the mainnet.

In Chapters 5 and 6, we have discussed the Proof-of-Work (PoW) algorithm, which implies that any participant on the network can become a miner competing in the block generation game. A miner must be the first one to solve a specific computational problem to confirm a new block and receive a reward. As we have shown in Chapters 5 and 6, economic considerations force miners to coalesce in highly centralized mining pools.

Due to the inability of the original PoW to scale and its prodigious use of electricity, it is natural to entertain other consensus algorithms. We mention several popular approaches:

(1) Proof-of-Stake (PoS);
(2) Proof-of-Capacity (PoC);
(3) Byzantine Fault Tolerance (BFT);
(4) Direct Acyclic Graph (DAG).

For additional insights, see Barreiro-Gomez and Tembine (2019); Houy (2014); Mazieres (2015); Nguyen et al. (2019); Seang and Torre (2018); Zhou et al. (2020) among many others.[5] Let us briefly discuss the corresponding algorithms.

PoS is based on the conventional capitalist assumption that only owners of coins or smart contracts can decide which transactions are legitimate. Stakeholders are randomly chosen to become validators for a new block, with the probability of being selected increasing with the holdings' size. Besides the straightforward PoS, Delegated Proof-of-Stake (DPoS) and Leased Proof-of-Stake (LPoS) have been tried. Some of the popular PoS-based protocols are Tezos, Steemit, and forthcoming Ethereum 2.0 upgrade. Several researchers pointed out that PoS has serious drawbacks as a distributed consensus protocol. The "nothing-at-stake" issue is especially tough to resolve; see Poelstra (2014). In the PoS framework, block generators can vote for multiple blockchain histories without incurring significant costs. If they decide to do so, the consensus could not be achieved.

PoC works similarly to PoW; however, rather than solving the PoW puzzle, participants need to allocate a large amount of memory or disk space required to solve a computational problem specified by the protocol. Some popular protocols using this consensus are Algorand, Filecoin, and Solana.

BFT-based consensus is an inspiration for numerous blockchain protocols, thus we describe it in detail. BFT-based consensus tolerates dishonest or faulty validators, provided that their number is less than one-third of all validators. BFT plays a pivotal role in computer science, as a key tool for dealing with hardware failures; see Lamport et al. (1982).

The BFT consensus is based on the classical Byzantine Generals Problem (BGP), which works as follows. Several generals attacking a city from different directions. Some generals are loyal; others are not. It is not known in advance who is faithful and who is treasonous. The generals communicate with each other via messengers. It is necessary to design an algorithm to ensure that all loyal generals reach a consensus and act in unison. Assuming

---

[5]A very detailed source of valuable information about consensus algorithms used by popular blockchain protocols is given by Cédric Walter in Blockchain Consensus Encyclopedia https://tokens-economy.com/.

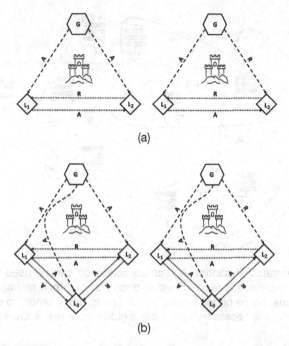

(a)

(b)

Figure 11.11 (a) The BGP with three generals, including one traitor, cannot be solved. The left figure shows faithful commanding general G and lieutenant L_1, and treacherous lieutenant L_2. Since lieutenant, L_1 receives conflicting messages, A (advance) from the general, and R (retreat) from lieutenant L_2, he cannot decide what to do. The right figure shows treacherous general G, and faithful lieutenants L_1, L_2. Lieutenants receive conflicting orders and cannot decide how to proceed. (b) The BGP with four generals, including one traitor, can be solved. The left figure shows faithful commanding general G and lieutenants L_1, L_3, and treacherous lieutenant L_2. Lieutenants L_1, L_3 receive two A messages, and one R message. Hence they advance, together with G. The right figure shows treacherous general G, and faithful lieutenants L_1, L_2, L_3. Lieutenants L_1, L_2, L_3 receive two A messages, and one R message. Hence they advance. Own graphics.

that there is a commanding general (not necessarily loyal), Lamport et al. (1982) formulate Interactive Consistency (IC) conditions for the BGP for n generals as follows:

A commanding general must send an order to his $n-1$ lieutenant generals such that

IC1. All loyal lieutenants obey the same order.

IC2. If the commanding general is loyal, then every loyal lieutenant obeys the order he sends.

Conditions IC1 and IC2 are called the interactive consistency conditions. Note that if the commander is loyal, then IC1 follows from IC2. However, the commander need not be loyal.

When there is a commanding general, solving the BGP problem and achieving BFT-based consensus is possible if the number of traitors is less than one-third of the total and impossible otherwise. Figure 11.11 illustrates both possibilities.

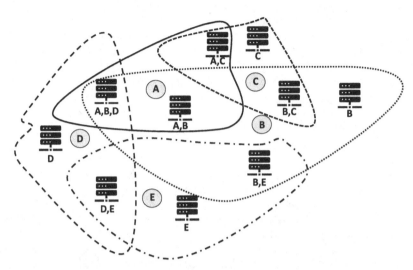

Figure 11.12 Federated Byzantine Agreement tolerance can be used to scale transactions. Initially, a consensus is reached within quorums, and then among them. The figure shows five quorums. Some nodes belong to just one quorum, others to several. The overlap between quorums is necessary to achieve a global consensus. Own graphics.

Extensions of the original BFT concept include Practical Byzantine Fault Tolerance (PBFT) and Federated Byzantine Agreement (FBA); see, e.g., Castro and Liskov (1999); Mazieres (2015). Major protocols based on FBA include Stellar and Ripple. The original BFT-based consensus is reachable by having a list of approved validators (commanding generals); accordingly, it is not decentralized. The FBA is a decentralized version of the original consensus. It starts with dividing nodes into quorums, comprised of nodes that trust each other. The consensus is reached by voting first inside quorums and then amongst them. As mentioned in Chapter 7, quorums have to overlap to make an agreement between them possible. The idea of FBA is illustrated in Figure 11.12.

Direct Acyclic Graph (DAG) algorithms reject a simple sequential topology of conventional blockchains, such as Bitcoin or Ethereum, in favor of the so-called tangle. Although DAG has not received a wide recognition, we view it as a promising tool in resolving the trilemma in future DLT designs. It presents an alternative consensus to much more popular PoW and PoS. DAG has no validators and uses no blocks, hence, potentially, it can be decentralized, scalable, and secure.

To add a new transaction to the tangle, a participant needs to validate two previous transactions of her choosing. She has to perform a small amount of work (much less than the PoW required to become a Bitcoin miner). The purpose is to suppress spam, as in the original proposal of Back (2002). The binary consensus (each transaction validates two previous ones) strengthens and potentially accelerates with additional transactions added to the tangle. It allows building low-fee, secure transaction protocols. A typical DAG is shown in Figure 11.13.

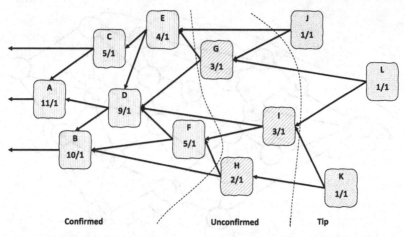

Figure 11.13 DAG is a scalable alternative to the linear blockchain topology. Own graphics.

This figure shows that transactions can receive both direct and indirect approval from their peers. For instance, TX A is directly approved by TX C, and indirectly by TX E (either via E– >C– >A or E– >D– >A). Of course, every transaction approves itself. The cumulative number of approvals determines whether a transaction is confirmed, unconfirmed, or at the tip of the tangle. TX A has 11 approvals in total and hence is confirmed; TX G has three approvals, and hence is unconfirmed; TX J has only one approval (from itself) and hence is a tip transaction. Similarly to the longest chain rule, shown in Figure 3.9 in the case of forking, TXs with the highest number of cumulative transactions survive, while others are disregarded, making a double-spend attempt impossible.

The best-known DAG-based protocols are IOTA and Hashgraph. IOTA claims to be capable of processing 250 TpS. However, as of this writing (January 2021) the actual number of TpS is less than 10 because of the relatively low network demand; see https://thetangle.org/.

Permissioned ledgers, such as Azur, Corda, Hyperledger Fabric, and Quorum rely on Proof-of-Authority (PoA) or PBFT algorithms; see Brown et al. (2016); Gupta (2017); Hearn (2016). They are scalable but certainly not decentralized. From a practical standpoint, the lack of decentralization might not be a severe issue.

11.4.4.4 Layer2 protocols

Lightning Network, Radien Network, and Plasma achieve scalability by executing most inter-member transactions off the blockchain and settling the total scores amongst members on the main blockchain infrequently; see Poon and Dryja (2016); Poon and Buterin (2017); Raiden (2016). Figure 11.14 shows a typical setup. Recall that in Section 11.3 we discussed how to use Lightning Network to achieve better privacy.

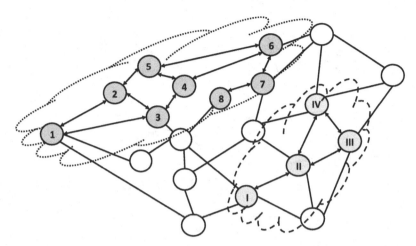

Figure 11.14 Lightning network can be used to scale transactions (and achieve additional privacy). Two lightning sets are shown: nodes 1 to 8, and nodes I to IV. Own graphics.

11.4.4.5 Conclusions

Comparison between varous consensus algorithms shows that successful transition to the decentralized + scalable + secure framework is only possible by combining several protocols described in this section. The PoS, sharding and DAG-based consensus look particularly promising due to their relatively low energy consumption, and ability to process transactions in parallel.

11.5 Summary

This chapter discussed some of the most critical and challenging technical problems that need to be solved before DLT goes mainstream. Given that these problems are inherent to the technology itself, we can hope that cryptographers and software engineers will find their satisfactory solution in the not too distant future. The archetypal Bitcoin and Ethereum protocols showed how to build the foundational DLT-based framework. Getting to a higher level requires applying additional techniques, some of which are covered in this chapter.

11.6 Exercises

1. What are the advantages and disadvantages of PoW as a mechanism for establishing distributed trust? Can you describe other potential approaches such as proof-of-stake (PoS), proof-of-activity (PoA), proof-of-burn (PoB), etc.? Why do we want to replace PoW with other mechanisms?

2. Why implementing PoS-based consensus is so hard?
3. What are the top pros and cons of permissioned blockchains?
4. What are the advantages and disadvantages of PoW compared with BFT? Do you feel that BFT is as robust as PoW?
5. Is DAG a viable alternative to DL in the long run?

12 | Present and Future of DLT

12.1 Introduction

As was discussed earlier, a distributed ledger is a system of records shared among participants, eliminating the need to reconcile disparate ledgers. It operates as an "append-only" distributed database, which securely and immutably stores transactions (changes) and allows for these transactions to be tracked in a highly efficient manner. By using distributed ledger technology (DLT), one can:

(1) Ensure data trustworthiness and accuracy;
(2) Increase transactional reliability, transparency, and speed;
(3) Reduce transaction costs;
(4) Simplify and automate complex business processes;
(5) Improve security against fraud and cybercrime.

Iansiti and Lakhani (2017) put it succinctly: "Blockchain is not a disruptive technology, which can attack a traditional business model with a lower-cost solution and overtake incumbent firms quickly. Blockchain is a foundational technology: It has the potential to create new foundations for our economic and social systems." Its power lies in its ability to offer standardized secured shared data. Even when blockchain-related activity is not central to the overall efforts, it creates the impetus for change across different industries and organizations.

The choice of an appropriate blockchain is crucial to the overall success of a particular application. While attempts to use public blockchains, such as Ethereum, are quite common, they face practical obstacles that are very hard to overcome, not least of which is the sheer cost of using CaaS, see Chapter 6. As such, private or permissioned blockchains are better suited to the development of most applications, since they usually serve preauthorized participants within a comparatively small private network. Some popular open-source blockchain platforms include Corda Enterprise, Hyperledger, Multichain, and Quorum. IBM, Microsoft Azur, SAP, and Vechain provide blockchain as a service (BaaS).

Central to the evolution of the crypto space is whether or not DLT can potentially, or even already, has real-life applications above and beyond cryptocurrencies. Figure 12.1 illustrates potential areas where blockchains can be used.

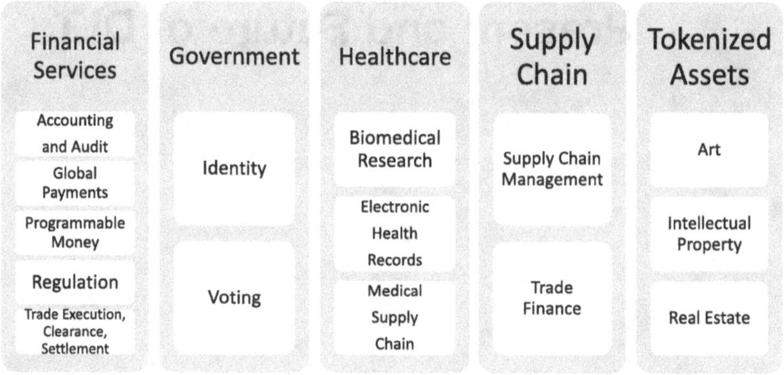

Financial Services	Government	Healthcare	Supply Chain	Tokenized Assets
Accounting and Audit	Identity	Biomedical Research	Supply Chain Management	Art
Global Payments		Electronic Health Records		Intellectual Property
Programmable Money				
Regulation	Voting	Medical Supply Chain	Trade Finance	Real Estate
Trade Execution, Clearance, Settlement				

Figure 12.1 Possible blockchain applications. Own graphics.

We emphasize that, outside of the narrow domain of cryptocurrencies and, somewhat more broadly, cryptoassets, the progress is not as fast as some have hoped for. So far, most of the projects are at the proof-of-concept stage.

However, the speed of development is not the only issue plaguing the broad adoption of DLT. Several high-profile projects, which launched to critical fanfare and grand promises from their founders, had to be curtailed and moved to a scaled-down version of original plans. Although not a problem by itself, it shows that the appeal of the sensationalist nature of some projects can outweigh that of realistic and knowledgeable proposals. Nevertheless, as long as these downsized projects serve their intended purpose, they are unquestionably welcomed into the club of successful DLT companies.

In this final chapter, we describe existing applications of DLT as they pertain to various activities, from financial services, with the exception of cryptoassets addressed in the previous chapters, to government services, to identity and security, to healthcare, and supply chain management. We also compare existing approaches and develop the outline for potential, forward-thinking solutions that are both novel and practical. The chapter is organized as follows. In Section 12.2, we discuss potential applications of DLT in financial services, which go above and beyond cryptocurrencies addressed in the previous chapters. There are several promising areas where DLT can be helpful, including accounting and audit, domestic and cross-border payments, programmable money, regulations, including Know Your Customer (KYC) and Anti-Money Laundering (AML), and trade execution, clearing, and settlement relying on distributed financial market infrastructure.[1] Section 12.3 covers government, emphasizing identity, and voting. Healthcare, covered in Section 12.4, is likewise an attractive application domain for DLT. Here, we discuss three areas of interest: (a) sharing of biomedical research data while preserving its integrity; (b) patient control over electronic health records; and (c) rationalization of medical supply chains. Section 12.5 describes trade finance and non-healthcare related supply chain management. In Section

[1]It is clear that the growing value of the cryptoassets, standing at 1 trillion as of this writing, makes it inevitable that a more stringent regulatory regime will be applied to them in the not too distant future. Strict regulations will diminish free-wheeling ways of the crypto market, and, as a result, increase its institutional appeal.

12.6, we cover approaches to tokenization of assets, with an emphasis real estate. Finally, in Section 12.7, we draw our conclusions.

12.2 Financial Services

12.2.1 Accounting and audit

Bookkeeping is as old as writing itself. For several millennia, single-entry accounting reigned supreme. However, in the 15th century, or possibly even earlier, double-entry bookkeeping and accounting were developed and popularized by authors such as Benedetto Cortiglia and Luca Pacioli, among others; see, e.g., a detailed discussion in De Roover (1956).

It is impossible to overstate the role of double-entry accounting in the development of capitalism in general, and banking in particular. In the novel "Wilhelm Meister's Apprenticeship" published in 1795–96, the great German writer Johann Wolfgang von Goethe (1749–1832) expressed his enthusiasm for double-entry methods: "I could not think of any man whose spirit was, or needed to be, more enlarged than the spirit of a genuine merchant. What a thing it is to see the order which prevails throughout his business! By means of this he can at any time survey the general whole, without needing to perplex himself in the details. What advantages does he derive from the system of bookkeeping by double entry? It is among the finest inventions of the human mind; every prudent master of a house should introduce it into his economy."

It is often said that double-entry bookkeeping and accounting allow the owners of an enterprise, along with auditors and authorities, to have a clear idea of its inner workings, and, equally important, to find both accidental and intentional errors in its books. While the first claim is understandable, the second is more difficult to accept. The fundamental accounting equation, or the balance sheet equation, states that debits are equal to credits for every transaction, so that:

$$Assets = Liabilities + Equity. \qquad (12.1)$$

This equation is useful for finding basic discrepancies, whether intentional or not, but fails to detect more sophisticated ones.[2]

The issue with auditing based on double-entry bookkeeping is that it is performed based on information obtained primarily from within the organization, while a proper and thorough audit requires information from outside of that organization. For example, if Alice buys equipment from Bob for $100, in a double-entry paradigm, she will make appropriate entries in her books. At the same time, Bob will record the transaction in his.

---

[2] Vladimir Lenin idolized "bookkeeping and control," and clearly stated: "The smart one is not the one who does not make mistakes. There are no such people and cannot be. The smart one makes mistakes that are not very significant and knows how to correct them easily and quickly." Luckily, we have since borrowed several sophisticated error correction tools from linear algebra and coding theory that are much more useful in detecting mistakes, both significant and otherwise; see Arya et al. (2004).

Finally, Charlie, the banker, has to move $100 from Alice's account to Bob's and change his books accordingly. Thus, a single transaction is reflected in three different ledgers, which makes auditing individual books challenging. The auditor needs to verify not only Alice's books and records but also the parts of Bob's and Charlie's that are pertinent to the transaction. Details are shown in Figure 12.2.

Audits are currently performed by outside organizations on a regular basis but by necessity, are only able to cover a small fraction of all transactions, namely those deemed representative of the overall nature of a business' finances. This backward-looking *modus operandi* inevitably misses a large share of both unintentional and deliberate fraudulent errors. For example, the Association of Certified Fraud Examiners estimates the overall accounting fraud in 2017 at US$4 trillion per year, of which audit captured only 4–5%; see Cai (2019). This is not entirely surprising — in most cases, the perpetrators of fraud attempt to conceal their actions using methods such as record altering or destruction, etc.[3]

Thus, the logical step is to switch from double-entry bookkeeping to triple-entry bookkeeping. The latter term was coined by Ijiri (1986). However, his triple-entry framework,

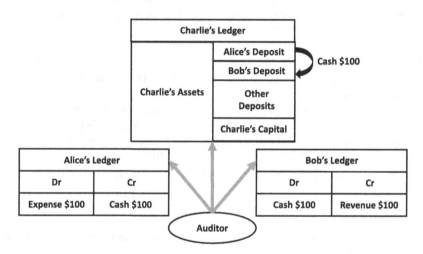

Figure 12.2 In the double-entry accounting framework, Alice's purchase of goods from Bob is reflected in three ledgers. To audit Alice's ledger, it is necessary to get information from her, Bob, and Charlie. Own graphics.

---

[3]Wirecard AG, a German payment processor and financial services provider, is a recent example of egregious accounting fraud. Before its bankruptcy, the company was the second-largest German tech company. For years, Wirecard told its auditors that it had US$2 billion in two Philippine banks, which they substantiated using fraudulent bank statements. However, in June 2020, it was forced to announce that the money did not exist and declare bankruptcy. Wirecard's ability to evade its auditors by forging bank statements is a clear consequence of the fact that there is no single truth source for its ledger.

while ingenious falls squarely outside of our current focus and shall not be discussed further.[4]

Confusingly, Grigg used the same term, triple-entry accounting (or bookkeeping) for an entirely different purpose; see Grigg (2004, 2005). He proposed to describe financial transactions in a shared way via the so-called Ricardian contracts, which are *both* human and machine-readable. Ibanez et al. (2020) provide an interesting discussion of Grigg's ideas, as well as earlier discoveries of McCarthy and Boyle; see McCarthy (1979, 1982); Boyle (2000). Grigg's vision was to create a single receipt for the above transaction between Alice and Bob, which will be shared between Alice, Bob, and trusted authority, Ivan. Initially, Bob sends Alice a cryptographically signed invoice. Provided that Alice finds the invoice acceptable, she also cryptographically signs and sends it to Ivan, the issuer. Ivan finally signs and time stamps the receipt and puts it in the common depository. This triple-signed receipt is kept by all three parties involved — Alice, Bob, and Ivan (hence, the name triple-entry bookkeeping).[5] Details are shown in Figure 12.3.

Assuming that Ivan is an honest broker in good standing, the transaction record cannot be altered or removed. In Grigg's view, this triple-signed receipt is the transaction itself. Rather than keeping books at all times, the corresponding accounting information can be created on the fly on an as-needed basis from the receipts. For example, if Alice wants to know her balance of trade with Bob, all she needs to do is sum all the receipts, having Bob as a counterparty. We emphasize that, as far as accounting proper is concerned, Grigg's

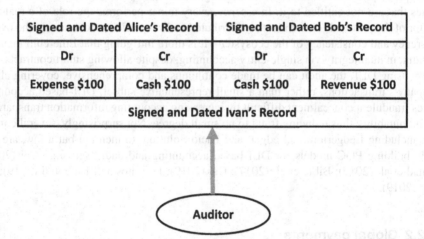

Signed and Dated Alice's Record		Signed and Dated Bob's Record	
Dr	Cr	Dr	Cr
Expense $100	Cash $100	Cash $100	Revenue $100
Signed and Dated Ivan's Record			

Auditor

Figure 12.3 In the triple-entry accounting framework, Alice's purchase of goods from Bob is reflected in three receipts, signed by Alice, Bob, and Ivan. To audit Alice's ledger, it is necessary to get information only from her. Own graphics.

---

[4]Ijiri aimed to add to stocks and flows measured in monetary units such as dollars and captured in the account statements as wealth and income, respectively, the rate of change of flows, which he called momentum measured in monetary units per time period, such as dollars per year or month.

[5]In reality, since Charlie executes payments, his participation is needed too, but Grigg does not go in this direction.

ideas boil down to triple-entry bookkeeping but single-entry accounting. Every transaction is entered into the database only once; its subsequent accounting processing is left to the relevant parties to handle the way they see fit.

Expanding the work of Boyle, as well as his own research, Grigg (2005) articulated several critical requirements for the practical implementation of triple-entry accounting:

(**1**) Strong Pseudonymity;
(**2**) Entry Signing;
(**3**) Message Passing;
(**4**) Entry Enlargement and Migration;
(**5**) Local Entry Storage and Reports;
(**6**) Integrated Hard Payments;
(**7**) Integrated Application-Level Messaging.

The astute reader can recognize that the Bitcoin protocol meets and exceeds these requirements, which might not be a mere coincidence. We emphasize that Bitcoin, Ethereum, and other protocols use single-entry accounting. In itself, it's not particularly remarkable given that the Bitcoin protocol simply records movements of unspent transaction outputs, UTXOs, from an old address to a new one.

In our example, the role of Ivan, the issuer, can be replaced with blockchain, thus streamlining bookkeeping, accounting, and audit. To implement this idea in practice, one needs to use cryptographic obfuscation, such as hushing, to hide the business information from those parties that are not entitled to it. In essence, every miner performs the role of a real-time auditor of the protocol. This collective distributed audit is sufficient to maintain the overall coherency and consistency of the ecosystem. It is more intriguing that Ethereum manages to maintain its integrity via single-entry accounting, despite allowing smart contracts.

By using DLT, the audit can be made continuous and comprehensive, covering all the company's transactions, rather than a small representative subset. DLT-based accounting makes fraudulent concealment difficult by increasing accounting information transparency and immutability thus reducing fraud to no small degree. Not surprisingly, several companies, including Ledgerium, zkLedger, and Pacio Solution, to mention but a few, are currently building POC models for DLT-based accounting and audit; see, e.g., Cai (2019); Ahmad et al. (2019); Bible et al. (2017); Cai (2019); Drosatos and Kaldoudi (2019); Liu et al. (2019).

12.2.2 Global payments

While the current payment system is not broken, it is significantly outdated; for instance, continuous linked settlement (CLS), which is used in FX trading to solve the delivery vs. payment problem, is certainly tried and true, but clunky and expensive, making it challenging to use, especially for retail clients. According to some estimates, about 40% of all banking revenue comes from payment and other fees. As such, it is paramount to find swifter and more streamlined solutions that are both easier and cheaper to implement and maintain, while reducing ancillary costs associated with payment processing that eat into revenues and cost companies billions of dollars every year.

In global payments, one can use BTC, or ETH, to change one currency into another but this is still fraught with many obstacles, such as speed, transaction costs (incurred twice), credit risk, and general apathy of the banking community. Many practitioners think that using gold deposited offshore instead of cryptocurrency would be much more straightforward. Figure 12.4 compares and contrasts current and future cross-border payment flows.

Figure 12.4 (a) shows a convoluted and costly way of how cross-border payments between an originator in Japan and a beneficiary in Brazil are executed at present. The transfers use the corresponding banking framework. Let's consider payment from a Japanese

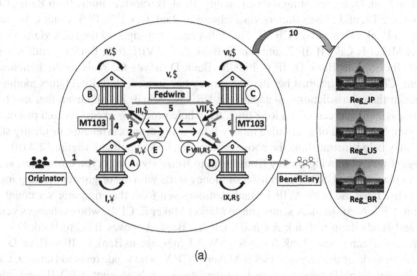

(a)

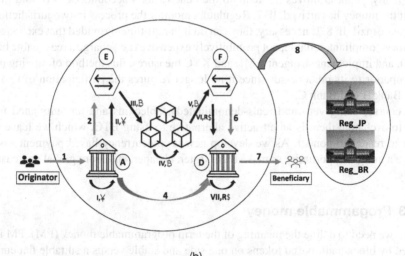

(b)

Figure 12.4 Global payments: (a) current, (b) future. Own graphics.

buyer to a Brazilian seller and follow the money step-by-step. Since direct exchange of the Japanese yen into the Brazilian real is uncommon, the US dollar is used as an intermediate currency. We use Arabic numerals to mark sequential information flows (IFs) and Roman numerals to mark cash flows (CFs). The transfer of money starts with the originator sending instructions to her Japanese bank A, IF 1. Bank A moves yen from the originator's account to its own account, CF I. Bank A sends yen to Market Maker E, CF II, IF 2, who exchanges yen into dollars and sends them to Bank A's account with an American bank B, CF III, IF 3. At the same time, Bank A sends SWIFT message MYT103, IF 4, to Bank B, instructing it to forward the dollars to an American bank C, which is the corresponding bank of the Brazilian bank D, representing the beneficiary. Bank B transfers funds from Bank A to its account, CF IV, and moves dollars via Fedwire to Bank C, CF V, IF 5. Bank C moves the dollars to Bank D's account, CF VI. The dollars are exchanged to the reals via the second Market Maker F, CF VII, IF 7, and sent to Bank D, CF VIII, IF 8. Bank C sends a SWIFT MT103 message to Bank D, IF 6. Finally, Bank D moves the reals to the beneficiary's account, CF IX, and informs her that the money has arrived, IF 9. Regulators monitor the process in three jurisdictions — Japan, the US, and Brazil, IF 10. Given that most steps require payment and can take a long time, the whole process is expensive and protracted.

A simple example of alternative money flows using cryptocurrencies, or ideally stablecoins, may help illustrate how the above process can be streamlined. Figure 12.4 (b) shows an elegant way of moving money from Japan to Brazil by using cryptocurrencies, such as BTC, in place of the USD. The transfer of money starts with the originator sending instructions to her Japanese bank A, IF 1. Bank A moves yen from the originator's account to its account, CF I. Next, Bank A sends yen to Market Maker E, CF II, who exchanges yen into BTC and sends them to Bank A's node, CF III. Bank A moves BTC to Bank D's node, CF IV. At the same time, Bank A sends a SWIFT message to Bank D, IF 4. Bank D sends BTC to the node of the second Market Maker F, CF V, who sends reals to Bank D, CF VI, IF 6. Finally, Bank D moves the reals to the beneficiary's account, CF VII, and informs her that the money has arrived, IF 7. Regulators monitor the process in two jurisdictions — Japan and Brazil, IF 8. If necessary, they can do it in real-time. Provided that exchanges are regulatory-compliant, reliable, not prohibitively expensive (i.e., charge a reasonable bid-ask spread), and implement stringent AML and KYC measures, this method of moving money is far superior to its predecessor since it no longer requires the participation of two other banks, Bank B and Bank C.

Of course, using yen- and reals-denominated stablecoins and an automated market maker to exchange them is an attractive alternative to using BTC, which we leave to the interested reader to ponder. As we develop new cryptocurrency-based payment systems, we have a unique opportunity to make them faster, cheaper, and more global and customer friendly.

12.2.3 Progammable money

To start, we need to define the meaning of the term programmable money (PM). PM is represented by blockchain-based tokens on one side and stable versus a suitable fiat currency or a basket thereof on the other. In addition, PM possesses additional functionalities, such as conditional payments, outright payments, and payment commitments.

We emphasize three points. First, as was mentioned in Chapter 2, currently, money exists predominantly in digital form and is held by commercial banks. Many of the existing banks provide payment processing automation, making life easier for their customers while simultaneously reducing their expenses. For instance, a client can instruct her bank via the Internet to make the monthly mortgage, utility payments, and other recurring payments or send her credit card issuer the minimum monthly amount not fixed in advance. While we appreciate the utility of this functionality, we do not think it makes banks' money programmable since it relies on programming the account, rather than the money directly, which could in theory be done using blockchain without a bank account.

Second, as discussed in Chapter 6, advanced blockchain protocols, such as Ethereum, support smart contracts, which can be used to define various tokens, including stablecoins. Once again, while fully appreciating the advantages of stablecoins, we do not view them as PM because they typically do not have the required built-in conditions. Third, central bank digital currency (CBDC) studied in Chapter 8 cannot be considered PM either because CBDC has to be the electronic version of cash with perfectly fungible units, while PM comes with restrictions and hence is non-fungible.

Given that all of the aforementioned discussions present useful and practical avenues for the evolution of digital money, it is natural to ask why we need PM in the first place. The answer is simple — we can think of PM as "narrowly tailored money." Put another way, it can be programmed to be used solely at specific locations or for particular purposes, thus allowing the issuer to tailor its use as it sees fit. PM could have numerous applications, some more exciting than others; see, e.g., Elsden et al. (2019); Deutsche Bundesbank (2020). Here are some potential business use cases:

(1) Bi-directional clearing resolving the Delivery vs. Payment problem and resulting simultaneous settlement of mutual obligations between the parties. For instance, two enterprises can automatically settle invoices using PM and triple-entry bookkeeping, thus creating an immutable audit trail; see Section 12.2.1.
(2) Automated securities clearing and settlement; see Section 12.2.5.
(3) Streamlining of supply chain management and trade finance. Settlement of letters of credit against the presentation of stipulated digital information; see Section 12.5.2.
(4) Economic rejuvenation and development schemas, relying on PM usable only in specific, economically disadvantaged areas.
(5) Supplementary currency in a particular economic circle, such as the celebrated Wirtschaftsring (WIR) of Switzerland.
(6) Delivery of humanitarian aid performed in a transparent and corruption-free manner.
(7) Machine-to-Machine (M2M) payments. For instance, an electric car can use PM to pay for electricity or parking in real-time.

We briefly discuss two possible applications of PM — humanitarian relief efforts and fighting climate change. Humanitarian relief efforts are often plagued by corruption. One option in reducing waste and increasing the direct impact of financial humanitarian aid is to use PM, which could encourage desirable outcomes and discourage undesirable ones by making them very difficult, if not outright impossible, to pay for; see, e.g., Rosenoer and Yong (2020). PM can be deposited directly to the phones of the people in need.[6] PM's features can be used to allow only the purchase of specific products at designated stores thus reducing opportunities for extortion and corruption. Moreover, given PM's enhanced traceability, even small diversions would be easier to identify and rectify by analyzing the purchasing patterns.

Let us turn our attention to the use of PM in alleviating climate change. While the innumerable ravages of climate change permeate into all aspects of society, we focus here on its profound financial component. According to the Bank of England: "Climate change poses significant risks to the economy and to the financial system, and while these risks may seem abstract and far away, they are in fact very real, fast approaching, and in need of swift and immediate action to stymie their potential destruction." Yet, it is clear that protestations alone are insufficient to convince people to make their behavior more environmentally friendly. Thus, we need to design suitable financial tools and incentives that push people in the right direction, gently, of course. One exciting possibility is the use of green programmable money (GPM), first introduced by Pentland et al. (2020) as the environmental digital trade coin (EDTC). GPM is a special case of the Digital Trade Coin (DTC) discussed in Chapter 8; see Lipton et al. (2018b).

We can structure GPM based on a coalition loyalty program, which amalgamates several single issuer loyalty programs that allow participants to accrue and redeem loyalty points at various participating businesses.[7] The GPM ecosystem consists of several components. At its core is an efficient, precise, and easy-to-use app that records program participants' "positive" or environmentally-friendly actions and plays the role of a "money printing press." The app is used by individual participants interested in fighting climate change and/or that are attracted to suitable financial initiatives. Their natural counterparts are corporate sponsors that are concerned about climate change and prepared to spend money to fight it. The ecosystem is held together by coalition members, such as utilities, stores, restaurants, gas and electric stations, etc., that accept GPM as partial payment for their goods and services.

The ability to program GPM is indispensable to the implementation of a robust, stable, sustainable monetary policy by making it possible to set an expiry date on the GPM earned by participants.[8] Suitable circulation rules identify "sources and sinks" for GPM. GPM is created at "source" when individual participants perform environmentally-friendly actions and destroyed at "sink" when they redeem it as partial payment for goods and services.

However, a more general approach might be better. GPM can be both earned by participants and borrowed by sponsors. In this scenario, end consumers still receive and redeem GPM based on their ecofriendly actions. However, the intermediate-level businesses that accept GPM can now also use them as partial payment for their own needs with

---

[6] The aid sponsors should furnish the phones and pay for their usage.

[7] American Airlines launched the first frequent flyer program (along with coining the term itself) in the seventies. Since then, loyalty programs have grown exponentially and gained considerable economic value. For instance, in January 2019, Air Canada, along with Toronto-Dominion Bank (TD), Canadian Imperial Bank of Commerce (CIBC), and Visa, acquired a loyalty program called Aeroplan, with about 5 million active members, from Aimia for CA$450 million in cash and liability for the unused Airplane points at an estimated value of CA$1.9 billion.

[8] All successful loyalty programs, including various frequent flyer programs, use demurrage so that loyalty points accumulated by participants expire and disappear after a certain period.

the participating coalition members. "Super sponsors," including tax authorities and major multinationals, spend part of their climate change budgets on promoting environmentally-friendly policies by accepting and destroying GPM, thus acting as a "sink" rather than sending it back into circulation.

12.2.4 Regulations

Bitcoin came into existence as a spontaneous force, unregulated, uncontrolled, and untamed. Arguably, that was the only way to break the existing boundaries and succeed in the beginning. Bitcoin became a lodestar among those who lost faith in governments and central banks, initially more of a cult-like trend than a rationale for a new asset class. However, as the movement gained recognition, the universe of cryptoassets and crypto-currencies has significantly expanded beyond Bitcoin. Thus, it became clear that without a proper regulatory framework, the crypto universe cannot achieve its full potential of going mainstream. This section discusses critical regulatory problems, which must be solved as crypto enters our lives. Most of them are inherent to the financial system as a whole, including legacy and crypto, while others are specific to the crypto universe. This area of research and development constitutes RegTech, although, for different people, this term might offer a slightly different meaning.

The Greek philosopher Aristotle articulated legal aspects of money, emphasizing that money and government are joined at the hip: "[M]oney has been introduced by convention as a kind of substitute for need or demand, and this is why we call it money ($\nu o\mu\iota\sigma\mu\alpha$) because its value is derived, not from nature, but from the law ($\nu o\mu o\zeta$), and can be altered or abolished at will" (Crisp, 2014).

When we talk about the banking system in general and financial transactions in particular, it is hard to overestimate the importance of the KYC and AML requirements for regulatory compliance. KYC measures deal with identity verification and asserting suitability and risks for maintaining a business relationship with a customer, whether an individual or business. In the current setting, the onboarding of a new customer can take up to 50 days, thereby hurting customer experience. AML administers the detection of money laundering and other illegal activities and criminal behaviors. The KYC/AML procedures are equally applicable to traditional financial transactions and those with cryptoassets, such as issuing new tokens and redeeming existing ones.

Davies (2010) looked at the situation more pragmatically: "Because of the anonymity and widespread acceptance of money, financial fraud is always likely to be among the most attractive and direct of the criminal's temptations, from forgery to money-laundering, but it must be emphasized that many of the most calamitous of financial failures began with the cleverest of people operating with the best of intentions. Incompetents and Nobel Laureates find themselves in the same basket."

KYC measures for cryptoassets are further simplified because exchanging fiat currencies into stable tokens and vice versa uses existing legacy bank accounts, whose owners have passed the standard KYC procedures performed by their banks. Such an ability to take advantage of existing bank accounts reduces KYC costs to less than a dollar per person. Digital Identity discussed in Section 12.3.1 can be used for the purposes of KYC.

AML policies can also be significantly simplified with the intelligent use of DLT. Given that stable tokens circulate on a blockchain — currently on Ethereum, subsequently on others — which keeps an immutable record of all transactions, AML can be performed indirectly by analyzing the corresponding transaction social graph. In this regard, transactions with stable tokens are more transparent than conventional physical cash and are on par with traditional banking transactions. Initially, criminals were enthusiastic about using cryptocurrencies, such as Bitcoin and Ethereum, for illegal actions. However, with time, they realized that the immutable record of transactions opens new avenues for law enforcement agencies to monitor and prosecute their activities. At present, they are reverting to using cash or cryptocurrencies with built-in privacy.

Moreover, for finding suspicious activities or fraudulent transactions, DLT opens a new domain to further analyze the transactional record at the level of an individual transaction. It enables chain analysis, including Know Your Transaction (KYT), to discover and follow all the chain links that led to questionable or illicit actions. KYT is offered as a service by several companies, such as Chainalysis, for constantly monitoring transactions for fraud detection in real-time. For additional insights, see Campbell-Verduyn (2018) and references therein.

While the use of stable coins can significantly facilitate all steps involved in KYC and AML procedures, it also bears a mark of a surveillance hazard when used by a repressive state. When high volumes of cash transactions are completed on a blockchain, the government can track a considerable portion of its citizens' payments and exert unprecedented control over their lives. Therefore, at least small cash-like financial transactions must be anonymized, except when the anonymity is lifted as part of criminal investigations or in similar situations. When there is a social imperative to override anonymity, it should rely on an adequately qualified judiciary system. Thus, developing a suitable and competent legal infrastructure is paramount for successfully using stable tokens as a medium of exchange.

12.2.5 Trade execution, clearing and settlement

The holy trinity of the stock market is trade execution, clearing, and settlement. The initial public offering of stock is an essential rite of passage for a new company. Its secondary trading is a mechanism for allocating ownership and control in a (somewhat) optimal fashion. In addition to stocks, many other products, such as equity derivatives, interest rate swaps, commodities, etc., are traded on public exchanges. Moving many over-the-counter (OTC) products to exchanges is an essential regulatory imperative; see Lipton (2018); Subramanian (2017), among many others.

Unfortunately, this system suffers from numerous shortcomings and outright failures. Let us briefly discuss how Trading, Clearing, and Settlement function at present. Figure 12.5 shows the corresponding steps with black and grey arrows indicating the information requests and securities transfers, respectively; see Lipton (2018); Pinna and Ruttenberg (2016).

Here are the necessary steps:

Seller (S) instructs her Broker (SB) to sell a security. At the same time, Buyer (B) instructs her Broker (BB) to buy this security.

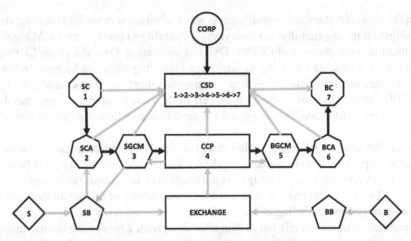

Figure 12.5 The current state of trading, clearing, and settlement is unnecessarily complicated. Numerous expensive intermediaries have to assist the sale of a security by a seller
and its purchase by a buyer. Own graphics.

(1) Sell and buy orders by the SB and BB are matched on an exchange, which can be
organized in various ways; for instance, it can function as a Limit Order Book or use
market makers.
(2) Matched orders are novated. The original bilateral trade is transformed into a pair of
trades: a sale of the security by Seller's General Clearing Member (SGCM) representing
the SB to the Central Counterparty (CCP), and a sale of this security by CCP to Buyer's
General Clearing Member (BGCM) representing the BB to the CCP.
(3) SGCM asks SB to deliver the corresponding security.
(4) SB sends this request to Seller's Clearing Agent (SCA).
(5) SCA forwards it to Seller's Custodian (SC).
(6) SC's request is sent to Central Securities Depository (CSD), where a series of transfers
of the security ownership take place in:

 (a) from SC to SCA (1– >2);
 (b) from SCA to SGCM (2– >3);
 (c) from SGCM to CCP (3– >4);
 (d) from CCP to BGCM (4– >5);
 (e) from BGCM to Buyer's Clearing Agent (BCA) (5– >6);
 (f) from BCA to Buyer's Custodial (BC) (6– >7);

As a result, the security in question originally held by SC is transferred to BC. Step 1
from the above sequence represents trading, step 2 — clearing, steps 3–6 — settlement.
Figure 12.5 does not show the flow of money, which naturally occurs in the opposite direction, i.e., from buyer to seller.

Vastly different time scales characterize these steps — trading often occurs in milliseconds, while clearing and settlement take 1–3 days! Although the proverbial T+2 days, T+3
days irritate many people, the push for a T+15 minute settlement may be misguided. Let us
briefly summarize the pros and cons of the current setup.

On the pros side, there are several aspects which it solves very well: (a) netting; (b) DvP risk mitigation more generally; (c) anonymity; (d) ability to borrow stocks. Many authors argue that the entire purpose of CCPs is DvP risk mitigation. Over the years, CCPs developed an arsenal of tools for doing so, including collecting Variation Margin, Initial Margin, and Guarantee Fund from its general clearing members; see Barker et al. (2017); Lipton (2019). On the cons side, numerous aspects of the current setup are somewhat disconcerting: (a) cost; (b) speed; (c) the need for reconciliation, which can fail, to mention but a few.

We can envision the following evolutionary path for the CCPs' infrastructure. A straightforward attempt to apply a blockchain to trading, clearing, and settlement cannot be successful. In particular, blockchain, by design, would result in instantaneous settlement (occasionally called T+15 min settlement), which would consequently obliterate all the advantages of the current system, including netting and trade compression (by having to move money at predetermined time intervals rather than after every trade), borrowing ability, anonymity (to some degree), etc. As a result, transactions would no longer be netted against each other to the extent they are now, thus increasing the amount of money moving around at any time by substantial amount.

The difficulty with redesigning the trading, clearing, and settlement paradigm is that the slow speed is not so much a consequence of the technological backwardness of exchanges (even if they often lag in technological innovation themselves), but rather a reflection of the nature of trading. However, by using a permissioned private ledger, one can certainly increase the speed of clearing and settlement and reduce the need for reconciliation and failures. In particular, smart contracts (provided they are legally enforceable) and PM can solve part of the DvP conundrum, digitally signed by both parties.

For DLT-based framework to operate, it is necessary to have both tokenized securities and stablecoins to coexist on the ledger. Of course, it is not a panacea, but nonetheless a necessary step in the right direction. Figure 12.6 illustrates a potential evolutionary path culminating with automated market makers operating on a blockchain.

12.3 Government

12.3.1 Identity

The United Nations' goal is to "provide [by 2030] legal identity for all, including birth registration." Given the ubiquity of the Internet, creating digital identities (DIDs) and developing sound identity and access management (IAM) systems can help achieve the above goal. Even in developed countries, where obtaining legal identity is less of a problem than in some developing countries, the implementation and usage of DIDs are well established, but still inadequate.

The prevalent centralized models of IAM are constantly under attack from regular data breaches resulting in loss of privacy, reputation damage, and identity theft; see, e.g., Dunphy and Petitcolas (2018) and references therein. Moreover, end-users currently have little or no control and ownership of their DIDs.

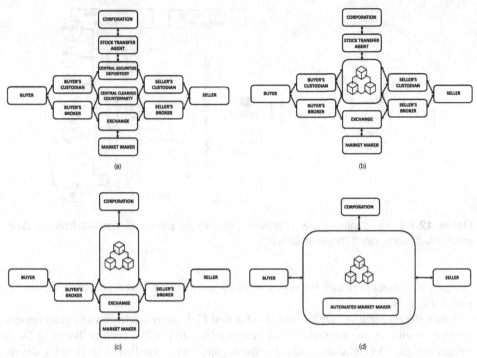

Figure 12.6 CCPs infrastructure: (a) current, (b) a near future, (c) a distant future, (d) a very distant future. Own graphics.

Before dealing with DLT-specific aspects of DID and IAM, it is useful to summarize Core ID concept, developed by Pentland and his coauthors; see Hardjono et al. (2019b). Figure 12.7 illustrates their ideas graphically.

At the root of Core ID concept, there lies core identity, which is represented by a set of characteristics making a person uniquely recognizable and generating a unique core identifier based on the assortment of relevant personal data, such as birth certificate, school and college diploma, driver's license, passport, financial information, credit history, etc. A core identifier, which has to be kept secret at all times, is private data (represented by a string) or a hash function uniquely specifying the person. A persona is defined by and derived from a subset of attributes used in a particular context, such as a work-persona, a home-persona, etc. A core identifier is used to derive as many personas as needed. Finally, various personas specify transaction identities and identifiers required to execute a particular transaction. Each transaction identity is derived via privacy-preserving verification from the corresponding persona, which, in turn, is obtained via a similar procedure from the core identifier.

The main emphasis of Core ID is privacy preservation. Core ID allows to reveal as little information as possible to complete a particular transaction and make different personas

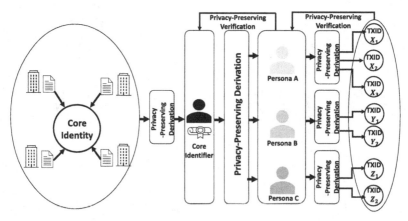

Figure 12.7 Core identity, core identifier, and various personas derived from it. Own graphics. Source: Hardjono et al. (2019b).

sufficiently distinct to reduce the opportunities to correlate various digital activities of the underlying person.[9]

Given the properties of DLT outlined in Section 12.1, many academics and practitioners have proposed it as an essential tool for bringing IAM to the 21st century. By using DLT, we can design a decentralized, tamper-resistant, and user-controlled DID. Besides, developing the blockchain-based DID is cost-efficient, compared to the existing technologies, making DIDs more inclusive, and achieving the UN's goal mentioned at the beginning of this section. According to Dunphy and Petitcolas (2018), all DLT-based schemas to date can be divided into two categories:

(1) Self-Sovereign Identity, owned and controlled by its owner, does not rely on any external administrative authority. It is created by a decentralized identity eco-system, allowing recording and exchanging identity attributes and building trust among participating entities. As the name suggests, no one can revoke such an identity.

(2) Decentralized Trusted Identity, provided by a centralized service, relies on existing trusted credentials, such as passports, birth certificates, and the likes. The centralized service records identity attestations on a DLT, which are subsequently validated by third parties.

The Core ID concept is particularly well-suited to implementing DLT-based DID. Some of the best-known private companies, such as Microsoft, and several startups, including Sovrin, uPort, OneName, ShoCard, BitID, ID.me, and IDchainZare, are developing complementary approaches to building DLT-based DID. Decentralized Trusted Identity is likely to be better suited than Self-Sovereign Identity to the real-world realities. One possible implementation, proposed by Microsoft (2018), is shown in Figure 12.8.

---

[9]For instance, instead of using a driver's license to prove that a person is over 21, it is much better to use privacy-preserving zero-knowledge proof, issued by the relevant authority, which ascertains so.

Microsoft (2018) design depends on seven key components:

(1) World Wide Web Consortium (W3C) DIDs created, owned, and controlled by users independently of any organization or government. These identifiers are globally unique and linked to Decentralized Public Key Infrastructure (DPKI) metadata containing public key material, authentication descriptors, and service endpoints.
(2) Blockchains supporting DIDs and providing the mechanism and features required for DPKI.
(3) Wallet-like DID User Agents empowering real people to create DIDs, manage data and permissions, and sign/validate DID-linked claims.

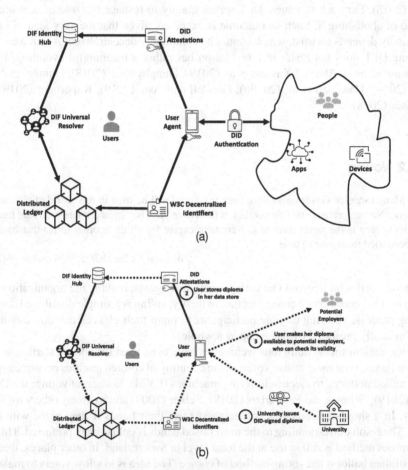

Figure 12.8 A DLT-based system for DID: (a) general architecture; (b) a particular example of a user sharing her university diploma with potential employees. Own graphics. Source: Microsoft (2018).

(4) Decentralized Identity Foundation (DIF) Universal Resolver (UR) — a server utilizing a collection of DID Drivers and returning the DID Document Object (DDO) encapsulating DPKI metadata associated with a DID.
(5) DIF Identity Hubs forming a replicated mesh of encrypted personal datastores, facilitating identity data storage and interactions.
(6) DID Attestations are DID-signed statements based on standard formats and protocols, enabling identity owners to generate, present, and verify claims and establish trust between the ecosystem members.
(7) Decentralized apps and services enable the creation of a new class of apps and services. They store data at the user's Identity Hub and utilize it according to their permissions.

In theory, DLT is supposed to be a remedy for centralized system architectures; however, in practice, it does not work for the DLT-based IAM schemes; see Dunphy and Petitcolas (2018). For such schemes, DLT serves mainly to reshape the role of centralization instead of abolishing it. Such an outcome is expected, given that the very idea of identity profoundly depends on trust, which cannot be completely decentralized. Again we see that applying DLT does not result in a revolution but rather a meaningful evolution. Further details are given in Alsayed Kassem et al. (2019); Dunphy et al. (2018); Dunphy and Petitcolas (2018); Hardjono et al. (2019b); Goodell and Aste (2019); Kuperberg (2019); Zhu and Badr (2018).

12.3.2 Voting

"Many forms of Government have been tried, and will be tried in this world of sin and woe. No one pretends that democracy is perfect or all-wise. Indeed it has been said that democracy is the worst form of Government except for all those other forms that have been tried from time to time. ..."

Winston S Churchill, 12 November 1947

Democracy relies on free and fair voting by eligible agents, making the organization of the voting process enormously consequential. In theory, voting is a simple distributed decision-making process, allowing eligible participants to input their choices and arrive at an outcome mutually preferred by the group as a whole.

Since ancient times, numerous techniques have been tried for voting, starting with the show of hands, *viva voce,* or live voice, to the counting of colored marbles or ostraca (pieces of discarded pottery), to electronic voting machines (EVM), to internet voting; see Dhillon et al. (2019); Grontas and Pagourtzis (2019); Selker (2004) among many others for further details. In a show of hands vote, participants raise their hands if they agree with a proposal. The result corresponding to the most raised hands is collectively preferred. This non-anonymous method is still in use at the local level in Switzerland. In other places, the secret or Australian ballot is the voting method of choice. The idea is to allow voters to make their electoral choices anonymously. On the election day, voters congregate at the polling place, fill their paper ballots in secret, put them inside an envelope, and cast the envelope in a sealed ballot box. When the election is finished, a magistrate opens the box. The envelopes are unsealed, and the votes are tallied publicly in front of various competing parties' representatives. Of course, by its very nature, the process is slow and error-prone. However, it does leave a paper trail, which is auditable if needed, though the audit of election outcomes poses several problems, which are not dissimilar to those discussed in Section 12.2.1.

Many researchers and inventors have tried their hand at improving the voting process. For instance, Thomas Edison patented the first electric voting machine in 1868. Lever machines were used in the US as early as 1892; mark-sense ballots were tried for the first time in 1962; devices utilizing Hollerith punch cards debuted in 1964; direct-recording electronic (DRE) voting machines in 1976; and Internet voting in 2000; for further details, see Selker (2004).

Here, we focus on electronic voting systems and, specifically, whether one can improve these systems by using DLT. Such systems must satisfy several requirements, which we now articulate following Grontas and Pagourtzis (2019) and references therein. These requirements include:

(1) availability and efficiency;
(2) authentication and authorization;
(3) privacy;
(4) correctness;
(5) end-to-end (E2E) verifiability;
(6) several other, less critical properties.

Availability and efficiency mean that the system is readily available, easy to implement, and easy to use, not least by those with varying levels of technical or language proficiency. This would ideally also include ways to address hurdles that may prevent voters from being physically present at the polls. Finally, ensuring that these systems are easily accessible to individuals with disabilities is also of paramount importance. Authentication and authorization refer to a system's ability to authenticate a voter's identity and eligibility. Privacy means that a voter's choice is not revealed to anyone. Correctness signifies that the voting system, when run as designed, produces the result that correctly reflects the actual votes cast, and anyone can check the outcome and the process by which it was reached. End-To-End (E2E) verifiability means that all the voting process stages — collection, accumulation, storage, and counting of votes — should be verifiable.

While relatively easy to articulate, these requirements are hard to satisfy. For many years, academics and practitioners have aimed at finding the best ways of doing so. The celebrated CALTECH/MIT Voting Technology Project — a multi-disciplinary, collaborative project between two preeminent academic institutions — is a case in point. Yet even today, several of their goals proved to be elusive, which is not particularly surprising given the contradictory nature of the requirements. For instance, authentication and authorization can make privacy challenging to achieve.

In principle, one can use cryptographic primitives introduced in Chapter 4 to build a cryptographic election system modeled after voting with a show of hands, with a snapshot maintained for tallying. The idea is to replace the paper trail with mathematical proofs based on such constructs as blind signatures discussed in Chapter 4, zero-knowledge proofs, covered in Chapter 12 below, and other similar tools; see Adida (2006). Specifically, the cryptographic voting paradigm is based on the Bulletin Board (BB) concept, defined as a broadcast channel with memory, which supports only the insertion of new data (append-only), does not allow any deletions or modifications, and retains data for verification.

Despite numerous attempts, there is no universally acceptable implementation of the BB. Since blockchains and BBs are databases operating in an append-only manner thus allowing participants to reach consensus on the validity of transactions, it is only natural to try blockchain-based voting. In the Bitcoin protocol, transactions represent the transfer of value, while in the elections protocol, transactions represent the record of preference.

All proposed DLT-based voting systems are conceptually similar. After choosing a particular blockchain protocol (both permissioned and permissionless blockchains have been suggested), all eligible voters are represented by pseudonymous addresses, while candidates are represented by known addresses. A small amount of tokens is sent to voters in advance by the election authority. There is room for PM discussed in Section 12.2.3, which can be explicitly designed for voting, instead of pocketing or spending these crypto tokens on unrelated matters. A vote for a specific candidate is cast by sending the received amount of tokens to the address of the candidate of the voter's choice. The corresponding transaction is recorded on the blockchain. When the election ends, the amount accumulated by each candidate is checked, and the winner is declared; see Nasser et al. (2018). This approach is the 21st-century version of the show of hands used in classical Greece with added pseudonymity.

Unfortunately, while superficially attractive, the proposed approach has many conceptual and logistical drawbacks. First, since only eligible voters should participate in the election, there must be a mapping between their real identities and the corresponding blockchain addresses. Since such a mapping has to be executed by a registration authority, a trusted third party is introduced to the voting system *nolens volens*. Second, blockchain-based voting does not provide proper vote secrecy (similar to its underlying blockchain protocol). Third, by its very nature, DLT-based voting is conceptually extremely complicated, making it potentially untrustworthy to the electorate. Hence it can disenfranchise many voters — a highly undesirable outcome.

One can still run small scale elections, such as board elections, among tech-savvy voters known to each other on a smart-contract-supporting blockchain; see, e.g., McCorry et al. (2017). However, we share Dhillon et al. (2019); Grontas and Pagourtzis (2019); Park et al. (2020) thinking that large-scale DLT-based voting is not ready for its prime time. Nevertheless, we feel that practical usage of DLT for less glamorous but vital tasks, such as running census, maintaining and updating lists of eligible voters, helping to tally results from different precincts, is a more promising short-term goal as we find a way to build a DLT framework for voting on a larger scale.

12.4 Healthcare

12.4.1 Electronic health records system

Not dissimilarly to the financial system, the existing healthcare system also suffers from several drawbacks. This fact is not particularly surprising since its informational backbone — the electronic health records (EHRs) system — was not designed to be multi-institutional and real-time. Instead, patient's EHRs are kept by numerous medical care providers, retaining *de facto* (and often *de jure*) primary stewardship of the data; see Ekblaw et al. (2016). As a result, patients cannot get a full picture of their health records, instead getting the information piecemeal, if at all; the same is true as far as healthcare providers are concerned. Thus, it is natural to see if DLT can rectify the situation; see, e.g., Ekblaw et al. (2016); Gordon and Catalini (2018); Hussein et al. (2019); Khezr et al. (2019); Kuo et al. (2017) among many others.

DLT-based system for handling EHRs achieves several key objectives, including securing sensitive patient data, allowing access to real-time data to authorized parties, and facilitating collaboration of several authorized healthcare providers on the same medical records. Such a system has the following advantages compared to the existing one:

(1) Increase of patients control over their EHR;
(2) Improves data provenance and integrity;
(3) Reduces the burden of data duplication and reconciliation;
(4) Allows for seamless clinical care by different healthcare providers;
(5) Enables effective management of emergency medical situations.

It also opens the possibility for patient-generated health data from wearables, sensors, smartphone applications, and other Internet of Things (IoT) devices to be integrated with information obtained during interactions with healthcare providers. All of these data are permanently stored in designated databases and made available to patients and authorized providers as needed. We illustrate one potential implementation of DLT-based EHRs system in Figure 12.9.

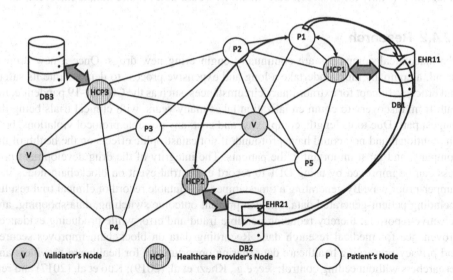

Figure 12.9 A DLT-based system for maintaining EHRs. Own graphics.

This figure shows three healthcare providers and several patients. Healthcare providers maintain their separate databases, which contain EHRs of their patients. A distributed ledger manages interactions between a patient and the healthcare provider ecosystem as a whole. Validators maintain the integrity of the distributed ledger. For example, consider a $Patient_1$ who visits HCP_1. As a result of this visit, an EHR_{11} is created and stored in the DB_1. The access privileges to EHR_{11} are stored on the distributed ledger node, which belongs to the $Patient_1$. Assume now that $Patient_1$ requires additional treatment from HCP_2. She can grant access privileges to HCP_2 by broadcasting an appropriate transaction to other distributed ledger nodes. DB_1 verifies that HCP_2 has the right to access (either to read or to copy) EHR_{11} and allows them to do so. As a result, HCP_2 has a holistic picture of $Patient_1$'s health. Depending on the circumstances, the access can be a one-off or permanent. However, there is one nuance of this paradigm that the reader needs to appreciate.

We tacitly assumed that DB_1 and DB_2 are compatible so that EHR_{11} can be read and understood by DB_2. We leave it to the reader to figure out what to do if it is not the case. To improve the integrity of the EHRs kept in the databases, HCPs can hash them periodically and post the corresponding Merkle roots on the distributed ledger.

Technically, the corresponding blockchain protocol is not as complicated as, say, the Bitcoin or Ethereum protocols. However, the real challenge lies in creating a functional infrastructure that overcomes the inertia of the existing ecosystem without disruption to high-quality patient care.

Given that efficient medical billing and health insurance claim management heavily rely on the availability and reliability of EHR, DLT can be just the right instrument in modernizing this industry. It can be used to implement smart health insurance contracts, thus allowing their automatic processing and self-execution as per pre-programmed terms and conditions. Such automation eliminates the need for intermediaries and results in overall cost reduction. It also reduces the number of claim disputes and speeds up claims processing — only the most difficult situations would now require human intervention.

12.4.2 Research

Pharmaceutical companies are continually engineering new drugs. Once a new drug is found, a company must undertake a long and expensive process to demonstrate its safety and efficacy. Except for extraordinary circumstances, such as the COVID-19 pandemic, the path from discovery to commercialization takes many years, with clinical trials being the longest part. Due to its length, complexity, and cost, any possible protocol violations, both unintentional and not, could have profound, if not catastrophic effects on the health of the company, and most importantly, the patients. The integrity of the drug development process can be improved by using DLT to record every trial event on blockchain nodes in a tamper-proof way. By generating a timestamped, immutable record of clinical trial results, including patient-generated data, one can prevent outcome switching, data snooping, and selective reporting, thereby reducing possible fraud and error while producing evidenced provenance for medical research data. Recording data on blockchain improves secured and privacy-preserving healthcare data sharing opportunities for healthcare providers and researchers without ceding control; see, e.g., Khezr et al. (2019); Kuo et al. (2017) and references therein.

12.4.3 Supply chain

The integrity of the pharmaceutical supply chain is of great importance especially because of the catastrophic consequences of counterfeit drugs; see, e.g., Khezr et al. (2019); Kuo et al. (2017). For instance, according to the World Health Organization, more than 100,000 people die in Africa every year because of counterfeit drugs delivered via compromised supply chains. DLT can be useful to address some of the weaknesses of the existing supply chains. We can utilize the corresponding distributed ledger to register information at each step of the supply chain and permit proper authentication of drugs delivered to the end-user. By doing so, we can improve drug traceability, ensure data integrity and transparency across the supply chain, and perform a continuous real-time audit of the entire process, thus reducing the possibility of counterfeit drugs entering the supply chain. As an extra benefit, DLT facilitates interactions between supply chain participants, thus increasing healthcare provision efficiency. We show the process of strengthening the pharmaceutical supply chain in Figure 12.10.

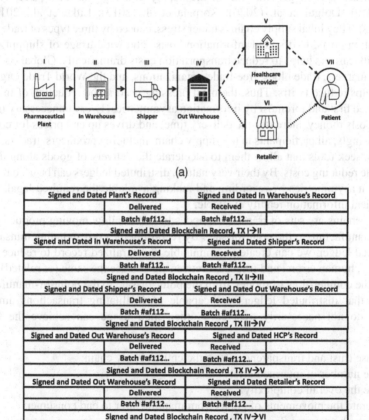

Figure 12.10 A DLT-based system for improving the integrity of PSCM: (a) physical shipments; (b) the corresponding blockchain TXs. Own graphics. Source: (a) Khezr et al. (2019).

This figure shows a DLT-based system for improving pharmaceutical supply chain management (PSCM). It is interesting to trace reflections of physical changes of the shipment's ownership on the distributed ledger. Grigg's triple-signed receipt introduced in Figure 12.3, is the key ingredient of the blockchain transaction chain presented in Figure 12.10 (b). It is imperative to understand that connections between the physical (analog) world, shown in Figure 12.10 (a), and its digital reflection shown in Figure 12.10 (b) is tenuous. Thus, the process makes sense only if participants also use proper tamper-proof containers for shipping the underlying drugs and other pharmaceutical products.

12.5 Supply chain

12.5.1 Supply chain management

Never before has the modern economy depended as much on international trade in goods and services in general and on the supply chain in particular; see, e.g., Allen et al. (2019); Dolgui et al. (2020); Korpela et al. (2017); Litke et al. (2019); Vyas et al. (2019). The global supply chain is under stress, caused by three types of trade frictions: transportation costs, tariffs, and information costs. The wide usage of shipping containers and bulk carriers helps to reduce transportation costs dramatically. Global coordination via single markets, trade blocs, free-trade organizations, and the World Trade Organization (WTO) helps reduce tariffs. Thus, the most pressing need is the reduction of information costs, caused first and foremost by information asymmetries. Every unnecessary inspection of goods costs money, increases the delivery time, and drives up end prices for consumers. Not surprisingly, all participants in the supply chain, including producers, traders, shippers, and banks, seek tools that allow them to accelerate the delivery of goods along the supply chain while reducing costs. By their very nature, distributed ledgers can be used to decrease information asymmetries by recording and making available the required legal, financial, and technical information readily available.

Supply chains are sets of transaction nodes linked to allow moving products from the point-of-manufacture to the point-of-sale. By recording the corresponding transactions on a distributed ledger, we can create an immutable decentralized record to reduce disputes, time delays, human errors, and fraud. Moreover, we can use the same (or related) ledger to facilitate the counter-flow of money from the point-of-sale to the point-of-manufacture.

Given that distributed ledgers are capable of facilitating transactions among parties, who do not necessarily trust each other, their usage can achieve the following benefits:

(1) increase trust and transparency along the entire supply chain;
(2) cut the number of intermediaries;
(3) reduce the overall complexity;
(4) accelerate the movements of goods from manufacturer to end consumer;
(5) facilitate the counter-flow of money;
(6) and, finally, make the entire process paperless.

According to Allen et al. (2019), blockchain will facilitate new forms of economic organization governing supply chain coordination, decrease information asymmetries, shift economic power towards primary producers and final consumers, differentiate goods, and decrease consumer reliance on quality proxies. Figure 12.11 shows the flow of goods from Seller to Buyer and the counter-flow of money in the opposite direction.

Numerous startups, public and private companies, and consortia are exploring applications of DLT to supply chain management. For instance, the container-shipper Maersk is part of a proof-of-concept initiative, using an IBM-designed blockchain to digitize cargo inventories, visualize information flow, and optimize supply chain real-time state. Similarly, several companies have built DLT-based software to trace goods moving across supply chains, while facilitating trade finance and other related tasks.

12.5.2 Trade finance

Trade finance is the lubricant on the supply chain wheels — in fact, international trade would be impossible without it. Trade finance is a hybrid activity involving both supply chain and financial services. International trade has experienced explosive growth over several decades due to intermediaries that have positioned themselves to finance transactions between the buyers (importers) and the sellers (exporters). These intermediaries offer a wide range of options, such as a bank letter of credit (LC), bank guarantee, lending, forfaiting, and export credit and financing. Thus, trade financing requires at least three parties — a buyer, a seller, a financier — and possibly several more. Pain points of trade finance include lack of transparency, copious bureaucracy, fraud, errors, among others. As such, DLT is well-suited to reduce payment risk and facilitate international trade.

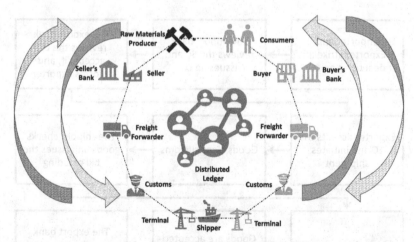

Figure 12.11 A typical supply chain and associated trade finance cycles. Goods flow in the counter-clockwise direction, money flow in the clockwise direction. Own graphics.

Letters of credit are a case in point. Recall that an LC is an obligation of the importer's bank to pay the exporter's bank upon presenting all the documents specified by the buyer and seller's purchase agreement. It is essential to understand that an LC deals exclusively with documentation. A typical LC requires a long time to negotiate, is quite expensive, and often challenging to obtain. Besides, it is often subject to many ambiguities, which can cause severe issues on the ground, including payment disputes, payment delays, and the need for costly amendments. Not surprisingly, these negative externalities, combined with the emergence of new financing options, has resulted in the decline in the usage of LCs in favor of open account trade. Still, if the pain points are adequately addressed, an LC is a beneficial instrument. DLT can be very useful to address the above-mentioned issues. We can use an SCs-supporting blockchain to simplify documentation preparation and processing and partially solve the DvP problem. Figure 12.12 shows a sequence of blockchain transactions associated with the successful execution of an LC.

Using DLT along the lines shown in Figure 12.12 provides several advantages compared to the existing process:

(1) Paperwork is processed in real-time, and its overall amount is reduced due to standardization.
(2) Forfeiting is greatly simplified since all the invoices are accessible in real-time to the authorized parties.
(3) Counterparty risk is reduced since all bills of lading are accessible, so that double spending is more difficult (though not impossible).
(4) Ownership of goods in transit is transparent.
(5) Settlement is automated, streamlined, and cheap.
(6) Regulations are applied in real-time.

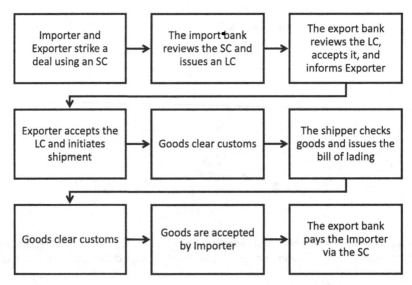

Figure 12.12 How DLT can streamline trade finance. Own graphics.

12.6 Tokenization of real assets

Tokenization of art, intellectual property, music, and real estate are areas where DLT can be beneficially deployed; see Pentland et al. (2020) and references therein.

For brevity, we shall restrict ourselves to real estate. The world's real estate (RE) is the most valuable asset class in the world. According to Savills Impact, at the end of 2017, its value reached US$280.6 trillion, with residential, commercial, and agricultural real estate (RRE, CRE, and ARE for brevity) worth US$220.6 trillion, US$33.3 trillion, and US$27.1 trillion, respectively. When combined, global RE value dwarfs the value of financial instruments, oil reserves, gold, and cryptocurrencies.

Yet, RE transactions tend to be lumpy, slow, expensive, and notoriously hard to execute. A further complication is that, in many countries, establishing a transparent chain of ownership is difficult, if not impossible. Given the nature of RE transactions, DLT can help address some of the aforementioned pain-points. As was discussed earlier, revolutionary changes in the way RE is transacted, financed, insured, and held are unlikely. However, even some of the seemingly less radical innovations can have profound and significant consequences in reshaping the inefficiencies and asymmetries in RE transactions.

Here is a particular example illustrating the above point succinctly. According to Davies et al. (2020), on March 6, 2019, it took 22 weeks for a residential house in Gillingham, UK, to change hands rather than initially estimated six weeks. It is important to note that there were no additional complications involving uncertainties related to house ownership.[10] To check whether DLT can fasten the process, a public/private consortium of Her Majesty's Land Registry's (HMLR) Digital Street, Mishcon de Reya, R3, and several other parties designed a DLT-based conveyance process. The consortium built a prototype enabling a digital transfer of the property, resulting in the Land Register's automatic updates. Their prototype distributed ledger was successful beyond expectations. It completed the transaction mentioned above, end to end, in less than 10 minutes, instead of 22 weeks!

There is a lot to be learned from this experience from a methodological standpoint. First, the interested parties created a consortium. Second, they created a flow chart of the existing conveyance process. Third, the consortium produced a distributed application built on the Corda DLT (also known as a CorDapp), modeling a primary end-to-end automated conveyance. Finally, they used the CorDapp to replicate the property transaction of interest. As a result, a prototype for executing a fast, trustworthy, secure, and transparent conveyancing process was successfully designed and implemented.

Of course, there are several applications of DLT in RE that go above and beyond the conveyancing process. These applications span the entire spectrum of RE-related activities, stretching from land acquisition and construction of new properties to their sale and subsequent maintenance.

As a starting point, authorities must begin by recording land ownership on a suitably chosen blockchain, in all likelihood, a permissioned one. Given that usage of blockchain does not solve pre-existing disputes over land ownership, ownership must be established

---

[10] A typical conveyancing transaction results in the transfer of RE's legal title from one person to another. It has two major phases: the exchange of contracts and completion, when legal title passes from the seller to the buyer, and equitable rights merge with the legal title.

with certainty before moving property rights of a particular parcel to the DL. The most dramatic step in this direction is to conduct a modern analog of the "Great Survey" of England and Wales completed in 1086 by King William the Conqueror, which resulted in the creation of the celebrated Domesday Book. An essential aspect of the book is that the assessors' reckoning of land possessions was dispositive and without appeal — hence the book's name. Of course, modern democracies might be less inclined to such steps than Medieval kings, so the blockchain-based land description can be created gradually — one parcel at a time, as part of the purchasing process, for example.

Next, the construction of RE can be facilitated by DLT, particularly in the rationalization of its supply chain management, as discussed in Section 12.5. Of course, the title for the finished property should be recorded on the blockchain.

DLT facilitates the buying and selling of newly constructed and existing properties on several levels. The property search can be harmonized and streamlined by replacing multiple listing services with posting everything on a commonly accessible distributed ledger. We already mentioned that DLT makes the conveyance process much faster, cheaper, and more reliable than it is at present. It is only natural to move the title of the newly acquired property to the underlying blockchain.

RE financing is also a promising target for DLT. PM, discussed in Section 12.2.3, is one direction to explore to solve the DvP problem. The other, more revolutionary in spirit, is RE tokenization, which allows crowd financing of mortgages, and opens new opportunities for small investors to invest in RRE, CRE, and ARE as an asset class. As such, using cryptocurrencies for buying RE is an exciting dimension to explore.

Finally, maintenance of the existing properties, particularly rental and commercial ones, can be significantly facilitated by PM and related instruments by simplifying financial arrangements.

Of course, deployment of DLT in the RE ecosystem is not without issues:

(1) As discussed above, the line of ownership of the existing or newly-built RE has to be established with finality outside of DLT.
(2) The level of awareness among decision-makers has to be raised to make educated choices regarding advantages and disadvantages of DLT.
(3) Considerable technical obstacles need to be overcome since distributed ledgers underpinning RE have to be extremely robust and fast, backward compatible, and satisfy several additional requirements.
(4) Regulatory uncertainty must be resolved before the mass adoption of DLT in RE, which is highly regulated. DLT-based RE processing raises many legal and regulatory questions, such as privacy and data protection, system governance rules, tax and property law.

12.7 Summary

Throughout the book we developed the underlying principles and mathematical apparatus that can guide creation of these new applications of blockchain and DLT. In this chapter we pivot to reflect on possible applications of DLT outside of cryptocurrencies. Although

DLT applications in finance are conceptually straightforward, they often remain difficult to implement in practice. These difficulties are only amplified when trying to broaden the application of DLT outside of finance, since many of these use cases stumble on problems detailed throughout the book that require resolution outside DLT. Moreover, these projects will take a long time to move from the proof-of-concept stage to the production-ready state. Still, given the broad, long-term advantages in many areas, including healthcare, supply chain management, and real estate, to mention but a few, it is entirely justifiable for governments, private companies, and academic institutions to pursue them. Our overall impression is that DLT holds a lot of promise, but it will require competence, effort, resources, and patience to achieve its full potential.

Many non-financial applications would represent a somewhat scaled-down version of the original grand vision of a fully distributed, decentralized ledger. As such, one should not undervalue the utility of DLT and its innumerable applications as a driver of significant change. That said, the real barrier to reshaping the backbone of our society using a globalized implementation of DLT remains our ability to practically translate its ideas and requirements into a full infrastructure and ecosystem that can be used by anyone and everyone. The enthusiasm expressed by multiple professionals shows that DLT is slowly making its way into various areas of human activity. To satisfy this interest, several universities and business schools are offering courses in DLT and all things crypto, including the Hebrew University of Jerusalem (HUJI), where one of the authors (AL) reads the course; the École Polytechnique Fédérale de Lausanne (EPFL), where both authors gave lectures and practical seminars on the subject; as well as Oxford, University College London (UCL), Imperial College, and many other schools.

In short, the overhaul of everyday life using DLT technology has already begun, but it will take time for the promised all-encompassing blockchain revolution to materialize.

12.8 Exercises

1. Compare and contrast double- and triple-entry accounting. Which aspects of triple-entry accounting are used in the Bitcoin protocol?
2. Describe how one can use DLT for central clearing. What are the obstacles to its wide adoption in this area?
3. Choose a particular healthcare-related topic, for instance, EHRs access, and design a DLT-based system that can remove some of its most egregious pain points.

Bibliography

Aas, K. and Hobaek Haff, I. (2005). *NIG and Skew Student's t: two special cases of the Generalised Hyperbolic Distribution.* Oslo: Norwegian Computing Center Preprint.

Abadi, J. and Brunnermeier, M. (2018). Blockchain Economics. In: *Working Paper No. w25407.* Cambridge: National Bureau of Economic Research.

Abramowitz, M. and Stegun, I.A., eds. (1948). *Handbook of mathematical functions with formulas, graphs, and mathematical tables.* Washington, D.C.: US Government Printing Office.

Adida, B. (2006). *Advances in Cryptographic Voting Systems.* PhD. Massachusetts Institute of Technology, Dept. of Electrical Engineering and Computer Science.

Admati, A. and Hellwig, M. (2014). *The bankers' new clothes: what's wrong with banking and what to do about it — updated edition.* Princeton: Princeton University Press.

Adrian, M.T. and Griffoli, M.T.M. (2019). *The rise of digital money.* Washington, D.C.: International Monetary Fund.

Ahmad, A., Saad, M. and Mohaisen, A. (2019). Secure and Transparent Audit Logs with BlockAudit. *Journal of Network and Computer Applications,* 145, p. 102406. doi:10.1016/j.jnca.2019.102406.

Akaike, H. (1974). A New Look at the Statistical Model Identification. *IEEE Transactions on Automatic Control,* 19, pp. 716–723.

Akcora, C.G., Dixon, M.F., Gel, Y.R. and Kantarcioglu, M. (2018). Bitcoin Risk Modeling with Blockchain Graphs. *Economics Letters,* 173, pp. 138–142.

Alharby, M., Aldweesh, A. and van Moorsel, A. (2018). Blockchain-Based Smart Contracts: A Systematic Mapping Study of Academic Research. In: *2018 International Conference on Cloud Computing, Big Data and Blockchain (ICCBB).* IEEE, pp. 1–6.

Ali, A., Latif, S., Qadir, J., Kanhere, S., Singh, J. and Crowcroft, J. (2019). Blockchain and the Future of the Internet: A Comprehensive Review. *arXiv preprint,* arXiv:1904.00733.

Al-Jaroodi, J. and Mohamed, N. (2019). Blockchain in Industries: A Survey. *IEEE Access,* 7, pp. 36500–36515.

Allen, D.W., Berg, A. and Markey-Towler, B. (2019). Blockchain and Supply Chains: V-form Organisations, Value Redistributions, De-commoditisation and Quality Proxies. *The Journal of the British Blockchain Association,* 2(1), pp. 1–8.

Al-Naji, N., Chen, J. and Diao, L. (2017). Basis: A Price-Stable Cryptocurrency with an Algorithmic Central Bank. In: *White Paper.* [online] Basis. Available at: https://www.basis.io/basis_whitepaper_en.pdf.

Alsayed Kassem, J., Sayeed, S., Marco-Gisbert, H., Pervez, Z. and Dahal, K. (2019). DNS-IdM: A Blockchain Identity Management System to Secure Personal Data Sharing in a Network. *Applied Sciences,* 9(15), p. 2953, doi:10.3390/app9152953.

Ammous, S. (2018). *The Bitcoin Standard: the decentralized alternative to central banking.* Hoboken: John Wiley & Sons.

Anderson, R.M., Anderson, B. and May, R.M. (1992). *Infectious diseases of humans: dynamics and control.* Oxford: Oxford University Press.

Anderson, T.W. and Darling D.A. (1954). A Test of Goodness of Fit. *Journal of the American Statistical Association,* 49, pp. 765–769.

Angeris, G., Kao, H.T., Chiang, R., Noyes, C. and Chitra, T. (2019). An Analysis of Uniswap Markets. *Cryptoeconomic Systems Journal*.

Antonakakis, N., Chatziantoniou, I. and Gabauer, D. (2019). Cryptocurrency Market Contagion: Market Uncertainty, Market Complexity, and Dynamic Portfolios. *Journal of International Financial Markets, Institutions and Money*, 61, pp. 37–51.

Antonopoulos, A.M. (2017). *Mastering Bitcoin: programming the Open Blockchain*. Sebastopol: O'Reilly Media.

Antonopoulos, A.M. and Wood, G. (2018). *Mastering Ethereum: building smart contracts and DApps*. Sebastopol: O'Reilly Media.

Archer, D.W., Bogdanov, D., Lindell, Y., Kamm, L., Nielsen, K., Pagter, J.I., Smart, N.P. and Wright, R.N. (2018). From Keys to Databases — Real-World Applications of Secure Multi-Party Computation. *The Computer Journal*, 61(12), pp. 1749–1771.

Ardaillon, È. (1897). *Les mines du Laurion dans l'antiquité*. Paris: Ancienne Librairie Thorin et Fils.

Arya, A., Fellingham, J., Schroeder, D. and Young, R. (2004). *Double Entry Bookkeeping and Error Correction*. Report No. 014. [online] Columbus: Ohio State University. Available at: https://cpb-us-w2.wpmucdn.com/u.osu.edu/dist/7/36891/files/2019/06/DoubleEntry.pdf.

Back, A. (2002). *Hashcash — A Denial of Service Counter-Measure*. In: *White Paper*.

Baird, L. (2016). *Hashgraph Consensus: Fair, Fast, Byzantine Fault Tolerance*. [online] Swirlds, pp. 1–28. Available at: https://www.swirlds.com/downloads/SWIRLDS-TR-2016-01.pdf.

Bank of England, (2020). *Central Bank Digital Currency. Opportunities, Challenges and Design*. London: BoE.

Barinov, I., Arasev, V., Fackler, A., Komendantskiy, V., Gross, A., Kolotov, A. and Isakova, D. (2019). Proof of Stake Decentralized Autonomous Organization.

Barker, E.B. (2002). *Secure Hash Standard (SHS) [includes Change Notice from 2/25/2004] (No. Federal Inf. Process. Stds. (NIST FIPS)-180-2)*. Gaithersburg: NIST.

Barker, R., Dickinson, A., Lipton, A. and Virmani, R. (2017). Systemic Risks in CCP Networks. Risk, 31(1), pp. 91–97.

Barndorff-Nielsen, O.E. (1977). Exponentially Decreasing Distributions for the Logarithm of Particle Size. In: *Proceedings of the Royal Society A*. London: The Royal Society, pp. 409–419.

Barndorff-Nielsen, O. E. (1997). Normal Inverse Gaussian Distributions and Stochastic Volatility Modelling. *Scandinavian Journal of Statistics*, 24, pp. 1–13.

Barreiro-Gomez, J. and Tembine, H. (2019). Blockchain Token Economics: A Mean-Field-Type Game Perspective. *IEEE Access*, 7, pp. 64603–64613.

Bartolucci, S. and Kirilenko, A.A. (2020). A Model of the Optimal Selection of Crypto Assets. Available at: SSRN 3578450.

Baumann, A., Fabian, B. and Lischke, M. (2014). Exploring the Bitcoin Network. In: *10th International Conference on Web Information Systems and Technologies (WEBIST)*. Berlin, Heidelberg: Springer, pp. 369–374.

Baur, D.G., Hong, K. and Lee, A.D. (2018). Bitcoin: Medium of Exchange or Speculative Asset? *Journal of International Financial Markets, Institutions and Money*, 54, pp. 177–189.

Beccuti, J. and Jaag, C. (2017). The Bitcoin Mining Game: On the Optimality of Honesty in Proof-of-Work Consensus Mechanism. In: *Swiss Economics Working Paper 0060*.

Bechtel, W. (2007). Biological mechanisms: organized to maintain autonomy. In: F. Boogerd, F. J. Bruggerman, J.-H. S. Hofmeyr and H. V. Westerhoff, eds., *Systems Biology: Philosophical Foundations*. Amsterdam: Elsevier, pp. 269–302.

Begenau, J. and Stafford, E. (2019). Do Banks Have an Edge? Available at: SSRN 3095550.

Belotti, M., Božic, N., Pujolle, G. and Secci, S. (2019). A Vademecum on Blockchain Technologies: When, Which, and How. *IEEE Communications Surveys & Tutorials*, 21(4), pp. 3796–3838.

Bene, J. and Kumhof, M. (2012). *The Chicago plan revisited*. Washington, D.C.: International Monetary Fund.

Beniiche, A. (2020). A Study of Blockchain Oracles. *arXiv preprint*, arXiv:2004.07140.

Berentsen, A. and Schär, F. (2017). *Bitcoin, blockchain und kryptoassets: eine umfassende einführung*. Hamburg: BoD, Norderstedt.

Bernanke, B.S. and Blinder, A.S. (1988). *Credit, money, and aggregate demand (No. w2534)*. Cambridge: National Bureau of Economic Research.

Bernstein, D.J., Chou, T., Chuengsatiansup, C., Hülsing, A., Lambooij, E., Lange, T., Niederhagen, R. and Van Vredendaal, C. (2015). How to Manipulate Curve Standards: A White Paper for the Black Hat. In: *International Conference on Research in Security Standardisation*. Cham: Springer, pp. 109–139.

Bernstein, D.J., Duif, N., Lange, T., Schwabe, P. and Yang, B.Y. (2012). High-Speed High-Security Signatures. *Journal of Cryptographic Engineering*, 2(2), pp. 77–89.

Bheemaiah, K. (2017). *The blockchain alternative: rethinking macroeconomic policy and economic theory*. New York: Apress.

Bhutoria, R. (2020). Addressing Persistent Bitcoin Criticisms. In: *Fidelity Digital Assets Working Paper*.

Biais, B., Bisiere, C., Bouvard, M. and Casamatta, C. (2019). The Blockchain Folk Theorem. *The Review of Financial Studies*, 32(5), pp. 1662–1715.

Bible, W., Raphael, J., Taylor, P. and Valiente, I.O. (2017). *Blockchain Technology and its Potential Impact on the Audit and Assurance Profession*. [online] Durham: American Institute of Certified Public Accountants. Available at: https://www.aicpa.org/content/ dam/aicpa/ interestareas/frc/assuranceadvisoryservices/downloadabledocuments/blockchain-technology-and-its-potential-impact-on-the-audit-and-assurance-profession.pdf.

Bissias, G., Levine, B.N. and Thibodeau, D. (2018). Using Economic Risk to Model Miner Hash Rate Allocation in Cryptocurrencies. In: *Data Privacy Management, Cryptocurrencies and Blockchain Technology*. Cham: Springer, pp. 155–172.

Bistarelli, S., Mazzante, G., Micheletti, M., Mostarda, L. and Tiezzi, F. (2019). Analysis of Ethereum Smart Contracts and Opcodes. In: *International Conference on Advanced Information Networking and Applications*. Cham: Springer, pp. 546–558.

Black, F. (1987). A gold standard with double feedback and near zero reserves. In: F. Black, ed., *Business Cycles and Equilibrium*. Hoboken: John Wiley & Sons, pp. 115–120.

Blandin, A., Cloots, A.S., Hussain, H., Rauchs, M., Saleuddin, R., Allen, J.G., Zhang, B.Z. and Cloud, K. (2019). Global Cryptoasset Regulatory Landscape Study. In: *University of Cambridge Faculty of Law Research Paper*.

Bodenhorn, H. (2000). *A history of banking in Antebellum America: financial markets and economic development in an era of nation-building*. Cambridge: Cambridge University Press.

Bodó, B. and Giannopoulou, A. (2019). The logics of technology decentralization: the case of distributed ledger technologies. In: M. Ragnedda and G. Destefanis, eds., *Blockchain and Web 3.0: Social, Economic, and Technological Challenges*. London: Routledge, 16 pp.

Bordo, M.D. and Levin, A.T. (2017). Central bank digital currency and the future of monetary policy. In: *Working Paper 23711*. Cambridge: National Bureau of Economic Research.

Borgonovo, E., Caselli, S., Cillo, A., Masciandaro, D. and Rabitti, G. (2018). Cryptocurrencies, Central Bank Digital Cash, Traditional Money: Does Privacy Matter? In: *Working Paper No. 1895*. Milano: Centre for Applied Research on International Markets Banking Finance and Regulation, Universita'Bocconi.

Bovet, A., Campajola, C., Lazo, J.F., Mottes, F., Pozzana, I., Restocchi, V., Saggese, P., Vallarano, N., Squartini, T. and Tessone, C.J. (2018). Network-Based Indicators of Bitcoin Bubbles. *arXiv preprint*, arXiv:1805.04460.

Bowman, A. W., and Azzalini, A. (1997). *Applied smoothing techniques for data analysis.* New York: Oxford University Press Inc.

Box, G.E., Jenkins, G.M. and Reinsel, G.C. (2011). *Time series analysis: forecasting and control.* Hoboken: John Wiley & Sons.

Boyle, T. (2000). The Public Transaction Repository (PTR) Project. In: *White Paper.* GL Dialtone.

Brands, S. (1993). Untraceable Off-line Cash in Wallet with Observers. In: *Annual International Cryptology Conference.* Berlin, Heidelberg: Springer, pp. 302–318.

Brauer, F., Driessche, P.D. and Wu, J. (2008). *Lecture notes in mathematical epidemiology.* Berlin: Springer.

Brown, E. (2000). Three Fermat Trails to Elliptic Curves. *The College Mathematics Journal,* 31(3), pp. 162–172.

Brown, R.G., Carlyle, J., Grigg, I. and Hearn, M. (2016). Corda: An Introduction. In: *Corda Introductory Whitepaper.*

Brown, R.G., Carlyle, J., Grigg, I. and Hearn, M. (2016). Corda: An introduction. In: *R3 CEV, August,* 1, p. 15.

Burniske, C. and Tatar, J. (2017). *Cryptoassets: the innovative investor's guide to Bitcoin and beyond.* New York: McGraw Hill Professional.

Buterin, V. (2013). A Next-Generation Smart Contract and Decentralized Application Platform. In: *White Paper.* Available at: https://github.com/ethereum/wiki/wiki.

Buterin, V. (2016). Chain Interoperability. In: *R3 Research Paper.*

Cachin, C. and Vukolić, M. (2017). Blockchain Consensus Protocols in the Wild. *arXiv preprint,* arXiv:1707.01873.

Cai, C.W. (2019). Triple Entry Accounting with Blockchain: How Far Have We Come? *Accounting & Finance,* doi:10.1111/acfi.12556.

Campbell-Verduyn, M. (2018). Bitcoin, Crypto-Coins, and Global Anti-Money Laundering Governance. *Crime, Law and Social Change,* 69(2), pp. 283–305.

Casino, F., Dasaklis, T.K. and Patsakis, C. (2019). A Systematic Literature Review of Blockchain-Based Applications: Current Status, Classification and Open Issues. *Telematics and Informatics,* 36, pp. 55–81.

Castro, M. and Liskov, B. (1999). Practical Byzantine Fault Tolerance. In: *Proceedings of the Third Symposium on Operating Systems Design and Implementation.* New York: Association for Computing Machinery, pp. 173–186.

Catalini, C. and Gans, J.S. (2016). Some Simple Economics of the Blockchain. In: *NBER Working Paper 22952.* Cambridge: National Bureau of Economic Research.

Certicom Research, (2009). *SEC 1: Elliptic Curve Cryptography.* [online] Mississauga: Certicom Corp. Available at: https://www.secg.org/sec1-v2.pdf.

Chan, S., Chu, J., Nadarajah, S. and Osterrieder, J. (2017). A Statistical Analysis of Cryptocurrencies. *Journal of Risk and Financial Management,* 10(2), pp. 12–34.

Chase, B. and MacBrough, E. (2018). Analysis of the XRP Ledger Consensus Protocol. *arXiv preprint,* arXiv:1802.07242.

Chaudhry, N. and Yousaf, M.M. (2018). Consensus Algorithms in Blockchain: Comparative Analysis, Challenges and Opportunities. In: *2018 12th International Conference on Open Source Systems and Technologies* (ICOSST). IEEE, pp. 54–63.

Chaum, D. (1983). Blind Signatures for Untraceable Payments. In: *Advances in Cryptology.* Boston: Springer, pp. 199–203.

Chaum, D., Fiat, A. and Naor, M. (1988). Untraceable Electronic Cash. In: *Conference on the Theory and Application of Cryptography.* New York: Springer, pp. 319–327.

Chen, T., Zhu, Y., Li, Z., Chen, J., Li, X., Luo, X., Lin, X. and Zhange, X. (2018). Understanding Ethereum via Graph Analysis. In: *IEEE INFOCOM 2018-IEEE Conference on Computer Communications*. IEEE, pp. 1484–1492.

Chu, J., Nadarajah, S. and Chan, S. (2015). Statistical Analysis of the Exchange Rate of Bitcoin. *PLOS ONE*, 10(7), e0133678.

Chu, S. and Wang, S. (2018). The Curses of Blockchain Decentralization. *arXiv preprint*, arXiv:181 0.02937.

Clanchy, M. (1979). *From memory to written record*. 3rd ed. Oxford: Blackwell.

Clark, D. (1988). The Design Philosophy of the DARPA Internet Protocols. In: *Symposium Proceedings on Communications Architectures and Protocols*. New York: Association for Computing Machinery, pp. 106–114.

Cochrane, J.H. (2014). Toward a run-free financial system. In: M.N. Baily and J.B. Taylor, eds., *Across the Great Divide: New Perspectives on the Financial Crisis*. Stanford: Hoover Institution, pp. 214–215.

Collomb, A. and Sok, K. (2016). Blockchain/Distributed Ledger Technology (DLT): What Impact on the Financial Sector? *Digiworld Economic Journal*, [online] Volume 103. Available at: https://search.proquest.com/openview/b5b5fa49be78d9d574a4c20bc94fc42f/1?pq-origsite=gs cholar&cbl=616298.

Conti, M., Kumar, E.S., Lal, C. and Ruj, S. (2018). A Survey on Security and Privacy Issues of Bitcoin. *IEEE Communications Surveys & Tutorials,* 20(4), pp. 3416–3452.

Cover, T. and King, R. (1978). A Convergent Gambling Estimate of the Entropy of English. *IEEE Transactions on Information Theory,* 24(4), pp. 413–421.

Crisp, R. (2014). *Aristotle: Nicomachean ethics*. Cambridge: Cambridge University Press.

D'Amiano, S. and Di Crescenzo, G. (1994). Methodology for Digital Money based on General Cryptographic Tools. In: *Workshop on the Theory and Application of Cryptographic Techniques*. Berlin, Heidelberg: Springer, pp. 156–170.

Damgård, I.B. (1988). Payment Systems and Credential Mechanisms with Provable Security against Abuse by Individuals. In: *Conference on the Theory and Application of Cryptography*. New York: Springer, pp. 328–335.

Danezis, G. and Meiklejohn, S. (2015). Centrally Banked Cryptocurrencies. *arXiv preprint*, arXiv:1 505.06895.

Davies, E., Kirby, N., Bond, J., Grogan, T, Moore, A., Roche, N., Rose, A., Tasca, P. and Vadgama, N. (2020). Towards a Distributed Ledger of Residential Title Deeds in the UK. In: *Mishcon de Reya White Paper*.

Davies, G. (2010). *History of money*. Cardiff: University of Wales Press.

De Filippi, P. and Wright, A. (2018). *Blockchain and the law: the rule of code*. Cambridge: Harvard University Press.

de Leon, D.C., Stalick, A.Q., Jillepalli, A.A., Haney, M.A. and Sheldon, F.T. (2017). Blockchain: Properties and Misconceptions. *Asia Pacific Journal of Innovation and Entrepreneurship*, 11(3), pp. 286–300.

De Roover, R. (1956). *The development of accounting prior to Luca Pacioli according to the account books of medieval merchants*. London: Sweet & Maxwell.

Decker, C. and Wattenhofer, R. (2013). Information Propagation in the Bitcoin Network. In: *IEEE P2P 2013 Proceedings*. IEEE, pp. 1–10.

Dembo, R.S., Lipton, A. and Burkov, S., Stronghold Labs LLC, (2018). *Account platform for a distributed network of nodes*. U.S. Patent Application 15/946,381.

Deutsche Bundesbank, (2020). Money in Programmable Applications: Cross-Sector Perspectives from the German Economy. In: *White Paper*.

Devaynes v Noble [1816] 35 ER 781.

Dhillon, A., Kotsialou, G., McBurney, P. and Riley, L. (2019). Introduction to Voting and the Blockchain: Some Open Questions for Economists. In: CAGE Online *Working Paper Series 416*. Competitive Advantage in the Global Economy (CAGE).

Di Crescenzo, G. (1994). A Non-Interactive Electronic Cash System. In: *Italian Conference on Algorithms and Complexity*. Berlin, Heidelberg: Springer, pp. 109–124.

Diffie, W. and Hellman, M. (1976). New Directions in Cryptography. *IEEE Transactions on Information Theory,* 22(6), pp. 644–654.

Diffie, W. and Hellman, M.E. (1977). Exhaustive Cryptanalysis of the NBS Data Encryption Standard. *Computer,* 10(6), pp. 74–84.

Dittmer, K. (2015). 100 Percent Reserve Banking: A Critical Review of Green Perspectives. *Ecological Economics,* 109, pp. 9–16.

Dobbertin, H., Bosselaers, A. and Preneel, B. (1996). RIPEMD-160: A Strengthened Version of RIPEMD. In: *International Workshop on Fast Software Encryption*, Berlin, Heidelberg: Springer, pp. 71–82.

Doherty, K. and Surane. J. (2020). Citi Asks Revlon Lenders to Return Mistaken $900 Million. *Bloomberg*, [online]. Available at: https://www.bloomberg.com/news/articles/2020-08-14/citi-asks-lenders-to-return-mistaken-900-million-revlon-payment.

Dolgui, A., Ivanov, D., Potryasaev, S., Sokolov, B., Ivanova, M. and Werner, F. (2020). Blockchain-Oriented Dynamic Modelling of Smart Contract Design and Execution in the Supply Chain. *International Journal of Production Research,* 58(7), pp. 2184–2199.

Douglas, P.H., Fisher, I., Graham, F.D., Hamilton, E.J., King, W.I. and Whittlesey, C.R. (1939). *A Program for Monetary Reform*.

Drosatos, G. and Kaldoudi, E. (2019). Blockchain Applications in the Biomedical Domain: A Scoping Review. *Computational and Structural Biotechnology Journal,* 17, pp. 229–240.

Druon, M. (2014). *Les rois maudits*. Paris: Plon.

Duffie, D. (2019). Digital Currencies and Fast Payment Systems: Disruption is Coming. In: *Asian Monetary Forum Mimeo*.

Dunphy, P. and Petitcolas, F.A. (2018). A First Look at Identity Management Schemes on the Blockchain. *IEEE Security & Privacy,* 16(4), pp. 20–29.

Dunphy, P., Garratt, L. and Petitcolas, F. (2018). Decentralizing Digital Identity: Open Challenges for Distributed Ledgers. In: *2018 IEEE European Symposium on Security and Privacy Workshops* (EuroS&PW). IEEE, pp. 75–78.

Egorov, M. (2019). StableSwap — Efficient Mechanism for Stablecoin Liquidity. In: *White Paper*.

Ekblaw, A., Azaria, A., Halamka, J.D. and Lippman, A. (2016). A Case Study for Blockchain in Healthcare: "MedRec" Prototype for Electronic Health Records and Medical Research Data. In: *Proceedings of IEEE Open & Big Data Conference*. IEEE, p. 13.

ElGamal, T. (1985). A Public Key Cryptosystem and a Signature Scheme based on Discrete Logarithms. *IEEE Transactions on Information Theory,* 31(4), pp. 469–472.

Ellis, S., Juels, A. and Nazarov, S. (2017). Chainlink a Decentralized Oracle Network. In: *White Paper*.

El-Qorchi, M. (2002). The Hawala System. *Finance and Development,* 39(4). Available at: https://www.imf.org/external/pubs/ft/fandd/2002/12/elqorchi.htm.

Elsden, C., Feltwell, T., Lawson, S. and Vines, J. (2019). Recipes for Programmable Money. In: *Proceedings of the 2019 CHI Conference on Human Factors in Computing Systems*. New York: Association for Computing Machinery, pp. 1–13.

Elton, H. (1998). *Warfare in Roman Europe, AD 350-425*. Oxford: Oxford University Press.

European Central Bank, (2012). *Virtual Currency Schemes. EC Bank, Virtual Currency Schemes* (h. 13-14). Frankfurt am Main: ECB.

European Central Bank, (2015). *Virtual Currency Schemes — A Further Analysis.* Frankfurt am Main: ECB.

European Central Bank, (2018). *The Role of Euro Banknotes as Legal Tender.* Speech by Yves Mersch, Member of the Executive Board of the ECB, at the 4th Bargeldsymposium of the Deutsche Bundesbank, Frankfurt am Main, 14 February 2018.

Eyal, I. and Sirer, E.G. (2014). Majority is not Enough: Bitcoin Mining is Vulnerable. In: *International Conference on Financial Cryptography and Data Security.* Berlin, Heidelberg: Springer, pp. 436–454.

Fan, C.I., Huang, V.S.M. and Yu, Y.C. (2013). User Efficient Recoverable Off-line E-cash Scheme with Fast Anonymity Revoking. *Mathematical and Computer Modelling,* 58(1-2), pp. 227–237.

Federal Financial Institutions Examination Council, (2016). *Retail Payment Systems. IT Examination Handbook.* FFIEC.

Federal Reserve Bank of San Francisco, (2015). *Cash Continues to Play a Key Role in Consumer Spending: Evidence from the Diary of Consumer Payment Choice.* San Francisco: FRBSF.

Ferguson, N., Schneier, B. and Kohno, T. (2010). *Cryptography engineering: design principles and practical applications.* Hoboken: Wiley.

Fisher, I. (1933). *Stamp scrip.* New York: Adelphi Company.

Fisher, I. (1935). *100% money.* New York: Adelphi Publication.

Frank, P., Goldstein, S., Kac, M., Prager, W., Szegö, G., Birkhoff, G., eds. (1964). *Selected Papers of Richard von Mises.* 2. Providence: American Mathematical Society.

Friedman, M. (1960). *A program for monetary stability (No. 3).* Ravenio Books.

Frost, J., Shin, H.S. and Wierts, P. (2020). An Early Stablecoin? The Bank of Amsterdam and the Governance of Money. In: *BIS Working Papers (No. 902).*

Galati, G. (2002). Settlement Risk in Foreign Exchange Markets and CLS Bank. *BIS Quarterly Review,* 4, pp. 55–65.

Gao, Y.L., Chen, X.B., Chen, Y.L., Sun, Y., Niu, X.X. and Yang, Y.X. (2018). A Secure Cryptocurrency Scheme based on Post-Quantum Blockchain. *IEEE Access,* 6, pp. 27205–27213.

Garay, J. and Kiayias, A. (2020). Sok: A Consensus Taxonomy in the Blockchain Era. In: *Cryptographers' Track at the RSA Conference.* Cham: Springer, pp. 284–318.

Gencer, A.E., Basu, S., Eyal, I., Van Renesse, R. and Sirer, E.G. (2018). Decentralization in Bitcoin and Ethereum Networks. In: *International Conference on Financial Cryptography and Data Security.* Berlin, Heidelberg: Springer, pp. 439–457.

Gershon, D., Lipton, A. and Levine, H. (2020). Managing COVID-19 Pandemic without Destroying the Economy. *arXiv preprint,* arXiv:2004.10324.

Gervais, A., Karame, G.O., Wüst, K., Glykantzis, V., Ritzdorf, H. and Capkun, S. (2016). On the Security and Performance of Proof of Work Blockchains. In: *Proceedings of the 2016 ACM SIGSAC Conference on Computer and Communications Security.* New York: Association for Computing Machinery, pp. 3–16.

Gesell, S. (1958). *The natural economic order.* London: Owen.

Gillilland, C.L.C. (1975). *The stone money of Yap: a numismatic survey.* Washington, D.C.: Smithsonian Institution.

Girasa, R. (2018). *Regulation of cryptocurrencies and blockchain technologies: national and international perspectives.* Berlin: Springer.

Gkillas, K. and Katsiampa, P. (2018). An Application of Extreme Value Theory to Cryptocurrencies. *Economics Letters,* 164, pp. 109–111.

Glaser, F., Hawlitschek, F. and Notheisen, B. (2019). Blockchain as a Platform. In: *Business Transformation through Blockchain.* Cham: Palgrave Macmillan, pp. 121–143.

Godley, W. and Lavoie, M. (2007). *Monetary economics: an integrated approach to credit, money, income, production and wealth.* London: Palgrave Macmillan.

Goetzmann, W.N. (2017). *Money changes everything: how finance made civilization possible.* Princeton: Princeton University Press.

Goldwasser, S. and Bellare, M. (2008). *Lecture Notes on Cryptography. Summer Course "Cryptography and Computer Security" at MIT.*

Goldwasser, S., Micali, S. and Rackoff, C. (1989). The Knowledge Complexity of Interactive Proof Systems. *SIAM Journal on Computing,* 18(1), pp. 186–208.

Goodell, G. and Aste, T. (2019). A Decentralised Digital Identity Architecture. *Frontiers in Blockchain,* doi:10.3389/fbloc.2019.00017.

Gordon, W.J. and Catalini, C. (2018). Blockchain Technology for Healthcare: Facilitating the Transition to Patient-Driven Interoperability. *Computational and Structural Biotechnology Journal,* 16, pp. 224–230.

Gorton, G. and Pennacchi, G. (1993). Money market funds and finance companies: are they the banks of the future? In: M. Klausner and L. White, eds., *Structural Change in Banking.* Homewood: Irwin Publishing, pp. 173–214.

Gosset, W.S. (1908). The Probable Error of a Mean. *Biometrika,* 6, pp. 1–25.

Graham, B. (1933). Stabilized Reflation. *Economic Forum,* 1, pp. 186–193.

Graham, F.D. (1940). The Primary Functions of Money and Their Consummation in Monetary Policy. *American Economic Review,* 30, pp. 1–16.

Grasselli, M.R. and Lipton, A. (2019). On the normality of negative interest rates. *Review of Keynesian Economics,* 7(2), pp. 201–219.

Gray, A., Greenhalgh, D., Hu, L., Mao, X. and Pan, J. (2011). A Stochastic Differential Equation SIS Epidemic Model. *SIAM Journal on Applied Mathematics,* 71(3), pp. 876–902.

Graziani, A. (2003). *The monetary theory of production.* Cambridge: Cambridge University Press.

Greer, R.J. (1997). What is an Asset Class, Anyway? *Journal of Portfolio Management,* 23(2), pp. 86–91.

Griffin, J.M. and Shams, A. (2019). Is Bitcoin Really Un-tethered? Available at: SSRN 3195066.

Griffoli, M.T.M., Peria, M.M.S.M., Agur, M.I., Ari, M.A., Kiff, M.J., Popescu, M.A. and Rochon, M.C. (2018). *Casting light on Central Bank Digital Currencies.* Washington, D.C.: International Monetary Fund.

Grigg, I. (2004). The Ricardian Contract. In: *Proceedings. First IEEE International Workshop on Electronic Contracting.* IEEE, pp. 25–31.

Grigg, I. (2005). Triple Entry Accounting. In: *White Paper.* Systemics Inc.

Grontas, P. and Pagourtzis, A. (2019). Blockchain, Consensus, and Cryptography in Electronic Voting. *Homo Virtualis,* 2(1), pp. 79–100.

Grym, A. (2018). The Great Illusion of Digital Currencies. *BoF Economics Review,* 1(2018), pp. 1–17.

Gupta, M. (2017). *Blockchain for DUMMIES.* Hoboken: John Wiley & Sons.

Haber, S. and Stornetta, W.S. (1990). How to Time-Stamp a Digital Document. In: *Conference on the Theory and Application of Cryptography.* Berlin: Springer, pp. 437–455.

Haber, S. and Stornetta, W.S. (1997). Secure Names for Bit-Strings. In: *Proceedings of the 4th ACM Conference on Computer and Communications Security,* pp. 28–35.

Hahn, A. (1920). *Volkswirtschaftliche theorie des bankkredits.* Tübingen: J.C.B. Mohr.

Hamilton, J.D. (2020). *Time series analysis.* Princeton: Princeton University Press.

Hannan, E.J. and Quinn, B.G. (1979). The Determination of the Order of an Autoregression. *Journal of the Royal Statistical Society, Series B (Methodological),* 41(2), pp. 190–195.

Hardjono, T., Lipton, A. and Pentland, A. (2019a). Toward an Interoperability Architecture for Blockchain Autonomous Systems. *IEEE Transactions on Engineering Management.*

Hardjono, T., Shrier, D.L. and Pentland, A. (2019b). *Trusted data: a new framework for identity and data sharing.* Cambridge: MIT Connection Science & Engineering.

Hardjono, T. and Smith, N. (2019). Decentralized Trusted Computing Base for Blockchain Infrastructure Security. *Frontiers in Blockchain,* 2, p. 24. doi:10.3389/fbloc.2019.00024.

Härdle, W.K., Harvey, C.R. and Reule, R.C. (2020). Understanding Cryptocurrencies. *Journal of Financial Econometrics,* 18(2), pp. 181–208.

Hart, A.G. (1935). A Proposal for Making Monetary Management Effective in the United States. *The Review of Economic Studies,* 2(2), pp. 104–116.

Hayek, F.A.V. (1943). A Commodity Reserve Currency. *The Economic Journal,* 53(210/211), pp. 176–184.

Hayek, F.A.V. (2009). *Denationalisation of money: the argument refined.* Auburn: Ludwig von Mises Institute.

Hearn, M. (2016). Corda: A Distributed Ledger. In: *Corda Technical White Paper.*

Hethcote, H.W. and Yorke, J.A. (2014). *Gonorrhea transmission dynamics and control.* Berlin, Heidelberg: Springer.

Hope-Bailie, A. and Thomas, S. (2016). Interledger: Creating a Standard for Payments. In: *Proceedings of the 25th International Conference Companion on World Wide Web.* New York: Association for Computing Machinery, pp. 281–282.

Houy, N. (2014). It Will Cost You Nothing to 'Kill' a Proof-of-Stake Cryptocurrency. Available at: SSRN 2393940.

Hsieh, Y.Y., Vergne, J.P., Anderson, P., Lakhani, K. and Reitzig, M. (2018). Bitcoin and the Rise of Decentralized Autonomous Organizations. *Journal of Organization Design,* 7(1), pp. 1–16.

Hussein, A.F., AIZubaidi, A.K., Habash, Q.A. and Jaber, M.M. (2019). An Adaptive Biomedical Data Managing Scheme based on the Blockchain Technique. *Applied Sciences,* 9(12), p. 2494. doi:10.3390/app9122494.

Iansiti, M. and Lakhani, K.R. (2017). The Truth About Blockchain. *Harvard Business Review,* (01).

Ibanez, J.I., Bayer, C.N., Tasca, P. and Xu, J. (2020). REA, Triple-Entry Accounting and Blockchain: Converging Paths to Shared Ledger Systems. Available at: SSRN 3602207.

Ijiri, Y. (1986). A Framework for Triple-Entry Bookkeeping. *The Accounting Review,* 61(4), pp. 745–759.

Ikeda, K. (2018). Security and privacy of blockchain and quantum computation. In: P. Raj and G. C. Deka, eds., *Advances in Computers.* Amsterdam: Elsevier, pp. 199–228.

Jentzsch, C. (2016). Decentralized Autonomous Organization to Automate Governance. In: *White Paper.*

Johnson, N.L., Kotz, S. and Balakrishnan, N. (1995). *Continuous univariate distributions.* Hoboken: John Wiley & Sons, Ltd.

Kahn, D. (1996). *The codebreakers.* New York: Macmillan.

Kahn, C.M., Rivadeneyra, F. and Wong, T.N. (2019). Should the Central Bank Issue E-money? In: *Working Papers 2019-3.* St. Louis: Central Bank of St. Louis.

Kajtazi, A. and Moro, A. (2019). The Role of Bitcoin in Well Diversified Portfolios: A Comparative Global Study. *International Review of Financial Analysis,* 61, pp. 143–157.

Kaldor, N. (2007). *Causes of growth and stagnation in the world economy.* Cambridge: Cambridge University Press.

Kalecki, M. (2007). Essays in the theory of economic fluctuations (1939). In: J. Osiatynski, ed., *Collected Works of Michael Kalecki,* Vol. I; *Capitalism, Business and Full Employment.* Oxford: Clarendon Press, pp. 235–252.

Katz, J. and Lindell, Y. (2014). *Introduction to modern cryptography.* Boca Raton: CRC Press.

Kavuri, A.S. and Milne, A. (2019). FinTech and the Future of Financial Services: What are the Research Gaps? Available at: SSRN 3215849.

Kay, J. (2010). *Should We Have Narrow Banking?* [Blog] johnkay.com. Available at: https://www.johnkay.com/2011/06/02/should-we-have-narrow-banking/.

Keen, S. (2001). *Debanking economics. the naked emperor of the social sciences*. London and New York: Zed Books.

Keen, S. (2013). A Monetary Minsky Model of the Great Moderation and the Great Recession. *Journal of Economic Behavior & Organization, 86*, pp. 221–235.

Keen, S. (2015). What is Money and How is It Created? *Forbes*, [online]. Available at: https://www.forbes.com/sites/stevekeen/2015/02/28/what-is-money-and-how-is-it-created/?sh=5ed29c717df4.

Kerckhoffs, A. (1883). La Cryptographie Militaire. *Journal des Sciences Militaires, IX*, pp. 5–83.

Keynes, J.M. (1930). *A treatise on money*. London: Macmillan.

Keynes, J.M. (1936). *The general theory of employment, interest, and money*. London: Macmillan.

Keynes, J.M. (1943). The Objective of International Price Stability. *The Economic Journal, 53*(210/211), pp. 185–187.

Khezr, S., Moniruzzaman, M., Yassine, A. and Benlamri, R. (2019). Blockchain Technology in Healthcare: A Comprehensive Review and Directions for Future Research. *Applied Sciences, 9*(9), p. 1736.

Khovratovich, D. and Law, J. (2017). BIP32-Ed25519: Hierarchical Deterministic Keys over a Non-Linear Keyspace. In: *2017 IEEE European Symposium on Security and Privacy Workshops (EuroS&PW)*. IEEE, pp. 27–31.

Kim, S. and Oh, H. (2001). Making Electronic Refunds Reusable. In: *Proceedings of the 2nd International Workshop on Information Security Applications*. Seoul, pp. 125–138.

Klusman, R. and Dijkhuizen, T. (2018). *Deanonymisation in Ethereum Using Existing Methods for Bitcoin*. [online] Amsterdam: Cees de Laat, pp. 1–8. Available at: https://delaat.net/rp/2017-2018/p61/report.pdf.

Knight, F., Cox, G.V., Director, A., Douglas, P.H., Fisher, I., Hart, A.G., Mints, L.W., Schultz, H. and Simons, H.C. (1933). *Memorandum on Banking Reform (President's Personal File, 431)*. Hyde Park: Franklin D. Roosevelt Presidential Library.

Knuth, D.E. (1997). *Volume 2: seminumerical algorithms. The art of computer programming*. Boston: Addison-Wesley Professional.

Koens, T. and Poll, E. (2019). Assessing Interoperability Solutions for Distributed Ledgers. *Pervasive and Mobile Computing, 59*, p. 101079. doi:10.1016/j.pmcj.2019.101079.

Kokoris-Kogias, E., Jovanovic, P., Gasser, L., Gailly, N., Syta, E. and Ford, B. (2018). Omniledger: A Secure, Scale-Out, Decentralized Ledger via Sharding. In: *2018 IEEE Symposium on Security and Privacy* (SP). IEEE, pp. 583–598.

Kolmogorov, A. (1933). Sulla Determinazione Empirica di Una Legge di Distribuzione. *G. Ist. Ital. Attuari., 4*, pp. 83–91.

Kontoyiannis, I. (1997). *The Complexity and Entropy of Literary Styles*. [online] Stanford: Department of Statistics, Stanford University, pp. 1–15. Available at: http://pages.cs.aueb.gr/~yiannisk/PAPERS/english.pdf.

Kopp, E., Kaffenberger, L. and Jenkinson, N. (2017). *Cyber risk, market failures, and financial stability*. Washington, D.C.: International Monetary Fund.

Korpela, K., Hallikas, J. and Dahlberg, T. (2017). Digital Supply Chain Transformation toward Blockchain Integration. In: *Proceedings of the 50th Hawaii International Conference on System Sciences*.

Kotlikoff, L.J. (2010). *Jimmy Stewart is dead: ending the world's ongoing financial plague with limited purpose banking*. Hoboken: John Wiley & Sons.

Kuo, T.T., Kim, H.E. and Ohno-Machado, L. (2017). Blockchain Distributed Ledger Technologies for Biomedical and Health Care Applications. *Journal of the American Medical Informatics Association, 24*(6), pp. 1211–1220.

Kuperberg, M. (2019). Blockchain-Based Identity Management: A Survey from the Enterprise and Ecosystem Perspective. *IEEE Transactions on Engineering Management*, 64(7), doi:10.1109/TE M.2019.2926471.

Kwon, Y., Liu, J., Kim, M., Song, D. and Kim, Y. (2019). Impossibility of Full Decentralization in Permissionless Blockchains. In: *Proceedings of the 1st ACM Conference on Advances in Financial Technologies*. New York: Association for Computing Machinery, pp. 110–123.

Lainá, P. (2018). *Full-Reserve Banking: Separating Money Creation from Bank Lending*. PhD. University of Helsinki.

Lamport, L., Shostak, R. and Pease, M. (1982). The Byzantine Generals Problem. *ACM Transactions on Programming Languages and Systems*, 4(3), pp. 382–401

Larsson, J.-K. (2018). *Cryptology. Linkoping Univesity Lecture Notes*. Available at: http://www.icg. isy.liu.se/en/courses/tsit03/.

Leising, M., Rojanasakul, M., Pogkas, D. and Kochkodin, B. (2018). Crypto Coin Tether Defies Logic on Kraken's Market, Raising Red Flags. *Bloomberg* [online]. Available at: https://www.bloomber g.com/graphics/2018-tether-kraken-trades/.

Lévy, P. (1925). *Calcul des probabilités*. Paris: Gauthier-Villars.

Lewis, A. (2018). *The basics of bitcoins and blockchains: an introduction to cryptocurrencies and the technology that powers them*. London: Mango Media Inc..

Li, Q.L., Ma, J.Y., Chang, Y.X., Ma, F.Q. and Yu, H.B. (2019). Markov Processes in Blockchain Systems. *Computational Social Networks*, 6(1), pp. 1–28.

Lima, C. (2018). Developing Open and Interoperable DLT/Blockchain Standards. *Computer,* 51(11), pp. 106–111.

Lindell, Y. and Pinkas, B. (2007). An Efficient Protocol for Secure Two-Party Computation in the Presence of Malicious Adversaries. In: *Annual International Conference on the Theory and Applications of Cryptographic Techniques*. Berlin, Heidelberg: Springer, pp. 52–78.

Lindell, Y., Pinkas, B., Smart, N.P. and Yanai, A. (2015). Efficient Constant Round Multi-Party Computation Combining BMR and SPDZ. In: *Annual Cryptology Conference*. Berlin, Heidelberg: Springer, pp. 319–338.

Lipton, A. (2001). *Mathematical methods for foreign exchange: a financial engineer's approach*. Singapore: World Scientific Publishing Co.

Lipton, A. (2016a). Modern Monetary Circuit Theory, Stability of Interconnected Banking Network, and Balance Sheet Optimization for Individual Banks. *International Journal of Theoretical and Applied Finance*, 19(6), doi:10.1142/S0219024916500345.

Lipton, A. (2016b). Macroeconomic Theories: Not Even Wrong. *Risk*, 30(9), p. 29.

Lipton, A. (2017). Review: Money Changes Everything: How Finance Made Civilization Possible. *Quantitative Finance*, 17, pp. 1319–1322.

Lipton, A. (2018). Blockchains and Distributed Ledgers in Retrospective and Perspective. *The Journal of Risk Finance*, 19(1), pp. 4–25.

Lipton, A. (2019). Don't Fear the Clearer. *Risk*, 33(4).

Lipton, A. (2020). *What the Fed's New Inflation Policy Means for Stablecoins*. [online] Coindesk. Available at https://www.coindesk.com/fed-inflation-stablecoins.

Lipton, A. and Hardjono, T. (2021). Blockchain intra- and interoperability. In: R. Boute, ed., *Innovative Technology at the Interface of Finance and Operations*. Springer.

Lipton, A., Hardjono, T. and Pentland, A. (2018a). Narrow Banks and Asset-Backed Digital Coins. *Capco Journal — Digitization*, 47, pp. 101–116.

Lipton, A., Hardjono, T. and Pentland, A. (2018b). Digital Trade Coin: Towards a More Stable Digital Currency. *Royal Society Open Science*, 5(7), p. 180155. doi:10.1098/rsos.180155.

Lipton, A., Hardjono, T. and Pentland, A. (2019). *Hannibal Ad Portas!* [online] MIT Media Lab. Available at https://www.media.mit.edu/posts/hannibal-ad-portas/.

Lipton, A. and Lopez de Prado, M. (2020). Mitigation Strategies for COVID-19: Lessons from the K-SEIR Model. Available at: SSRN 3623544.

Lipton, A. and Pentland, A. (2018). Breaking the Bank. *Scientific American,* 318(1), pp. 26–31.

Lipton, A., Sardon, A., Schär, F. and Schüpbach, C. (2020). From Tether to Libra: Stablecoins, Digital Currency and the Future of Money. *arXiv preprint,* arXiv:2005.12949.

Lipton, A., Shrier, D. and Pentland, A. (2016). *Digital Banking Manifesto: The End of Banks?* [Blog] getsmarter.com. Available at: https://www.getsmarter.com/blog/market-trends/digital-banking-m anifesto-the-end-of-banks-mit-report/.

Litke, A., Anagnostopoulos, D. and Varvarigou, T. (2019). Blockchains for Supply Chain Management: Architectural Elements and Challenges towards a Global Scale Deployment. *Logistics,* 3(1), p. 5. doi:10.3390/logistics3010005.

Liu, M., Wu, K. and Xu, J.J. (2019). How Will Blockchain Technology Impact Auditing and Accounting: Permissionless versus Permissioned Blockchain. *Current Issues in Auditing,* 13(2), pp. A19-A29.

Luu, L., Narayanan, V., Zheng, C., Baweja, K., Gilbert, S. and Saxena, P. (2016). A Secure Sharding Protocol for Open Blockchains. In: *Proceedings of the 2016 ACM SIGSAC Conference on Computer and Communications Security.* New York: Association for Computing Machinery, pp. 17–30.

Macleod, H.D. (1905). *The theory and practice of banking, in 2 volumes (1855–1856).* London: Longman, Greens and Co.

Maker Team, (2017). The Dai Stablecoin System. In: *White Paper.* [online] MakerDAO. Available at: https://makerdao.com/whitepaper/DaiDec17WP.pdf.

Mandelbrot, B.B. (1997). The variation of certain speculative prices. In B.B. Mandelbrot, ed., *Fractals and Scaling in Finance.* New York: Springer, pp. 371–418.

Marshall, A. (1887). *Report by the Gold and Silver Commission of 1887.* London: H. M. Stationery Office.

Maxwell, G. (2013). *CoinJoin: Bitcoin Privacy for the Real World.* [online] Bitcoin Forum. Available at: https://bitcointalk.org/index.php?topic=279249.0.

Mazieres, D. (2015). *The Stellar Consensus Protocol: A Federated Model for Internet-Level Consensus.* Stellar Development Foundation.

McCarthy, W.E. (1979). An Entity-Relationship View of Accounting Models. *The Accounting Review,* 54(4), pp. 667–686.

McCarthy, W.E. (1982). The REA Accounting Model: A Generalized Framework for Accounting Systems in a Shared Data Environment. *The Accounting Review,* 57(3), pp. 554–578.

McCorry, P., Shahandashti, S.F. and Hao, F. (2017). A Smart Contract for Boardroom Voting with Maximum Voter Privacy. In: *International Conference on Financial Cryptography and Data Security.* Cham: Springer, pp. 357–375.

McGinn, D., Birch, D., Akroyd, D., Molina-Solana, M., Guo, Y. and Knottenbelt, W.J. (2016). Visualizing Dynamic Bitcoin Transaction Patterns. *Big Data,* 4(2), pp. 109–119.

McLeay, M., Radia, A. and Thomas, R. (2014). Money Creation in the Modern Economy. *Bank of England Quarterly Bulletin,* 2014(Q1). Available at: SSRN 2416234.

Meiklejohn, S. and Mercer, R. (2018). Möbius: Trustless Tumbling for Transaction Privacy. *Proceedings on Privacy Enhancing Technologies,* 2018(2), pp. 105–121.

Merkle, R.C. (1978). Secure Communications over Insecure Channels. *Communications of the ACM,* 21(4), pp. 294–299.

Merkle, R.C. (1987). A Digital Signature based on a Conventional Encryption Function. In: *Conference on the Theory and Application of Cryptographic Techniques.* Berlin, Heidelberg: Springer, pp. 369–378.

Microsoft, (2018). Decentralized Identity. Own and Control Your Identity. In: *White Paper.*

Miers, I., Garman, C., Green, M. and Rubin, A.D. (2013). Zerocoin: Anonymous Distributed E-cash from Bitcoin. In: *2013 IEEE Symposium on Security and Privacy*. IEEE, pp. 397–411.

Miller, F. (1882). *Telegraphic code to insure privacy and secrecy in the transmission of telegrams*. C.M. Cornwell.

Mises, L.V. (1998). *Human action: A treatise on economics. The scholar's edition*. Auburn: Ludwig von Mises Institute.

Mitchell-Innes, A. (1914). The Credit Theory of Money. *Banking Law Journal*, 31, pp. 151–168.

Moreno-Sanchez, P., Modi, N., Songhela, R., Kate, A. and Fahmy, S. (2018). Mind Your Credit: Assessing the Health of the Ripple Credit Network. In: *Proceedings of the 2018 World Wide Web Conference*. New York: Association for Computing Machinery, pp. 329–338.

Morhaim, L. (2019). Blockchain and Cryptocurrencies Technologies and Network Structures: Applications, Implications and Beyond.

Möser, M. and Böhme, R. (2017). Anonymous Alone? Measuring Bitcoin's Second-Generation Anonymization Techniques. In: *2017 IEEE European Symposium on Security and Privacy Workshops* (EuroS&PW). IEEE, pp. 32–41.

Mulligan, C., Scott, J.Z., Warren, S. and Rangaswami, J.P. (2018). Blockchain beyond the Hype: A Practical Framework for Business Leaders. In: *White Paper of the World Economic Forum*.

Nakamoto, S. (2008). Bitcoin: A Peer-to-Peer Electronic Cash System.

Narayanan, A., Bonneau, J., Felten, E., Miller, A. and Goldfeder, S. (2016). *Bitcoin and cryptocurrency technologies: a comprehensive introduction*. Princeton: Princeton University Press.

Narayanan, A. and Clark, J. (2017). Bitcoin's Academic Pedigree. *Communications of the ACM*, 60(12), pp. 36–45.

Nasser, Y., Okoye, C., Clark, J. and Ryan, P.Y. (2018). Blockchains and Voting: Somewhere between Hype and a Panacea. In: *White Paper*.

National Institute of Standards and Technology, (1999). *Recommended Elliptic Curves for Federal Government Use*. Gaithersburg: NIST. Available at: http://csrc.nist.gov/CryptoToolkit/dss/ecdsa/NISTReCur.pdf.

Naumenko, G., Maxwell, G., Wuille, P., Fedorova, A. and Beschastnikh, I. (2019). Bandwidth-Efficient Transaction Relay for Bitcoin. *arXiv preprint*, arXiv:1905.10518.

Nguyen, C.T., Hoang, D.T., Nguyen, D.N., Niyato, D., Nguyen, H.T. and Dutkiewicz, E. (2019). Proof-of-Stake Consensus Mechanisms for Future Blockchain Networks: Fundamentals, Applications and Opportunities. *IEEE Access*, 7, pp. 85727–85745.

Noether, S. (2015). Ring SIgnature Confidential Transactions for Monero. *IACR Cryptol. ePrint Arch.*, 2015, p. 1098.

Oh, J.H. and Nguyen, K. (2018). The Growing Role of Cryptocurrency: What Does It Mean for Central Banks and Governments? *International Telecommunications Policy Review*, 25(1), pp. 33–55.

Okamoto, T. and Ohta, K. (1991). Universal Electronic Cash. In: Annual *International Cryptology Conference*, Berlin and Heidelberg: Springer, pp. 324–337.

Oresme, N. (1956). *The de moneta of Nicholas Oresme, and English Mint documents: translated from the Latin with introduction and notes by Charles Johnson*. London: Nelson.

Pappalardo, G., Di Matteo, T., Caldarelli, G. and Aste, T. (2018). Blockchain Inefficiency in the Bitcoin Peers Network. *EPJ Data Science*, 7(1), p. 30.

Park, S., Specter, M., Narula, N. and Rivest, R.L. (2020). Going from Bad to Worse: From Internet Voting to Blockchain Voting. In: *MIT White Paper*.

Pass, R., Seeman, L. and Shelat, A. (2017). Analysis of the Blockchain Protocol in Asynchronous Networks. In: *Annual International Conference on the Theory and Applications of Cryptographic Techniques*. Berlin, Heidelberg: Springer, pp. 643–673.

Payments Risk Committee, (2016). *Intraday Liquidity Flows. Report*. PRC.

Peng, L., Feng, W., Yan, Z., Li, Y., Zhou, X. and Shimizu, S. (2020). Privacy Preservation in Permissionless Blockchain: A Survey. *Digital Communications and Networks,* doi:10.1016/j.dcan.2020.05.008.

Pennacchi, G. (2012). Narrow Banking. *Annual Review of Financial Economics,* 4(1), pp. 141–159.

Pentland, A., Lipton, A. and Hardjono, T. (2020). *Building the new economy.* Cambridge: MIT Press.

Peterson, J., Krug, J., Zoltu, M., Williams, A.K. and Alexander, S. (2015). Augur: A Decentralized Oracle and Prediction Market Platform. *arXiv preprint,* arXiv:1501.01042.

Petty, W. (1899). *The economic writings of Sir William Petty, Vol. I.* Cambridge: Cambridge University Press.

Phillips, R.J. (1996). The 'Chicago plan' and New Deal banking reform. In D.B. Papadimitriou, ed., *Stability in the Financial System.* London: Palgrave Macmillan, pp. 94–114.

Phillips, R.J. and Roselli, A. (2011). How to avoid the next taxpayer bailout of the financial system: the narrow banking proposal. In J.A. Tatom, ed., *Financial Market Regulation.* New York: Springer, pp. 149–161.

Pillai, B., Biswas, K. and Muthukkumarasamy, V. (2019). Blockchain Interoperable Digital Objects. In: *International Conference on Blockchain.* Cham: Springer, pp. 80–94.

Pinna, A. and Ruttenberg, W. (2016). Distributed Ledger Technologies in Securities Post-Trading Revolution or Evolution? In: *ECB Occasional Paper Series 172.* Frankfurt: European Central Bank.

PlanB. (2019). Modeling Bitcoin's Value with Scarcity. [online] Available at: https://medium.com/@100trillionUSD/modeling-bitcoins-value-with-scarcity-91fa0fc03e25.

Poelstra, A. (2014). Distributed Consensus from Proof of Stake is Impossible. In: *Self-Published Paper.*

Ponemon, L. (2017). *Cost of data breach study.* North Traverse City: Ponemon Institute.

Poon, J. and Buterin, V. (2017). Plasma: Scalable Autonomous Smart Contracts. In: *White Paper.*

Poon, J. and Dryja, T. (2016). The Bitcoin Lightning Network: Scalable Off-Chain Instant Payments. In: *White Paper.*

Popper, N. (2015). *Digital gold: the untold story of Bitcoin.* London: Penguin UK.

Prat, J. and Walter, B. (2018). An Equilibrium Model of the Market for Bitcoin Mining. In: *CESifo Working Paper Series No. 6865.*

Press, W.H., Teukolsky, S.A., Vetterling, W.T. and Flannery, B.P. (2007). *Numerical recipes. The art of scientific computing.* 3rd ed. Cambridge: Cambridge University Press.

Qin, R., Yuan, Y., Wang, S. and Wang, F.Y. (2018). Economic Issues in Bitcoin Mining and Blockchain Research. In: *2018 IEEE Intelligent Vehicles Symposium* (IV). IEEE, pp. 268–273.

Quesnay, F. (1991). *Physiocratie (1759). Edition de J. Cartellier.* Paris: Flammarion.

Raiden, (2016). Raiden Network Documentation.

Raskin, M. and Yermack, D. (2018). Digital currencies, decentralized ledgers and the future of central banking. In: P. Conti-Brown and R. M. Lastra, eds., *Research Handbook on Central Banking.* Cheltenham: Edward Elgar Publishing.

Reitwiessner, C. and Wood, G. (2015). *Solidity.* Available at: http://solidity.readthedocs.org.

Rice, A. and Brown, E. (2012). Why Ellipses are not Elliptic Curves. *Mathematics Magazine,* 85(3), pp. 163–176.

Rigby, D. (2014). Online Shopping isn't as Profitable as You Think. *Harvard Business Review,* (08).

Rijmen, V. and Daemen, J. (2001). Advanced Encryption Standard. In: *Proceedings of Federal Information Processing Standards Publications.* Gaithersburg: National Institute of Standards and Technology, pp. 19–22.

Rivest, R. and Dusse, S. (1992). The MD5 Message-Digest Algorithm. In: MIT Laboratory for Computer Science and RSA Data Security, Inc. Memo.

Rivest, R.L., Shamir, A. and Adleman, L. (1978). A Method for Obtaining Digital Signatures and Public-Key Cryptosystems. *Communications of the ACM,* 21(2), pp. 120–126.

Roberds, W. and Velde, F.R. (2014). Early Public Banks. In: *FRB Atlanta Working Paper No. 2014-9.*

Romiti, M., Judmayer, A., Zamyatin, A. and Haslhofer, B. (2019). A Deep Dive into Bitcoin Mining Pools: An Empirical Analysis of Mining Shares. *arXiv preprint,* arXiv:1905.05999.

Rosenoer, J.M. and Yong, S.W., International Business Machines Corp, (2020). *Bespoke programmable crypto token.* U.S. Patent 10,742,398.

Ross, S. (2018), *What's Normal for Profit Margin in Retail Sector?* [online] Investopedia. Available at: https://www.investopedia.com/ask/answers/071615/what-profit-margin-usual-company-retail-sector.asp.

Rudebusch, G. (2018). A Review of the Fed's Unconventional Monetary Policy. In: *FRBSF Economic Letter 2018-27.* San Francisco: Federal Reserve Bank of San Francisco.

Saga, (2018). Saga. In: *White Paper.*

Saleh, F. (2020). Blockchain without Waste: Proof-of-Stake. *Review of Financial Studies.* Available at: SSRN 3183935.

Saltzer, J.H., Reed, D.P. and Clark, D.D. (1984). End-to-End Arguments in System Design. *ACM Transactions on Computer Systems (TOCS),* 2(4), pp. 277–288.

Sams, R. (2015). *A Note on Cryptocurrency Stabilization: Seigniorage Shares.* [online] Auckland: Brave New Coin, pp. 1–8. Available at: https://blog.bitmex.com/wp-content/uploads/2018/06/A-Note-on-Cryptocurrency-Stabilisation-Seigniorage-Shares.pdf.

Samuelson, P. and Nordhaus, W. (1995). *Economics.* New York: McGraw-Hill.

Sanadhya, S.K. and Sarkar, P. (2008). Attacking Reduced Round SHA-256. In: *International Conference on Applied Cryptography and Network Security.* Berlin, Heidelberg: Springer, pp. 130–143.

Sasson, E.B., Chiesa, A., Garman, C., Green, M., Miers, I., Tromer, E. and Virza, M. (2014). Zerocash: Decentralized Anonymous Payments from Bitcoin. In: *2014 IEEE Symposium on Security and Privacy.* IEEE, pp. 459–474.

Scaillet, O., Treccani, A. and Trevisan, C. (2020). High-Frequency Jump Analysis of the Bitcoin Market. *Journal of Financial Econometrics,* 18(2), pp. 209–232.

Schär, F. (2020). Decentralized Finance: On Blockchain-and Smart Contract-Based Financial Markets. Available at: SSRN 3571335.

Schär, F. and Berentsen, A. (2020). *Bitcoin, blockchain, and cryptoassets: a comprehensive introduction.* Cambridge: MIT Press.

Schnorr, C.P. (1989). Efficient Identification and Signatures for Smart Cards. In: *Conference on the Theory and Application of Cryptology.* New York: Springer, pp. 239–252.

Schoof, R. (1985). Elliptic Curves over Finite Fields and the Computation of Square Roots mod p. *Mathematics of Computation,* 44(170), pp. 483–494.

Schwartz, D., Youngs, N. and Britto, A. (2014). The Ripple Protocol Consensus Algorithm. In: *Ripple Labs Inc White Paper.*

Schwarz, G.E. (1978). Estimating the Dimension of a Model. *Annals of Statistics,* 6(2), pp. 461–464.

Seang, S. and Torre, D. (2018). Proof of Work and Proof of Stake Consensus Protocols: A Blockchain Application for Local Complementary Currencies. In: *GREDEG Working Papers* 2019-24. Nice: Groupe de REcherche en Droit, Economie, Gestion.

Selker, T. (2004). Fixing the Vote: What Electronic Ballots can Do for Democracy. *Scientific American,* pp. 92–97.

Senner, R. and Sornette, D. (2019). The Holy Grail of Crypto Currencies: Ready to Replace Fiat Money? *Journal of Economic Issues,* 53(4), pp. 966–1000.

Shahaab, A., Lidgey, B., Hewage, C. and Khan, I. (2019). Applicability and Appropriateness of Distributed Ledgers Consensus Protocols in Public and Private Sectors: A Systematic Review. *IEEE Access,* 7, pp. 43622–43636.

Shalini, S. and Santhi, H. (2019). A Survey on Various Attacks in Bitcoin and Cryptocurrency. In: *2019 International Conference on Communication and Signal Processing* (ICCSP). IEEE, pp. 0220–0224.

Shannon, C.E. (1949). Communication Theory of Secrecy Systems. *Bell System Technical Journal,* 28, pp. 656–715.

Shannon, C.E. (1951). Prediction and Entropy of Printed English. *Bell System Technical Journal,* 30(1), pp. 50–64.

Sharma, G.D., Jain, M., Mahendru, M. and Bansal, S. (2019). Emergence of Bitcoin as an Investment Alternative. *International Journal of Business and Information,* 14(1), pp. 47–84.

Sherman, A.T., Javani, F., Zhang, H. and Golaszewski, E. (2019). On the Origins and Variations of Blockchain Technologies. *IEEE Security & Privacy,* 17(1), pp. 72–77.

Shrier, D. (2020). *Basic blockchain: what it is and how it will transform the way we work and live. Robinson.*

Sila (2018). *Sila:* A Stable Fiat-Backed Digital Tokenized Means of Exchange. In: *White Paper.*

Singh, S. (2000). *The code book: the science of secrecy from ancient Egypt to quantum cryptography. Anchor.*

Smirnov, N. (1948). Table for Estimating the Goodness of Fit of Empirical Distributions. *Annals of Mathematical Statistics,* 19(2), pp. 279–281.

Smith, A. (1977). *An inquiry into the nature and causes of the wealth of nations, 1776.* Chicago: The University of Chicago Press.

Soddy, F. (1933). *Wealth, virtual wealth and debt: the solution of the economic paradox.* London: George Allen and Unwin Ltd.

Sompolinsky, Y. and Zohar, A. (2013). Accelerating Bitcoin's Transaction Processing. Fast Money Grows on Trees, Not Chains. *IACR Cryptol. ePrint Arch.,* [online] 2013, p. 881. Available at: ia.cr/2013/881.

Stewart, I., Ilie, D., Zamyatin, A., Werner, S., Torshizi, M.F. and Knottenbelt, W.J. (2018). Committing to Quantum Resistance: A Slow Defence for Bitcoin against a Fast Quantum Computing Attack. *Royal Society Open Science,* 5(6), doi:10.1098/rsos.180410.

Subramanian, H. (2017). Decentralized Blockchain-Based Electronic Marketplaces. *Communications of the ACM,* 61(1), pp. 78–84.

Swan, M., Potts, J., Takagi, S., Witte, F. and Tasca, P. (2019). *Blockchain economics: implications of distributed ledgers: markets, communications networks, and algorithmic reality.* Singapore: World Scientific Publishing Co. Pte. Ltd.

Szabo, N. (1997). Formalizing and Securing Relationships on Public Networks. *First Monday,* 2(9).

Szabo, N. (2017). *Money, Blockchains, and Social Scalability.* [Blog] Unenumerated. Available at: h ttp://unenumerated.blogspot.com/2017/02/money-blockchains-and-social-scalability.html.

Tapscott, D. and Tapscott, A. (2016). *Blockchain revolution: how the technology behind bitcoin is changing money, business, and the world.* London: Penguin.

Tariq, P. and Jamison, M. (2019). Never Trust Bitcoin: Blockchain Technology — The Misnomer of a 'Trustless' System. In: *Workshop on the Ostrom Workshop 6.* Bloomington: Indiana University.

Tasca, P. and Tessone, C.J. (2017). Taxonomy of Blockchain Technologies: Principles of Identification and Classification. *arXiv preprint,* arXiv:1708.04872.

Taskinsoy, J. (2019). Is Facebook's Libra Project Already a Miscarriage? Available at: SSRN 3437857.

Taylor, J. (1955). Copernicus on the Evils of Inflation and the Establishment of a Sound Currency. *Journal of the History of Ideas,* 16(4), pp. 540–547.

Teahan, W.J. and Cleary, J.G. (1996). The Entropy of English Using PPM-Based Models. In: *Proceedings of Data Compression Conference-DCC'96.* IEEE, pp. 53–62.

Tether (2016). Tether: Fiat Currencies on the Bitcoin Blockchain. In: *White Paper.*

Thomas, S. and Schwartz, E. (2015). *A Protocol for Interledger Payments*. [online] Interledger, pp. 1–25. Available at: https://interledger.org/interledger.pdf.

Tiberiuscoin (2017). Tiberiuscoin. In: *White Paper*.

Tlustý, P. and Šulista, M. (2017). The Algorithm Used for Numbering German Banknotes: What Counterfeiters Might Not Have Known. In: *INPROFORUM* 2016.

Tobin, J. (1969). A General Equilibrium Approach to Monetary Theory. *Journal of Money, Credit and Banking*, 1, pp. 15–29.

Tobin, J., (1985). Financial Innovation and Deregulation in Perspective. *Bank of Japan Monetary and Economic Studies*, 3, pp. 19–29.

Tobin, J. and Golub, S. (1998). *Money, credit, and capital*. Boston: Irwin McGraw-Hill.

Tov, E. (2012). The scribal and textual transmission of the Torah analyzed in light of its sanctity. In: A. Moriya and G. Hata, eds., *Pentateuchal Traditions in the Late Second Temple Period*. Leiden: Brill, pp. 57–72.

Traiger, I.L., Gray, J., Galtieri, C.A. and Lindsay, B.G. (1982). Transactions and Consistency in Distributed Database Systems. *ACM Transactions on Database Systems (TODS)*, 7(3), pp. 323–342.

Trappe, W. and Washington, L.C. (2006). *Introduction to cryptography with coding theory*. 2nd ed. London: Pearson.

Tschoegl, A.E. (2001). Maria Theresa's Thaler: A Case of International Money. *Eastern Economic Journal*, 27(4), pp. 443–462.

Tschorsch, F. and Scheuermann, B. (2016). Bitcoin and beyond: A Technical Survey on Decentralized Digital Currencies. *IEEE Communications Surveys & Tutorials*, 18(3), pp. 2084–2123.

Vasek, M., Bonneau, J., Castellucci, R., Keith, C. and Moore, T. (2016). The Bitcoin Brain Drain: A Short Paper on the Use and Abuse of Bitcoin Brain Wallets. In: *International Conference on Financial Cryptography and Data Security*. Berlin, Heidelberg: Springer.

Vigna, P. and Casey, M.J. (2016). *The age of cryptocurrency: how bitcoin and the blockchain are challenging the global economic order*. New York: Macmillan.

Vigna, P. and Casey, M.J. (2019). *The truth machine: the blockchain and the future of everything*. London: Picador.

Vyas, N., Beije, A. and Krishnamachari, B. (2019). *Blockchain and the supply chain: concepts, strategies and practical applications*. London: Kogan Page Publishers.

Walch, A. (2015). The Bitcoin Blockchain as Financial Market Infrastructure: A Consideration of Operational Risk. *NYU Journal of Legislation & Public Policy*, 18, p. 837.

Wang, L., Shen, X., Li, J., Shao, J. and Yang, Y. (2019a). Cryptographic Primitives in Blockchains. *Journal of Network and Computer Applications*, 127, pp. 43–58.

Wang, W., Hoang, D.T., Hu, P., Xiong, Z., Niyato, D., Wang, P., Wen, Y. and Kim, D.I. (2019b). A Survey on Consensus Mechanisms and Mining Strategy Management in Blockchain Networks. *IEEE Access*, 7, pp. 22328–22370.

Wang, X. and Yu, H. (2005). How to Break MD5 and Other Hash Functions. In: *Annual International Conference on the Theory and Applications of Cryptographic Techniques*. Berlin, Heidelberg: Springer, pp. 19–35.

Wei, D. (1998). *b-money, 1998*. [online] weidai. Available at http://www.weidai.com/bmoney.txt.

Werner, R.A. (2005). *New paradigm in macroeconomics*. Basingstoke: Palgrave Macmillan.

Werner, R.A. (2014). Can Banks Individually Create Money out of Nothing? — The Theories and the Empirical Evidence. *International Review of Financial Analysis*, 36, pp. 1–19.

Wicksell, K. (1913). *Vorlesungen über nationalökonomie auf grundlage des marginalprinzipes*. Jena: Gustav Fischer.

Wicksteed, P.H. (1910). *The common sense of political economy, including a study of the human basis of economic law*. London: Macmillan.

Williamson, S. (2019). Central Bank Digital Currency: Welfare and Policy Implications. In: *2019 Meeting Papers (No. 386)*. Society for Economic Dynamics.

Wolf, M. (2014). Strip Private Banks of their Power to Create Money. *The Financial Times,* [online]. Available at: https://www.ft.com/content/7f000b18-ca44-11e3-bb92-00144feabdc0.

Wood, G. (2015). Ethereum: A Secure Decentralised Generalised Transaction Ledger Homestead Revision. In: *Yellow Paper.*

Wright, R.E. (2008). *One nation under debt: Hamilton, Jefferson, and the history of what we owe.* New York: McGraw-Hill.

Wüst, K. and Gervais, A. (2018). Do You Need a Blockchain? In: *2018 Crypto Valley Conference on Blockchain Technology* (CVCBT). IEEE, pp. 45–54.

Yaga, D., Mell, P., Roby, N. and Scarfone, K. (2019). Blockchain Technology Overview. *arXiv preprint*, arXiv:1906.11078.

Yang, W., Aghasian, E., Garg, S., Herbert, D., Disiuta, L. and Kang, B. (2019). A Survey on Blockchain-Based Internet Service Architecture: Requirements, Challenges, Trends, and Future. *IEEE Access,* 7, pp. 75845–75872.

Yellen, J. (2017). A Speech at the Commonwealth Club.

Zezza, G. and Dos Santos, C.H. (2004). The role of monetary policy in post-Keynesian SFC macroeconomic growth models. In: M. Lavoie and M. Seccareccia, eds., *Central Banking in the Modern World: Alternative Perspectives.* Cheltenham: Edward Elgar, pp. 181–208.

Zhang, Y., Chen, X. and Park, D. (2018). Formal Specification of Constant Product (xy = k) Market Maker Model and Implementation.

Zheng, Z., Xie, S., Dai, H., Chen, X. and Wang, H. (2017). An Overview of Blockchain Technology: Architecture, Consensus, and Future Trends. In: *2017 IEEE International Congress on Big Data* (BigData Congress), pp. 557–564.

Zhou, Q., Huang, H., Zheng, Z. and Bian, J. (2020). Solutions to Scalability of Blockchain: A Survey. *IEEE Access,* 8, pp. 16440–16455.

Zhou, X. (2009). Reform the International Monetary System. *BIS Review,* 41, pp. 1–3.

Zhu, X. and Badr, Y. (2018). Identity Management Systems for the Internet of Things: A Survey towards Blockchain Solutions. *Sensors*, 18(12), p. 4215. doi:10.3390/s18124215.

Zilliqa, (2018). The Zilliqa Project: A Secure Scalable Blockchain Platform. In: *Position Paper.*

Zolotarev, V.M. (1986). *One-dimensional stable distributions.* Providence: American Mathematical Society.

Name Index

A

Abel, Niels Henrik, 108
Adleman, Leonard, 80, 94

B

Babbage, Charles, 79, 84
Back, Adam, 11, 50, 372
Bellaso, Giovan Battista, 83
Bezout, Erienne, 113
Bosselaers, Antoon, 137
Boyle, Todd, 381
Buterin, Vitalik, 9, 73, 208–213, 220, 231

C

Caesar, Julius, 82
Cauchy, Augustin-Louis, 299
Chaum, David, 50, 97, 365
Chernomyrdin, Viktor, 249
Churchill, Winston, 394
Cocks, Clifford, 94
Copernicus, Nicolas, 250
Cotten, Gerald, 285

D

Dai, Wei, 11, 50
Diffie, Whitfield, 86, 89, 98, 105, 121, 126
Dobbertin, Hans, 137
Durer, Albrecht, 8

E

ElGamal, Taher, 82
Ellis, James, 94
Euler, Leonhard, 108

F

Feistel, Horst, 86
Finney, Hal, 64, 161

G

Gauss, Karl Friedrich, 299
Goethe, Wolfgang v., 379
Graham, Benjamin, 257
Graham, Frank, 257
Graziani, Augusto, 16
Grigg, Ian, 381, 382, 400

H

Haber, Stuart, 11, 50, 175
Hanyecz, Laszlo, 71
Hapsburg, House of, 7
Hapsburg, Maria Teresa, 7
Hardjono, Thomas, 249, 356, 357
Hayek, Friedrich August von, 1, 7
Hellman, Martin, 86, 89, 98, 105, 121, 126
Huygens, Christian, 189

J

Jacobi, Carl Gustav, 108

K

Kaldor, Nicholas, 257
Kasiski, Frederich, 83
Keen, Steve, 15
Kepler, Johannes, 111
Kerckhoffs, Auguste, 81, 82
Keynes, John Maynard, 255
Koblitz, Niel, 148

Subject Index

Printed in the United States
by Baker & Taylor Publisher Services